Lecture Notes in Computer Science 8326

Commenced Publication in 1973
Founding and Former Series Editors:
Gerhard Goos, Juris Hartmanis, and Jan van Leeuwen

T0183476

Cathal Gurrin   Frank Hopfgartner
Wolfgang Hurst   Håvard Johansen
Hyowon Lee   Noel O'Connor (Eds.)

MultiMedia Modeling

20th Anniversary International Conference, MMM 2014
Dublin, Ireland, January 6-10, 2014
Proceedings, Part II

 Springer

Volume Editors

Cathal Gurrin
Dublin City University, Ireland
E-mail: cgurrin@computing.dcu.ie

Frank Hopfgartner
Technische Universität Berlin / DAI-Labor, Germany
E-mail: frank.hopfgartner@dai-labor.de

Wolfgang Hurst
Universiteit Utrecht, The Netherlands
E-mail: huerst@uu.nl

Håvard Johansen
UiT The Arctic University of Norway, Norway
E-mail: haavardj@cs.uit.no

Hyowon Lee
Singapore University of Technology and Design, Singapore
E-mail: hlee@sutd.edu.sg

Noel O'Connor
Dublin City University, Ireland
E-mail: oconnor2n@eeng.dcu.ie

ISSN 0302-9743   e-ISSN 1611-3349
ISBN 978-3-319-04116-2   e-ISBN 978-3-319-04117-9
DOI 10.1007/978-3-319-04117-9
Springer Cham Heidelberg New York Dordrecht London

Library of Congress Control Number: 2013955783
CR Subject Classification (1998): H.3, H.5, I.5, H.2.8, H.4, I.4, I.2
LNCS Sublibrary: SL 3 – Information Systems and Application,
incl. Internet/Web and HCI

Typesetting: Camera-ready by author, data conversion by Scientific Publishing Services, Chennai, India

Printed on acid-free paper

Springer is part of Springer Science+Business Media (www.springer.com)

International Liaisons

USA: Alex Hauptmann   Carnegie Mellon University, USA
Europe: Susanne Boll   University of Oldenburg, Germany
Asia: Jialie Shen   Singapore Management University, Singapore

Local Organizing Co-chairs

Rami Albatal   Dublin City University, Ireland
Lijuan Zhou   Dublin City University, Ireland

Website

Yang Yang   Dublin City University, Ireland
David Scott   Dublin City University, Ireland

Program Committee

Amin Ahmadi   Dublin City University, Ireland
Rami Albatal   Dublin City University, Ireland
Laurent Amsaleg   CNRS-IRISA, France
Noboru Babaguchi   Osaka University, Japan
Jenny Benois-Pineau   LABRI/University of Bordeaux, France
Laszlo Boeszocrmenyi   Klagenfurt University, Austria
Susanne Boll   University of Oldenburg, Germany
Vincent Charvillat   University of Toulouse, France
Gene Cheung   National Institute of Informatics, Japan
Liang-Tien Chia   Nanyang Technological University, Singapore
Insook Choi   Columbia College Chicago, USA
Konstantinos Chorianopoulos Ionian University, Greece
Wei-Ta Chu   National Chung Cheng University, Taiwan
Tat-Seng Chua   National University of Singapore, Singapore
Kathy M. Clawson   University of Ulster, UK
Matthew Cooper   FX Palo Alto Laboratory, USA
W. Bas de Haas   Utrecht University, The Netherlands
Francois Destelle   Dublin City University, Ireland
Cem Direkoglu   Dublin City University, Ireland
Ajay Divakaran   SRI International, USA
Lingyu Duan   Peking University, China
Stéphane Dupont   University of Mons, Belgium
Thierry Dutoit   University of Mons, Belgium
Maria Eskevich   Dublin City University, Ireland
Jianping Fan   University of North Carolina, USA
Gerald Friedland   ICSI Berkeley, USA
Yue Gao   National University of Singapore, Singapore

Organization

MMM 2014 was organized by Dublin City University, Ireland.

Organizing Committee

General Chair

Cathal Gurrin   Dublin City University, Ireland

Program Co-chairs

Noel O'Connor   Dublin City University, Ireland
Wolfgang Hürst   University of Utrecht, The Netherlands
Hyowon Lee   Singapore University of Technology and Design, Singapore

Special Session Co-chairs

Frank Hopfgartner   Technische Universität Berlin, Germany
Håvard Johansen   University of Tromsø, Norway

Short Paper Co-chairs

Neil O'Hare   Yahoo Labs, Spain
Richang Hong   Hefei University of Technology, China

Demonstration Co-chairs

Hideo Joho   University of Tsukuba, Japan
Udo Kruschwitz   University of Essex, UK

Student Support Chair

Rami Albatal   Dublin City University, Ireland

Advertising/Sponsorship Chair

Yantao Zheng   Google, USA

three members of the Program Committee. The following five special sessions were selected for inclusion in MMM 2014:

- Social Geo-Media Analytics and Retrieval
- Multimedia Hyperlinking and Retrieval
- 3D Multimedia Computing and Modeling
- Multimedia Analysis for Surveillance Video and Security
- Mediadrom: Artful Post-TV Scenarios

We would like to thank our invited keynote speakers (Anil Kokaram from Google and Narrative/Memoto representative) for their stimulating contributions to the conference. Special thanks go to the short papers co-chairs, Neil O'Hare and Richang Hong, and the demonstrations co-chairs, Udo Kruschwitz and Hideo Joho. We are also fortunate to have worked with wonderful supporting chairs, such as Yantao Zheng (sponsorship chair and publicity chair), Alex Hauptmann, Susanne Boll, and Jialie Shen (international liaisons). We also acknowledge the commitment of our local organization team including, Rami Albatal (student sponsorship and local organization chair), and our two designers and webmasters, David Scott and Yang Yang. Special mention goes to Alan Smeaton for his constant availability to provide support and advice. Finally, we wish to thank the wonderful local organization team, Lijuan Marissa Zhou, ZhenXing Zhang, Zhengwei Qiu, Brian Moynagh, Stefan Terziyski, Teng Qi He, Na Li and Zaher Hinbarji. In addition, we wish to thank all authors who spent their time and effort to submit their work to MMM 2014, and all of the participants and student volunteers for their contributions and valuable support. Our gratitude also goes to the MMM 2014 Program Committee members, the Award Committee members, and the other invited reviewers for the 500 reviews required for MMM 2014.

We are grateful to the sponsors for generously providing financial support for the conference, including Dublin City University, Science Foundation Ireland, Fáilte Ireland, Google, ISCA and the Insight Centre for Data Analytics. We would also like to thank the School of Computing at Dublin City University, in particular David Sinclair, Mark Roantree, and Rory O'Connor for their support as the heads of the School of Computing. Our special thanks go to the Insight admin team, Margaret Malone, Deirdre Sheridan, Barbara Flynn, and Anne Troy from the DCU Finance Department, Ana Terres from the DCU Office of the Vice-President for Research, Harald Weinreich from ConfTool, and our contacts in the Guinness Storehouse and the Smock Alley Theatre. Finally, we thank all of the attendees at MMM 2014 who made the trip to Dublin to attend MMM 2014, VBS 2014, and WMPA 2014.

January 2014

Cathal Gurrin
Noel O'Connor
Wolfgang Hurst
Hyowon Lee
Frank Hopfgartner
Håvard Johansen

Preface

These proceedings contain the papers presented at MMM 2014, the 20th Anniversary International Conference on MultiMedia Modeling. The conference was organized by Inisght Centre for Data Analytics, Dublin City University, and was held during January 6-10, 2014, at the wonderful venue of the Guinness Storehouse in Dublin, Ireland. We greeted the attendees at MMM 2014 with the following address: "Táimid an-bhrodúil fáilte a chur romhaibh chuig Baile Átha Cliath agus chuig an fichiú Comhdháil Idirnáisiúnta bliain ar Samhaltú Ilmheán. Tá súil againn go mbeidh am iontach agaibh anseo in Éirinn agus go mbeidh bhur gcuairt taitneamhnach agus sásúil. Táimid an-bhrodúil go háirithe fáilte a chur roimh na daoine ón oiread sin tíortha difriúla agus na daoine a tháinig as i bhfad i gcéin. Tá an oiread sin páipéar curtha isteach chuigh an chomhdháil seo go bhfuil caighdeán na bpáipéar, na bpóstaer agus na léiriú an-ard ar fad agus táimid ag súil go mór le hócaid iontach. We are delighted to welcome you to Dublin for the 20th Anniversary International Conference on Multimedia Modeling. We hope that the attendees have a wonderful stay in Ireland and that their visits are both enjoyable and rewarding. We are very proud to welcome visitors from both Ireland and abroad and we are delighted to be able to include in the proceedings such high-quality papers, posters, and demonstrations."

MMM 2014 received a total 176 submissions across four categories; 103 full-paper submissions, 24 short paper submissions, and 12 demonstration submissions. Of these submissions, 55% were from Europe, 41% from Asia, 3% from the Americas, and 1% from the Middle East. All full paper submissions were reviewed by at least three members of the 110-person Program Committee, for whom we owe a debt of gratitude for providing their valuable time to MMM 2014. Of the 103 full papers submitted, 30 were selected for oral presentation, which equates to a 29% acceptance rate. A further 16 papers were chosen for poster presentation. For short papers, a total of 11 were accepted for poster presentation, representing a 45% acceptance rate. In addition, nine demonstrations from a total of 12 submissions were accepted for MMM 2014. We accepted 28 special session submissions across the five special sessions and six Video Browser Showdown (VBS 2014) submissions. The accepted contributions represent the state of the art in multimedia modeling research and cover a diverse range of topics including: applications of multimedia modelling, interactive retrieval, image and video collections, 3D and augmented reality, temporal analysis of multimedia content, compression and streaming.

As in recent years, MMM 2014 included VBS 2014. This year we made the VBS a half-day workshop, which took place on January 7, 2014. For the first time in 2014, we also co-located the WinterSchool on Multimedia Processing and Applications (WMPA 2014), which ran during January 6-7, 2014.

As is usual for MMM, there were a number of special sessions accepted for inclusion in MMM 2014. Each special session paper was also reviewed by at least

William Grosky	University of Michigan, USA
Cathal Gurrin	Dublin City University, Ireland
Martin Halvey	Glasgow Caledonian University, UK
Allan Hanbury	Vienna University of Technology, Austria
Andreas Henrich	University of Bamberg, Germany
Richang Hong	Hefei University of Technology, China
Frank Hopfgartner	Technische Universität Berlin, Germany
Jun-Wei Hsieh	National Taiwan Ocean University, Taiwan
Winston Hsu	National Taiwan University, Taiwan
Benoit Huet	EURECOM, France
Wolfgang Hürst	University of Utrecht, The Netherlands
Ichiro Ide	Nagoya University, Japan
Rongrong Ji	Xiamen University, China
Yu-Gang Jiang	Fudan University, China
Håvard Johansen	University of Tromsø, Norway
Mohan Kankanhalli	National University of Singapore, Singapore
Yoshihiko Kawai	NHK, Japan
Yiannis Kompatsiaris	Information Technologies Institute, Greece
Udo Kruschwitz	University of Essex, UK
Martha Larson	Delft University of Technology, The Netherlands
Duy-Dinh Le	National Institute of Informatics, Japan
Hyowon Lee	Singapore University of Technology and Design, Singapore
Michael Lew	Leiden University, The Netherlands
Haojie Li	Dalian University of Technology, China
Ke Liang	Advanced Digital Sciences Center, Singapore
Suzanne Little	Dublin City University, Ireland
Dong Liu	Columbia University, USA
Xiaobai Liu	University of California at Los Angeles, USA
Yan Liu	The Hong Kong Polytechnic University, Hong Kong
Yuan Liu	Ricoh Software Research Center, China
Guojun Lu	Monash University, Australia
Nadia Magnenat-Thalmann	University of Geneva (MIRALab), Switzerland
Jose M. Martinez	Universidad Autónoma de Madrid, Spain
Davide Andrea Mauro	Institut Mines-Télécom, TÉLÉCOM ParisTech, CNRS-LTCI, France
Kevin McGuinness	Dublin City University, Ireland
Robert Mertens	Hochschule Weserbergland, Germany
Florian Metze	Carnegie Mellon University, USA
David Monaghan	Dublin City University, Ireland
Henning Müller	HES-SO Valais, Switzerland
Chong-Wah Ngo	City University of Hong Kong, Hong Kong
Naoko Nitta	Osaka University, Japan

Cha Zhang   Microsoft Research, USA
Lijuan Zhou   Dublin City University, Ireland
Roger Zimmermann   National University of Singapore, Singapore

Additional Reviewers

Zheng Song   National University of Singapore
Hanwang Zhang   National University of Singapore

Sponsoring Institutions

In Cooperation with

Bernauer-Budiman Inc., Reading, Mass.
The Hofmann-International Company, San Louis Obispo, Cal.
Kramer Industries, Heidelberg, Germany

Table of Contents – Part II

Special Session: 3D Multimedia Computing and Modeling

Special Session: Social Geo-Media Analytics and Retrieval

Special Session: Multimedia Hyperlinking and Retrieval

Short Papers

Demonstrations

Video Browser Showdown

Table of Contents – Part I

Interactive Indexing and Retrieval

Multimedia Collections

Applications

Temporal Analysis

3D and Augmented Reality

Compression, Transcoding and Streaming

Organising Crowd-Sourced Media Content
via a Tangible Desktop Application

Sema Alaçam[1], Yekta İpek[1], Özgün Balaban[1], and Ceren Kayalar

[1] Istanbul Technical University,
Architectural Design Computing Graduate Program, Turkey
{semosphere,yektaipek,ozgunbalaban,cerenk}@gmail.com

Abstract. This paper aims to present a framework for collaborative design environments during the manipulation of multimedia content. The proposed desktop system can be assumed as an imitation of physical desktop including three additional properties: a tangible timeline layer, a projection of 2D plan and 3D perspective views which can be controlled through specified augmented reality markers. The potentials of this desktop application, such as unfoldable experience of time via a timeline, multi-touch manipulation and organisation of 2D and 3D visual data, were explored for the further augmented reality studies.

Keywords: Tangible desktop, data organisation, multi-touch, collaborative study, spatial interaction.

1   Introduction

Taking the roots in the Gestalt school of psychology, holistic approach provides a perspective for understanding the dynamics of multisensory perception. In the previous theories of perception it is accepted that information from different senses is initially processed in different unisensory areas of the brain and subsequently integrated in higher-order multisensory areas [1]. Similarly, regarding the development of interfaces in the field of human-computer interaction, the sensory modalities have been examined as discrete operations. However recent studies indicate that information from different senses are received as a continuous flow [1]. In other words, the nature of sensory experience is more complicated than previous assumptions and holistic approach will be more important within understanding the role of multisensory and cross-sensory interactions. Considering human-computer interaction, phenomenological perspective concept was proposed for underlining the requirement for holistic point of view in terms of an additional theory in user-centered design of context-aware systems [2]. In phenomenological perspective the context-aware system is assumed as "sensing" information instead of a static artifact, thus user's bodily nature can be taken into consideration [2].Therefore we focused on the samples which allow the integration of different sensory modalities. In particular we looked at the applications which might provide tactile, haptic, kinesthetic feedback or gestural interaction in the space/with the spatial attributes of 3D virtual models.

C. Gurrin et al. (Eds.): MMM 2014, Part II, LNCS 8326, pp. 1–10, 2014.
© Springer International Publishing Switzerland 2014

Physical desktop environment provides direct interaction with the 2D and 3D representations. The direct interaction includes tactile, haptic and kinesthetic feedback which also constitutes a spatial and multi-sensory experience. Being a continuous multi-sensory experience source, physical desktop metaphor/media, carries potential of being an intersection plane for different types of representation such as drawings, sketches, notes, physical 3D models. However it is difficult to present time based on changes in physical environment. On the other hand, digital environment provides a limited and reduced experience. As a superposition of physical and digital media, we propose a Tangible Desktop Application (TDA) for manipulating multi-content data during collaborative studies.

In this paper, we focused on the outcomes of student workshop sessions entitled #gezidocumentARy[1].While the augmentation process of #occupygezi movement data consists of city-cape, plans, maps, photos, sketches, 3D architectural models, twitter messages, we decided to improve our collaborative study environment in terms of (i) providing a time-based experience; (ii) organisation and (iii) manipulation of 2D and 3D data. We selected three scenarios among others, for exploring the possible implications of our TDA:

- Time-based visualisation of twitter data of Gezi Park on a 3D architectural model;
- Visualisation of the day by day changes on the façade of one specific building, the Atatürk Cultural Center (AKM);
- Augmentation of street art around the Gezi Park;

The common properties of four scenarios can be listed as:

- The scenarios are site-specific;
- The type of data consist both location and time information;
- Visual and verbal data should be matched with location information on 3D architectural model of Gezi Park.

In Section 2, we introduce the state of art about tangible and augmented tabletop environments. In Section 3, we present some crowd-sourced (dis)content about #occupygezi. In Section 4, we introduce three scenarios on how we deal with the large amount of data "belonging" to the Gezi Park. In Section 5, the experimental setup and framework for our tangible desktop application is presented. In Section 6, we review which complementary hypermedia technologies could contribute to our project. In Section 7, we summarize our contributions and open research questions towards possible future work.

2   Related Tangible Work

In this section we discuss collaborative work environments consisting of tangible interaction in relation with kinesthetic and spatial interaction. There are plenty of

---

[1] http://gezidocumentary.wordpress.com

studies on developing a framework for multisensory interaction on tabletops via supporting tactile, haptic, gestural, or integrated interaction.

"Tangible User Interface" (TUI) term is presented via metaDESK Platform [3] which can be assumed one of the earlier examples of tactile interaction. The metaDESK Platform inherits the desktop metaphor and provides graspable interaction with graphical input devices such as windows, icons, menus and widgets. The technical requirements of metaDESK consist of infrared lamps; active lens for navigating 3D model, various sensors for tracking spatial position and orientation. However this setup requires comprehensive knowledge and expensive equipment.

Another application "URP and shadow" is developed for urban planning and architectural design process [4]. Providing tangible feedback through direct manipulation of the 2D/3D objects, URP allows spatial interaction. With the display of additional physical information such as shadow reflections of light, and pedestrian level wind flow; URP can be considered an augmented reality application.

Including a tabletop surface a camera for detecting the coordinates of AR cards, 3D virtual representation in computer environment and displaying the augmented visuals through head mounted display devices, Table-Top AR Environment, ARToolKit technology can be mentioned as a Tangible Augmented Reality (TAR) application [5]. On one hand Table-Top AR Environment provides direct manipulation of virtual objects in a natural and intuitive manner. On the other hand, as being an additional layer, the AR interface might limit the perception of physical environment.

Some tangible interfaces consist of direct manipulation of 2D and 3D data such as SandScape and Illuminating Clay [6]. Hybrid User Interface (HYUI) provides intuitive deictic and tangible interaction via arbitrary combination of 2D and 3D data [7]. However, in these systems the input devices are considered as static artefacts instead of dealing with the context-aware system as a whole.

Marquardt et al. presents a cross-device interaction via f-formations which is developed for face to face, side to side and corner to corner interactions during collaborative studies [8]. Another study of Marquardt et al. focuses on tangible continuous interaction space which allows both tactile and gestural interaction [9].

Developed a decade ago, HandVu[2] is a simple hand gesture recognition user interface. Its authors later proposed a survey on the challenge and innovations of computer-vision techniques for hand gestures applications [10]. More recently, XKin[3]is an alternative that can be trained for hand gesture recognition with Hidden Markov Models [11]. Ludique's Kinect Bundle (LKB)[4], formerly simple-kinect-touch, is a cost-effective solution to turn different surface types into multi-touch surfaces, by the simple means of a Kinect camera. It aims to provide a simple user interface for calibration. However, while its target users are non-experts in computer vision techniques, it requires a careful setup with cues on the surface types. The temporal precision feels acceptable, but the spatial precision highly depends on the quality of the calibration.

---

[2] http://www.movesinstitute.org/~kolsch/HandVu/

[3] https://github.com/fpeder/XKin

[4] https://code.google.com/p/lkb-kinect-bundle/

3   Sources of (Dis)content

Gezi Park Protests in Istanbul can be dated back to May 28th 2013. The process was dynamic. On one hand, actions and events were happening in real-time; on the other hand the process itself was producing its own accumulative and public knowledge. In the end of second week of 2013 June, numerous songs were produced[5]. In the first three weeks, the number of twitter users increased from 1.800.000 to 9.500.000[6] in Turkey. Thousands of pictures and videos are shared through social media specifically in form of graffiti, street art, poster, illustration and records[7, 8].

In the beginning, the protests were location based. However, in a short time the characteristics of protests evolved and spread out to different locations. As a consequence, the events produced large amount of multi-content data as well as different aspects of abstraction and interpretation[9, 10, 11]. Considering previous examples of AR applications for such public protests[12,13] and their impact on people in terms of perception, we are particularly interested in developing an AR-based framework from Gezi Park protests' data.

4   Scenarios

In order to limit the context and the scope, we focus on three sub-scenarios.

4.1   Subplot 1 - Tweets around Gezi Park

During the demonstrations, the information flow between people became a vital thing because of the intention of "being connected". People, who do not know each other in person, gathered together via Twitter public figures and hashtags. Twitter became an essential connection channel naturally.

In this scenario, location enabled tweets sent with hashtags related to Gezi Park (for instance #occupygezi) are going to be mapped onto their physical locations using Twitter API. Tweets can be analysed according to different criteria and visualised in different ways by extracting the data. With the help of the physical timeline on the table, people can track the tweets around Gezi Park and find out different information patterns regarding to the criteria, as in [12].

---

[5] http://muzikoloji.co/capulcu-sarkilari
[6] http://budahabaslangicmucadeleyedevam.com
[7] http://readlists.com/919501ec/
[8] http://scriptogr.am/geziparkarsiv
[9] http://gdtjam.com/jamgezi/
[10] http://elmaaltshift.com/2013/06/24/occupy-gezi-illustrations/
[11] http://direnduvar.com
[12] http://aroccupywallstreet.wordpress.com/
[13] http://protestars.wordpress.com

4.2   Subplot 2 –Changing Façade of Atatürk Cultural Center (AKM)

The protesters hang banners on the front face of AKM, which is one of the significant buildings for Istanbul and has been subjected to massive debates on the demolishment. The banners on the AKM façade had been changing day by day during demonstrations. After about two weeks, the banners had been removed from the façade.

In this scenario, we are aiming to show the changes on the façade, which had been carved on people's memory. After collecting pictures showing daily changes of the façade, we are going to have the opportunity to project the pictures on the physical model. Temporal changes on the façade can be observed in-situ by controlling the timeline projected on the table.

4.3   Subplot 3 - Street Art around Gezi Park

People expressed themselves by creating street art in the demonstration area. The documentation of this data is highly chaotic because of its massive amount. Every document regarding to street art should be classified according to their properties and their location data.

In this scenario, three or four people can gather around the table, and work simultaneously to classify the documents as well as to find out their location. Because of the huge amount of the documents, it is going to be necessary to correlate the documents with surrounding environmental data. Thus, the table offers vast efficiency to organize the documents of street art within Gezi Park region.

The organized documents can be streamed online and used in mobile applications. The documents can be displayed on mobile AR interfaces mobile devices in their original locations.

5   Experiment Setup

Our experiment setup consists of a tabletop as the main interaction environment, a Kinect for gathering depth and position data from the environment, a projector for projecting the multimedia onto the environment, and a computer.

The working plane has dimensions of 100cm x 70cm. The projector and Kinect resides 170cm on top of the working plane, facing the tabletop (Figure 1). The tabletop can be divided into three areas; timeline area, 2D planar view display and 3D perspective display. The timeline is for controlling the temporal parameters of the events and multimedia. The timeline area represents the actual dates and it is controlled via a physical fiducial marker. If the user wants to scroll through time, he just simply scrolls the marker in the environment. The position of the marker represents the date, for example the leftmost position of the marker represents 28th May which is the start of the Gezi protests, and the rightmost position is the last date that we plan to observe, in between dates are divided into segments in the timeline area. 2D planar view area is for projecting the map of the Gezi Park, and also the position data of the multimedia on top of the Gezi Park. Also in this area our physical control markers which are rectangular prisms with different heights, resides. By being in different height Kinect can distinguish each one, and each of them controls different parameters. One of them controls the position and the angle of the camera that is used in the render of the area it is facing. One of them is used to determine the height of the camera. And the last one controls the position of the 2D objects. And finally in the 3D perspective area, multimedia and 3D render of the Gezi Park area is displayed.

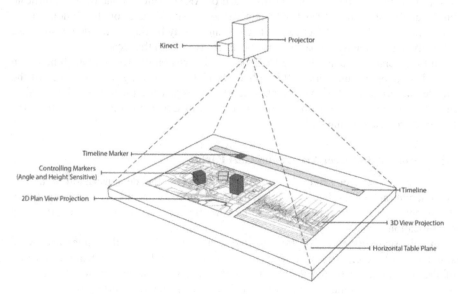

Fig. 1. Sketch of the architecture of our tangible user interface

The configuration of the setup is chosen with the aim of being cost-effective, easy to install, with parts easy to source so that everyone can build similar setup and use it. Also the reason of using projector instead of dozens of other possibilities is that we can project on top of an uneven surface which in our case is the physical model representation of Gezi Park.

The system works in 3 stages; scanning, analysis and projection. The desktop is constantly scanned for changes. In this stage the timeline marker, and control markers are controlled for any position change. These markers are paper based print-outs which are named as fiducial markers. Whenever there is a change in the position of the timeline marker, the system shows the resulting situation of the Gezi Park in the selected date. Besides these features control markers are used for sorting the images and controlling the angle of the projection plane. To access this data, the position of the timeline marker and control markers are scanned. In addition to the position data, Kinect provides the depth map of the desktop plane, so it is possible to access the height data of the control markers. This data is then used to control the height of the camera.

To realize this scenario we are using Rhinoceros 5.0 modeling software for creating virtual model of the Gezi Park[14], Grasshopper - a plug-in for graphical programming in Rhinoceros environment[15], Slingshot - a plug-in for Grasshopper for manipulating databases, MySQL for storing the multimedia and Processing[16], for

---

[14] http://www.rhino3d.com

[15] http://www.grasshopper3d.com

[16] http://www.processing.org

creating the environment. The Processing part and Rhinoceros works in parallel with via User Datagram Protocol (UDP).

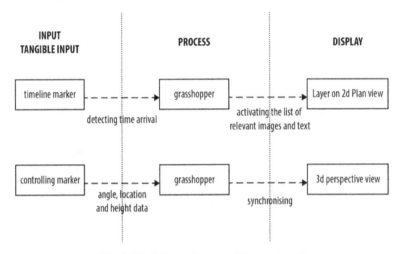

Fig. 2. Workflow of our tangible user interface

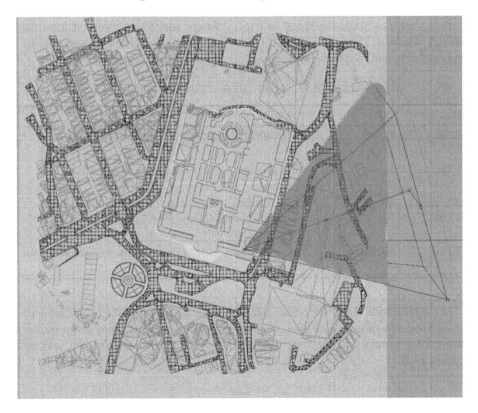

Fig. 3. Camera angle in Rhino model

Fig. 4. Street art from actual feeds overlaid on the 3D model

In the analysis stage the visual and text data are sorted and retrieved from the MySQL database. Data is stored first by the date field and all data related with the chosen date is retrieved from the database.

Table 1. Types and sources of media content

Data Source	Data Type	Properties
Photo	**Visual**	1)ID, (2)Name, (3)Keywords, (4) Location, (5)Date
Video	**Visual**	1)ID, (2)Name, (3)Keywords, (4) Location, (5)Date, (6) Duration
Processed Video	**Visual**	(1)ID, (2)Name, (3)Keywords, (4) Location1, Location2, LocationN, (5)Date1, Date2, Date3, (6) Duration1, Duration2, DurationN
Text	**Verbal**	(1)ID, (2)Keywords, (3)LocationDefault, LocationReal, (4) Content

In the projection stage, the selected data is projected on the desktop. The selected time arranges the images according to their places at that time and the control marker controls the camera and the final outcome is rendered in Rhino with the selected camera angle.

6   Evaluation of Possible Complementary Technologies

In this section, we brainstorm on technologies that are not of primary objective for our project, but that could complement it.

6.1   Data Collection and Hypermedia Linking

Crowd-sourced #occupygezi content reached giga- to terabytes over the peer to peer networks. It is hard to assess the amount of data accurately. Social media content providers such as YouTube and Twitter host as well tremendous quantities of information about it. We may require a digital asset management to store/link this with the information we would extract out of it. The first content to information links we would like to get is matching Tweets with videos through their location, which is sometimes mentioned on the metadata of the videos (title, tags, and comments). We would produce a map as in [12].

6.2   Content-Based Analysis

Having reviewed some of the video content ourselves, we believe that the following concepts and features may be extracted from the video content:

- color, that is also dependent on the time of the day and the presence of natural scenes, of smoke/fog [13];
- camera motion, that indicates whether the movie makers have smartphones or more professional stabilized equipment and/or their technique [14] also if there are some violent bursts;
- violent passages, that Acar et al. have recently been detecting through audiovisual cues [15], in #occupygezi videos people often chant, alarms and horns are quite loud and trebly;
- the presence of people or crowds, what would help to filter the content, plus their overall behavior, that Mancas et al. analyse through computational attention [16].

7   Discussion and Future Work

We have proposed scenarios that drove the design of a tangible desktop application for the organisation and understanding of crowd-sourced content (images, videos, Twitter feeds...) after the Gezi Park Protest in Turkey from Spring 2013. An experimental setup is in progress. Complementary technologies, hypermedia linking and content-based analysis, may support this project.

Acknowledgements. The authors of this paper would like to thank Christian Frisson from the numediart Institute of the University of Mons (UMONS) for his encouragement and intellectual contributions.

References

1. Liang, M., Mouraux, A., Hu, L., Iannetti, G.D.: Primary sensory cortices contain distinguishable spatial patterns of activity for each sense. Nature Communications 4, Article number: 1979 (2013), doi:10.1038/ncomms2979
2. Svanæs, D.: Context-Aware Technology: A Phenomenological Perspective. Human Computer Interaction 16(2, 3 & 4), 379–400 (2001)
3. Ullmer, B., Ishii, H.: The metaDESK: Models and prototypes for tangible user interface. In: Proceedings of the 10th Annual ACM Symposium on User Interface Software and Technology, pp. 223–232. ACM, New York (1997)
4. Underkoffler, J., Ishii, H.: Urp: A Lumious-Tangible Workbench for Urban Planning and Design. In: CHI 1999 Proceedings of the SIGHI Conference on Human Factors in Computing Systems, pp. 386–393. ACM, New York (1999)
5. Kato, H., Billinghurst, M., Poupyrev, I., Imamoto, K., Tachibana, K.: Virtual Object Manipulation on a Table-Top AR Environment. In: Proceedings of the International Symposium on Augmented Reality (ISAR 2000). Munich, Germany, pp. 111–119 (2000)
6. Ishii, H.: The tangible user interface and its evolution. Communications of the ACM 51(6) (2008)
7. Geiger, C., Fritze, R., Lehmann, A., Stöcklein, J.: HYUI: a visual framework for prototyping hybrid user interfaces. In: TEI 2008 Proceedings of the 2nd International Conference on Tangible and Embedded Interaction, New York, pp. 63–70 (2008)
8. Marquardt, N., Hinckley, K., Greenberg, S.: Cross-device interaction via micro-mobility and f-formations. In: UIST 2012 Proceedings of the 25th Annual ACM Symposium on User Interface Software and Technology, pp. 13–22 (2012)
9. Marquardt, N., Jota, R., Greenberg, S., Jorge, J.A.: The Continuous Interaction Space: Interaction Techniques Unifying Touch and Gesture On and Above a Digital Surface. In: Campos, P., Graham, N., Jorge, J., Nunes, N., Palanque, P., Winckler, M. (eds.) INTERACT 2011, Part III. LNCS, vol. 6948, pp. 461–476. Springer, Heidelberg (2011)
10. Wachs, J., Kölsch, M., Stern, H., Edan, Y.: Vision-Based Hand Gesture Applications: Challenges and Innovations. Communications of the ACM, Cover Article 54(2), 60–71 (2011)
11. Pedersoli, F., Adami, N., Benini, S., Leonardi, R.: XKin -: eXtendable hand pose and gesture recognition library for kinect. In: Proceedings of the 20th ACM International Conference on Multimedia (MM 2012), pp. 1465–1468. ACM, New York (2012), doi:10.1145/2393347.2396521
12. Leetaru, K., Wang, S., Cao, G., Padmanabhan, A., Shook, E.: Mapping the global Twitter heartbeat: The geography of Twitter. First Monday 18(5) (2013), doi:10.5210/fm.v18i5.4366
13. Calderara, S., Piccinini, P., Cucchiara, R.: Smoke detection in video surveillance: A MoG model in the wavelet domain. In: Gasteratos, A., Vincze, M., Tsotsos, J.K. (eds.) ICVS 2008. LNCS, vol. 5008, pp. 119–128. Springer, Heidelberg (2008)
14. Guo, J., Gurrin, C., Hopfgartner, F., Zhang, Z., Lao, S.: Quality Assessment of User-Generated Video Using Camera Motion. In: Li, S., El Saddik, A., Wang, M., Mei, T., Sebe, N., Yan, S., Hong, R., Gurrin, C. (eds.) MMM 2013, Part I. LNCS, vol. 7732, pp. 479–489. Springer, Heidelberg (2013)
15. Acar, E., Hopfgartner, F., Albayrak, S.: Violence detection in hollywood movies by the fusion of visual and mid-level audio cues. In: ACM MM 2013: Proceedings of the 21st ACM International Conference on Multimedia. ACM (2013)
16. Mancas, M., Gosselin, B.: Dense crowd analysis through bottom-up and top-down attention. In: Proceedings of BICS 2010, Madrid, Spain (2010)

Scenarizing Metropolitan Views: FlanoGraphing the Urban Spaces

Bénédicte Jacobs[1], Laure-Anne Jacobs[2], Christian Frisson[3],
Willy Yvart[4], Thierry Dutoit[3], and Sylvie Leleu-Merviel[4]

[1] Free University of Brussels (ULB), Centre de Recherche en Philosophie,
Avenue F.D. Roosevelt 50, 1050 Brussels, Belgium
[2] Larbits Lab, Avenue Clémentine 31, 1000 Brussels, Belgium
[3] University of Mons (UMONS), Numediart Institute, Boulevard Dolez 31,
7000 Mons, Belgium
[4] University of Valenciennes, DeVisu, UVHC – Campus Mont Houy,
59313 Valenciennes Cedex 9, France
benedicte.jacobs@larbitslab.be

Abstract. The recent decade has seen a rapid evolution in the field of digital media. Mobile devices are now being integrated into every aspect of urban life. GPS, sensor technologies and augmented reality have transformed the new generation of mobile devices from a communication and information platform into a navigational tool, fostering new ways of perceiving reality and image building. Touch sensor technology has changed the screen into a joint input and display device. In this paper we present the *FlanoGraph*, an application for smartphones and tablets designed to take benefit of the changes induced by mobile devices. We first briefly outline the conceptual background, evoking the work of some researchers in the fields of 'Non Representational Theory', mobile media, and computational data processing. We then present and describe the *FlanoGraph* through a set of use cases. Finally, we conclude discussing some techniques necessary for the development of the application.

Keywords: *FlanoGraph*, GPS, sensing technologies, database management, information retrieval, abstracting technologies, summarizing technologies, navigation, user interface design, gestural interaction, data visualization, timeline.

1 Introduction

The recent decade has seen a rapid evolution in the field of digital media. Smartphones and tablets are now being integrated into every aspect of urban life. Navigation and positioning, sensor-technologies and augmented reality have transformed mobile devices from a communication and information platform into a navigational tool, fostering new ways of perceiving reality and image building. Touch sensor technology has changed the screen into a joint input and visualization device.

In this paper we describe a series of usage scenarios for the *FlanoGraph*, an application for smartphones and tablets that is designed to take benefit of the changes

C. Gurrin et al. (Eds.): MMM 2014, Part II, LNCS 8326, pp. 11–21, 2014.
© Springer International Publishing Switzerland 2014

induced by mobile devices. The goal is to develop an application that creates choreographies with the metric of urban strolls taking as observation standard digital tracks gathered by the sensors of the smartphones of the inhabitants.

The *FlanoGraph* is based on the observation of a change in the processes of the sensing and processing of the environment by mobile technologies. The observation standard isn't anymore the eye or camera lens, but uses the sensors of smartphones, reproducing a reality based on computational analyses, carried out on datasets. A shift seems to occur at the level of cognitive representations, being increasingly augmented or enhanced by a layer of data quantifying the reality rather than representing it in an analogical way.

In section 2 we will return to this statement by briefly outlining the conceptual background, evoking the work of some researchers in the fields of 'Non Representational Theory', mobile media, and computational data processing. Then we will outline in section 3, the core of the paper, a set of use cases, with the aim to define the technical challenges that the project will have to solve. In section 4 we will discuss and describe some techniques that are necessary for the development of the application.

2   Background

The *FlanoGraph* is based on the following thoughts and intuitions:

(1) Change of paradigm induced by computational treatments of reality

As we noted in the introduction, the *FlanoGraph* starts from the assumption that mobile devices are rooted in a different approach on the reproduction of reality. What is basically different here is that we are no longer in a reproduction mode that could be described as 'sensitive' to the environment, rather it is calculative. Unlike the luminous flux captured by the camera lens and fixed on film, the signals – whether it's a luminous flux, a sound wave… – gathered by the sensors of the smartphone, are encoded and digitized, in short they are translated into a finite set of numbers.

As Lev Manovich has shown in *The Language of New Media* [1], this gives an infinite plasticity to the encoded material. The encrypted data keep no longer a link of resemblance, similarity or even causality to the received signals: data can be uncoupled, stored in databases and put through computational processes.

According to Nigel Thrift, these treatments are based on the principle of calculation and an indefeasible faith in the objectification of the world through numbers. Thrift argues in his book *Non-Representational Theory: Space, Politics, Affect* [2] that Euro-American cultures are poised to usher in a new era, a world of calculation which he called '*qualculation*'. This regime unfolds itself as a computational rationality in a constantly evolving environment that is continually calculating. Space is perceived as relative, even if it is 'riding back on the most absolute of absolute spaces' (Thrift, 2007: 106) in which each element is potentially locatable and traceable. Movement becomes eminently calculable. It is as a dynamic concept for determining spatiality. Thrift argues that this new *qualculative* world produce new kinds of spatial awareness.

(2) New apprehensions of space and time produced by mobile media

In her book *Mobile screens, the Visual Regime of Navigation* [3], media theorist Nanna Verhoeff argues that the use of smartphones, in a mobile surrounding, engages the user in practices and behaviors that change fundamentally the perception of space and time of the environment. Verhoeff stresses that the features of the smartphone – positioning, sensors, touch-screen – have transformed the mobile devices from a platform for communication and information into a tool for navigation, fostering new ways of perceiving reality which are based on a new regime of visualization. Verhoeff describes these new ways of experimenting space as 'performative cartography'. It would be a new procedural form of navigation in the environment, integrating the moving body in the process of observation. Mobile devices are creating a fundamentally different perception where mobility is an essential part of the practice of registration of spatiality. Mobile technologies become, from this point of view, a tool for recording movement.

(3) Visualizing urban life: movement, space and time in the age of computational analysis

Much has been done in the field of visual analytics of urban life. In this point we would like to focus attention on the computational data processing of visual representations of time, space and events. In *Cartographies of Time, A History of the Timeline*, Daniel Rosenberg and Anthony Grafton [4] provide an in-depth overview of visualizations and maps of events with a strong emphasis on time in the graphical representation. In short, the following items summarize the history of timelines: The earliest forms of timelines were timetables, with time and events as columns, providing a textual matrix-like representation. Throughout the 17th and 18th centuries, the advent of cartography mainly by Mercator gradually shifted timelines towards a better use of the 2D space provided by sheets of papers. These time maps make use of a visual vocabulary "to clearly communicate the uniformity, directionality and irreversibility of historical time". From the mid-nineteenth, the advent of chronography lead to possibly obtain objective representations of timed events through recording, less biased by human interpretation. Marey and Muybridge invented cameras initially for observing natural phenomena. The science of metrology would allow to analyze really small and really large scales. Since then, visualization artists and scientists seized such recording tools and crunched massive datasets to propose a visual understanding of time-based events, sometimes with derision. More recently, Powerpoints and Prezis have been aiding a wider range of people to visualize, sequence and present information, in a multimedia experience, so as to support dialogues or expressing ideas. With the rising access of multimedia databases, complementary tools are appearing, allowing to create pathways in information. Recent projects such as Timeline from Vérité.co/Northwestern University and ChronoZoom from Microsoft/Berkeley allow users to quickly craft timelines on open topics using data from online repositories of media data such as YouTube, Vimeo, SoundCloud, Flickr.

3   The FlanoGraph: Use Cases

3.1   Description

The *FlanoGraph* is a mobile application for smartphones, based on augmented reality and sensor technologies, which captures, stores and reproduces patterns of movement – digital tracks – by registering a number of sensor data. Hereby, the urban environment is not only perceived through the eye of the camera, but by the embedded sensors of the smartphone that record a wide range of physical and social processes – including sound, luminosity, temperature, speed, breathing, heartbeat. These processes are analyzed and converted into visual narratives. In this vision the smartphone is transformed into a measurement-tool. Measurability is conceived as a dynamic concept. The *FlanoGraph* is as much a tool for capturing and analyzing data, as a visualizing and editing tool. The aim is to generate new perceptions of spatiality, drawn from the kinetic properties of the movement within the geographical and social fabric. For the interaction, the *FlanoGraph* works on the HUD (Head-Up display) principle, which coordinates the gestures for accessing and using the key features of the application.

3.2   Use Cases

In this section we will discuss five scenarios. The first three examine some use cases of the *FlanoGraph* configured to act as a measuring tool. Scenarios four and five explore use cases of the *FlanoGraph* used as a visualizing and editing tool.

Fig. 1. The user accesses the sensor-interfaces by selecting one of them on the interactive wheel

3.2.1   Scenarios Exploring the Process of Sensing Urban Space by User Interface Design. The *FlanoGraph* as a Measurement Tool

As a tool for capturing and analyzing data, the smartphone is reduced to an instrument for measuring that allows the user to collect signals that are both environmental and behavioral, using a set of functions and navigation tools. To do this, the user accesses a range of 'sensor-interfaces' that appear superimposed on the smartphone screen that displays continuously the surrounding landscape, caught by the camera of the smartphone (Fig. 1). The 'sensor-interfaces' allow various recordings such as scanning the mood, measuring the heartbeat, the rhythm of movements, the frequency of travel, pollution levels, brightness and temperature. The user is free to use or ignore the sensors.

Persona

Name and occupation: Apolline, radiologist. Age: 27. Nationality and place of residence: Paris, Puteaux. Search behavior: explorative. Digital literacy: medium.

Use case 1: login, emoticon, the river and the heartbeat

Apolline is for a few days in Ljubljana. She decides to go into town and download the application *FlanoGraph*. Some of her friends had already tried and described it as a futuristic and strange "app that choreographies urban strolls". Once the app downloaded on her smartphone, Apolline starts up towards the center of Ljubljana. To activate the application, she touches the thumb of the application. A host interface pops up. It shows superimposed, the symbol of a very basic face represented by a circle, two dots for the eyes, a small horizontal dash for the nose and another dash more thicker and wider for the mouth. Behind this symbol also appears, shown superimposed, the imprint of a finger (Fig. 2).

Fig. 2. High-fidelity prototype of the login user interface

Apolline intuitively understands she is asked to login and generate an ID. So she puts her finger on the display. The scan is activated and reports an intravenous imprint of her finger. Once the task is completed, the two small pellets that are the eyes of the face are blinking and the scanner is activated again. Does the app scan the facial expression to infer her mood? Just in case, Apolline puts the smartphone against her face and smiles (Fig. 2.). Better to start the journey in a good mood, she thinks. Once the task is completed, an emoticon appears on the display. At the bottom of the screen a thin strip of the same color appears across the screen. Apolline reads the word 'surprise!' Exclamation point.

The emoticon vibrates. It look like as surprised as Apolline to found itself there. She pats on the emoticon, without really noticing, Apolline has passed successfully the formalities of identification. In all, it has taken a few minutes. She accesses the host interface of the application.

In the center of the screen she sees an action button in the form of an interactive wheel. Strange little signs or icons adorn the action button. She touches it with a finger (Fig. 3). It illuminates and the words 'light intensity and temperature activated' appear. Amused she touches another button. This time she sees the sensor icon 'heartbeat' animating a red color. Then nothing. How to continue? She taps on the red triangle. It displays the selected interface-sensor. This is great, Apolline is a cardiologist. Intrigued by this strange interface, shaped in a vermilion triangle, grafted to the left side of the screen and whose red nose is refined and lengthens and flattens, she try it out.

Fig. 3. High-fidelity test of the user interface display for heartbeat capturing

A small arrow points to the outside of the smartphone camera and shows she has to put her finger for a few seconds on the lens of the camera (Fig. 3). With the flash and the camera of the smartphone the application measures the heart rate. The measuring principle is based on an old eighties patent[1] for calculating the saturation of oxygen in the blood with each beat of the heart. A wealth of oxygen causes, in fact, a slight change in skin color that the camera analyses. The red triangle displayed on the interface is flat, it indicates a bmp 68. This correspond to the beat of her heart. Apolline restarts the operation while heading towards the road along the river Ljublanica; the finger still attached to the camera lens of the smartphone.

Use case 2: at the market light, colour, child, teddy

Arriving at the market, Apolline plunges herself into the crowd. It is very hot. It is nearly noon. A truck unloads its cargo, someone shouts. Lots of back and forth. Lots of people. Apolline feels a little unease. But regardless, she is desperate to capture the atmosphere of the market. Could the *FlanoGraph* relive the atmosphere of coming and going of people shouting, talking and walking in all directions? She selects the interface of the sound sensor to capture this sweet and slightly oppressive atmosphere that she feels uncomfortable with. The surrounding noise generates strange little green polygons on the display (Fig. 4). Apolline buries herself a little more in the crowd, noticing a young mother having her children at each hand. One of them, the one with the fuchsia t-shirt, drops his teddy. Instinctively, Apolline advances the smartphone in its direction that produces unwittingly an instant color, the 'color pix' sensor has been activated by the unusual movement of the arm (Fig. 4).

Apolline is leaving the market. Relieved and happy to let the crowd behind her, she is making some nonchalant arm gestures and then picks up speed. This sudden change of situation is soon detected and captured by the motion and orientation sensors. The screen adjusts accordingly and displays the interface of the corresponding sensor. Apolline notices a strange graffiti on the screen that shows abrupt changes of arm movements and steps (Fig. 4). She moves again her arms. The graffiti has changed. Taken by the game she had not noticed that the dog and passers-by also have left the market.

---

[1] http://patents.com/us-4258719.html

Fig. 4. High-fidelity mockups of the sound, color, motion, air quality, user interfaces

Use case 3: waiting for the bus, nitrogen dioxide

After having a drink to lessen her thirst, sitting on a bench, she goes to the bus, leaving the pedestrian area in the old center of Ljubljana and heads to the commercial area. She's struck by the noise of the traffic. The air is hardly breathable; the exhausts are everywhere, the hum of motorcycles, the passing of a bus. Apolline activates the interface sensor of the air quality to view the amount of nitrogen dioxide (Fig. 4). *Oops*, the index of air quality marks level 3. She taps on the interface to display the nitrogen dioxide, ozone and fine dust index. All the measures are expressed by a unique symbol. The level of nitrogen dioxide and ozone isn't good anymore. It exceeds the normal value, she noticed. How hot it is. Surprising for a pretty little green city like Ljubljana. Arriving at the bus stop, Apolline consults her smartphone for the schedule of the next bus.

3.2.2   Scenarios Exploring the Features of the FlanoGraph as Visualization and Editing Tool

The smartphone as visualization tool allows different viewing modes. The scenarios are focused on essentially two types of visualization. First, the mode 'synoptic view', which intends to restore snapshots, in real-time, of the collected data during the stroll.

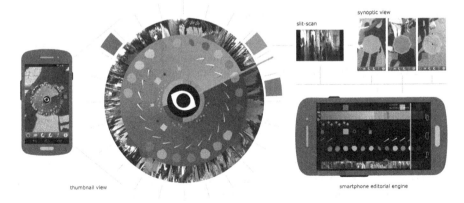

Fig. 5. Scheme of the *FlanoGraph* as visualization and editing tool

Second, the mode 'thumbnail view', which aims is to render the stroll in a more analytical and exploratory way. The user has at any time an immediate access to both viewing options. They are part of the basic functions and are included in the menu bar and are present on all interfaces of the application. The thumbnail mode allows also some editorial options. Below, we will return to these possibilities.

Use case 1: synoptic view, benches, tropical fruit juice

Arriving at the Square of the three bridges, Apolline decides to stop for a moment. She walks to a bench, sits, lays her smartphone and purse next to her and drinks a tropical fruit juice to lessen her thirst. This inactivity has been detected by the application: Apolline sitting on a bench and the smartphone, inert, laying next to her. After a while the smartphone becomes active. It shows a visualization in augmented reality. It is a synoptic view that is situating the entire journey (Fig. 6).

Synoptic views are in some ways exotic postcards. It allows having multi-dimensional, layered, visualizations of the walk. For Apolline, this feature acts primarily as a tool to compare different ambient environments based on the experiences she had during the walk.

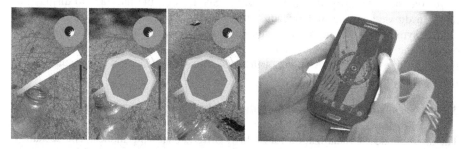

Fig. 6. High-fidelity mockups of the synoptic view interface and the thumbnail view interface

Use case 2: thumbnail view as editing mode

The purpose of the thumbnail view is to visualize the stroll in a more analytical and exploratory way. The thumbnail function displays superimposed on a smartphone or a tablet a circular graph, a disk indexed with exact time and space coordinates. The graph is composed of different concentric circles that fit together. Each circle corresponds with a data track. The outer circle visualizes the walk in a slit-scan mode. It emphasizes the movement and rhythm of the walk in the area. It shows in a loop the stroll. The inner concentric circles show the tracks of the sensor-data (Fig. 5).

Apolline hovers over the slit-scan with her finger. Small thumbs pop up. These are excerpts from the summary of the stroll. Oh, she shouts, that's what this thumbnail view visualization is meant for: to get quickly a more in-depth view of my data tracks. But I can take it further. I can use this thumbnail interface as an exploratory tool to edit my data. Apolline zooms in on one of the segments. Select. Taps. The segment now appears in a timeline. The timeline unfolds on the whole screen. Now the different tracks stack up, layered on top of each other, starting with the track of the slit-scan. They move across the screen from left to right. Apolline identifies the data of the heartbeat. They are shown superimposed on the track of the slit-scan at the bottom of the screen. And above, the track of sound and light, then the track of air quality, then the track of movement and finally the track of the mood. The timeline is interactive! Great, she thinks, this will allow me to even further explore the walk.

4   Analysis of Technologies

4.1   Intelligent Hypervideo Analysis

FlanoGraphs serve two cases in information visualization: monitoring the multimodal pathway at recording time, and assisting the browsing and comparison of multiple pathways at any time for data mining purposes. So as to compute such a representation and visualization, feature extraction techniques may help: at a low level colors, textures, contours or shapes; at a higher-level the detection of scenes (particularly natural versus urban settings as in [5]), people, the orientation of the camera (for instance to discard shots when the camera is facing the ground unintentionally). Dynamic saliency models proposed in [6] might help to compute these circular slit-scans in real-time by helping to segment the most spatial and temporal fragments. Regarding implementation, we plan to base our real-time analysis on OpenCV (running on Android and iOS platforms).

4.2   Linking Hypervideo to Web Content

The media content of which *FlanoGraphs* emerge is stored in a database of long videos and sensor signals. We plan to have *FlanoGraphers* upload their source content through our application, to worldwide hosts such as YouTube and Vimeo, under Creative Commons licenses. We aim at publishing a complementary website accessing such services through a digital asset management solution to be determined, that also stores the sensor signals data. For instance a similar project, *Walking The Edit* uses a commercial solution for that purpose; *MemoWays* [2]. We would rather choose an opensource solution for better sustainability.

Waypoints from GPS coordinates would allows maps to be reproduced and related geographical content to be accessed (information on city, neighborhoods...). Tag clouds may cluster terms of localization and mood. Linking *FlanoGraphs* to social media services such as Facebook is planned.

4.3   Interfaces and Presentation Engines

While several high-fidelity mockups have been illustrating the previous section on scenarios, Borgo et al. in [7] present extensive guidelines on how to design glyph-based visualizations. In [8] Viégas and Wattenberg illustrate how such techniques can be seized by artists for artful designs, including project *Last Clock*, by Jussi Angesleva and Ross Cooper that is inspiring for the *FlanoGraphs*. In [9] Hopfgartner et al. discuss user interface design cues for lifelogging applications. Such mobile augmented reality

---

[2] http://walking-the-edit.net and http://memoways.com

(MAR) applications yield at providing sensing, logging and displaying on the same device. Outdoor mobile augmented reality view management scenarios are surveyed and discussed in [10]. New companion devices to mobile phones for data logging are emerging: the *Autographer*, the Vicon *Revue*, both [3] costing half a thousand Euros, fitted with on-board memory and sensors, packed in a nice design. Their difference with mobile phones: these are carried on necklaces or shirt pockets, rather than held by hand. Alternatively Google Glasses also offer first-person views for *FlanoGraphs* with least effort. The implementation of the user interface is to be discussed, a choice to make among established solutions born with desktops (OpenSceneGraph), newer MAR solutions (Junaio, Metaio) and the rising HTML5 trend.

4.4   Contextualisation and Personalisation

As illustrated in the scenarios from the previous section, *cartographing* the mood of the *FlanoGrapher* is one important aspect of this project. In [11] Christian Nold had *wearers* record their Galvanic Skin Response (GSR) for an extensive period of time and in multiple cities, providing arousal- and geo-based maps. In [12], Rocca et al. propose a solution for marker-less face tracking, hence suitable for current mobile devices, waiting for depth sensing cameras and electrooculographical solutions. Facial emotion cues may be extracted as follow-up work. User-defined mood annotation through tags or avatars might be more robust and at least complementary.

5   Sources of Web and Media Content and Their Related Copyright/Copyleft Issues

All the crowd-sourced content made and recorded by the users (Larbits Lab being the inaugurating users), and all *FlanoGraphs* are released under Creative Commons licenses left for choice to the users' discretion among all that are available (http://creativecommons.org/choose/).

6   Conclusion

Smartphones, tablets are now everywhere. They have been integrated into every aspect of urban life. In this paper we introduce the *FlanoGraph* as a mobile application for mobile devices, based on augmented reality and sensor technologies, designed to take benefit of the changes induced by handheld devices. The *FlanoGraph* aims to capture, store and reproduce patterns of urban life by registering a number of sensor data. We first outlined briefly the conceptual background of the assumption on which the app is based, evoking some research in the fields of 'Non Representational Theory', mobile media and computational analysis. Further we describe the *FlanoGraph* by presenting use cases that demonstrate different usage scenarios. Finally we discussed and analyzed some techniques necessary for the development of the application.

This project has been chosen as scenario for the annual project for bachelor students in their second year at the computer science faculty of the University of Mons led by Prof.

---

[3] http://www.autographer.com and http://www.viconrevue.com

Tom Mens. In the upcoming academic year 2013-2014, these students will design the software architecture of prototype Android applications aiming at validating the usefulness and efficiency of each possible sensor type through user testing and data mining. So as to realize the *FlanoGraph*, a call for internships as qualification for Master students to obtain their degree will be opened, particularly on the design, development and evaluation aspects of mobile augmented reality and information visualization. So as to finalize this work, a project will be submitted to the eNTERFACE 2015 summer workshop on multimodal interfaces (http://www.enterface.net) so as to reach Master and PhD students from all over Europe. Subsequent consequences might be to help bridging Mons and Plzen as 2015 European Capitals of Culture.

Acknowledgements. This project is partly supported by a grand for research and development from the Belgium French and Flemish commissions. Christian Frisson's doctoral studies have been funded by the NUMEDIART research project financed through a grant of the Walloon Region of Belgium up to the end of summer 2013.

References

1. Manovich, L.: The Language of New Media. The MIT Press (2000)
2. Thrift, N.: Non-Representational Theory: Space, Politics, Affect. Routledge (2007)
3. Verhoeff, N.: Mobile Screens: The Visual Regime of Navigation. Amsterdam University Press (2012)
4. Rosenberg, D., Grafton, A.: Cartographies of Time: A History of the Timeline. Princeton Architectural Press (2010)
5. Mancas, M., Le Meur, O.: Memorability of Natural Scenes: the Role of Attention. In: Proceedings of the International Conference on Image Processing (IEEE ICIP 2013), Melbourne, Australia, September 15-18 (2013)
6. Riche, N., Mancas, M., Culibrk, D., Crnojevic, V., Gosselin, B., Dutoit, T.: Dynamic saliency models and human attention: A comparative study on videos. In: Lee, K.M., Matsushita, Y., Rehg, J.M., Hu, Z. (eds.) ACCV 2012, Part III. LNCS, vol. 7726, pp. 586–598. Springer, Heidelberg (2013)
7. Borgo, R., Kehrer, J., Chung, D.H., Maguire, E., Laramee, R.S., Hauser, H., Ward, M., Chen, M.: Glyph-based visualization: Foundations, design guidelines, techniques and applications. In: Eurographics State of the Art Reports, EG STARs, pp. 39–63. Eurographics Association (May 2013)
8. Viégas, F.B., Wattenberg, M.: Artistic data visualization: Beyond visual analytics. In: Schuler, D. (ed.) HCII 2007 and OCSC 2007. LNCS, vol. 4564, pp. 182–191. Springer, Heidelberg (2007)
9. Hopfgartner, F., Yang, Y., Zhou, L., Gurrin, C.: User interaction templates for the design of lifelogging systems. In: Hussein, T., Paulheim, H., Lukosch, S., Ziegler, J., Calvary, G. (eds.) Semantic Models for Adaptive Interactive Systems. Human- Computer Interaction Series, pp. 187–204. Springer, London (2013)
10. Kayalar, C., Balcisoy, S.: Look at the same screen, see differently: A discussion of view management methods in mobile augmented reality scenarios. In: Mobile Augmented Reality: Design Issues & Opportunities Workshop of Mobile HCI 2011 (2011)
11. Nold, C. (ed.): Emotional Cartography - Technologies of the Self. Creative Commons Attribution-NonCommercial-ShareAlike 3.0 license (2009), http://emotionalcartography.net
12. Rocca, F., Ravet, T., Tilmanne, J.: Humaface: Human to machine facial animation. In: Dutoit, T. (ed.) QPSR of the Numediart Research Program, UMONS/numediart, vol. 5, pp. 1–5 (March 2012)

Scenarizing *CADastre Exquisse*: A Crossover between Snoezeling in Hospitals/Domes, and Authoring/Experiencing Soundful Comic Strips

Cédric Sabato[1], Aurélien Giraudet[1], Virginie Delattre[1], Yves Desnos[2],
Christian Frisson[3], Rudi Giot[4], Willy Yvart[5], François Rocca[3],
Stéphane Dupont[3], Guy Vandem Bemden[3],
Sylvie Leleu-Merviel[5], and Thierry Dutoit[3]

[1] L'Art-Chétype ASBL, 10 Cours de l'âne barré, B-7000 Mons
[2] Université de Strasbourg, Laboratoire de Psychologie des Cognitions,
12 Rue Goethe F-67000 Strasbourg
[3] University of Mons (UMONS), numediart Institute, Bd Dolez 31, B-7000 Mons
[4] LARAS lab, ISIB, Rue Royale 100 B-1000 Brussels
[5] Université de Valenciennes et du Haut Cambrésis, Laboratoire DeVisu,
Campus Mont Houy, F-59313 Valenciennes
cedric@lart-chetype.eu, christian.frisson@umons.ac.be

Abstract. This paper aims at providing scenarios for the design of authoring and experiencing environments for interactive soundful comic strips. One setting would be a virtual immersive environment made of a dome including spherical projection, surround sound, where visitors comfortably lying down on an interactive mattress can appreciate exquisite corpses floating on the ceiling of the dome, animated, with sound, dependent of the overall behavior of the visitors. On tabletops, creators can generate comic-strip-like creatures by collage or sketching, and associate audiovisual behaviors and soundscapes to these. This creation system will be used in hospitals towards a living lab comforting patients in accepting their health trip. Both settings are inspired by snoezelen methods. These crossover scenarios associate a project by L'Art-Chétype retained to be featured for Mons 2015 EU Capital of Culture and other partners aiming at designing an environment for experiencing/authoring interactive comic-strips augmented with sound.

Keywords: virtual environments, collaborative media authoring, interactive comic strips, snoezelen.

1 Introduction and Context

CADastre Exquisse is a mashup name combining *"CAD"* as in computer aided design (the technology inherent in this project), *"astre"* reminiscent of sky sceneries in domes (one of the settings of this project), put together as *cadastre* or land registry, *exquisse* (French for "sketch", the creative practise gathering users for this project), also "cadavre exquis" (French for "exquisite corpses", a playful practice inviting for collaborative creations).

C. Gurrin et al. (Eds.): MMM 2014, Part II, LNCS 8326, pp. 22–33, 2014.
© Springer International Publishing Switzerland 2014

We will start by introducing the context: comic-strips (1.1), the relation of sound to image (1.2), snoezelen and immersive environments (1.3). Scenarios will follow (2) in two settings: in domes (2.1) then hospitals (2.2). We will assess related technologies 3: comic-strip analysis (3.1), hypercomics and web content (3.2), required user interfaces (3.3) and personalization (3.4). We will then conclude and open with future work (4).

1.1 Anatomy of Comic Strips

Scott McCloud [24] is one of the most revered authors of comic strips having theorized the analysis of these and their methods of creation, that he presented throughout three comic strip books: *Understanding* (1993), *Reinventing* (2000) and *Making* (2006) *Comics*. In the second book, he perspectives for the evolution of comic strips, notably the concept of "infinite canvas" which pushed comic strips beyond their frames, enabled at the time of writing of the book by the technologies of Internet, then more recently by mobile technologies.

Research works around comic strips in the field of information technologies come in majority from Asian countries (Japan, Korea, China, Singapore) then the US, incidentally countries where comics strips and equivalents (mangas, ...) have a strong cultural imprint. In Belgium, Pascal Lefèvre obtained the first PhD thesis around comic strips in the Flemish Region at the Faculty of Social Sciences at KU Leuven. He evaluated the current status of comic strips in Belgium, verbose in statistics [21], published by Smart ASBL, a non-profit association for creative freelance people. A recent reference [25] by the CETIC research center elaborates on methods for the digitization and recognition of handwritten letters and words. Walloon Belgian research project *SonixTrip* shared by some of the authors of the current paper aims at sustaining practitioners of comic books in Belgium and beyond and their craft, by shifting its paradigm to a new practice, now paper-less, interactive, and augmented with sound.

1.2 Augmenting Image with Sound

In one of his books [10], electroacoustic music composer and professor Michel Chion studies the relationship between sound and image. For movie analysis he coined new terms, for instance: *synchresis [is] an acronym formed by the telescoping together of the two words synchronism and synthesis: "The spontaneous and irresistible mental fusion, completely free of any logic, that happens between a sound and a visual when these occur at exactly the same time"*. Blesser and Salter provide an in-depth overview on how the human audition through age and evolution respond to spatialized sound [4]. New Zealand creators of the application Booktrack[1] for iOS filed a patent of a vast scope on the addition of sound to a narrative content. French company Byook[2] aims at providing a similar solution.

---

[1] http://www.booktrack.com

[2] http://www.byook.fr

1.3 Snoezelen and Immersive/Participatory Environments

Van der Borght's book [12] is a manual on how exquisite corpses, among other crafts born with surrealists, can benefit to children education. Porteous et al. proposed a social network interface for interactive narratives using a medical drama 3D animation as testbed [26]. Cavazza et al. provide feedback on how virtual narratives can facilitate the communication between patients and doctors [9]. Some of the current authors have used sensory integration devices for autistic people therapy [13]. Coming from the Netherlands in the 1970's, snoezelen is a controlled multisensory environment allowing therapies for people suffering from autism or disabilities. Kronenburg has studied all along his career the story of flexible architecture [19], including mobile housings such as domes. Grau elaborates on how these can provide artful immersion [15]. Some of the current authors produced in the past an evaluation of dome technologies to produce behavioral installations [14]. The current authors are influenced by Ernesto Neto's *Circleproteitemple* (2010) and Yayoi Kusama's *Metamorphosis* (2006).

2 Scenarios of CADastre Exquisse

Two sub-scenarios illustrate different systems of the *CADastre Exquisse* environment, in different settings. Scenario 2.1 focuses on an entertainment setting for participatory enjoyment of exquisite corpses. Scenario 2.2 focuses on the authoring system, that is also used in the former, but here applied to collaborative creation in living labs.

2.1 Snoezeling in the Dome

Malvina and Yves have both been studying psychology in France as part of their graduate studies. In 2015, they come to Brussels to pay a visit to their expatriate childhood friends. These bring them by train to Mons, 60 kilometers away. The town has been elected as one of this year's European Capital of Culture. The city is full with artful installations, games, performances. At Place du Parc, near the UMONS headquarters, a huge dome is installed. Some friendly people at the entrance invited them to enter, somehow as illustrated in Figure 1.

They enter the dome. Through the small entrance corridor, they first see people laying down on some mattress. The mattress looks very comfortable, like a cocoon. They approach and progress up to the inside of the dome. Malvina notices that the mattress is moving in a kind of living way, wobbling. They hear sounds that move around the dome. Soundscapes and sudden but soothing sound effects. The atmosphere, even if dim, looks inviting.

Influenced by the people laying down staring at the ceiling, they check what can draw their attention. An ocean of stars rotates very slowly around a center point of the dome. It reminds them of specific renderings of night sceneries, when "star trails" are mesmerized using a camera set with an aperture of several hours. Figure. 2 illustrates it. They decide to lay down as well so as to better enjoy it.

Fig. 1. Side/top views of the dome: physical smaller scale rendition for brainstormings

Some kind of monsters float as constellation overlaid on the night-like sky. They look funny, like impossible animals. Contours fit, but the association of heads, limbs and bodies is uncanny. They are animated, comic sounds burst every now and then. It seems that the floor vibration is connected to it as well. Yves points to one of the monster, and waves his hand to mimic the absurd movement of the creature. The monster suddenly changes its trajectory. How is it possible? Malvina waves at another, same effect.

They look at the periphery of the dome: some people are gathered around cardinal points. Monsters seem to climb to the ceiling from these points. Yves decides to go there take a look. He sees a kind of screen where the people gathered there create the monsters, as in Figure 3.

2.2   Snoezeling in Hospitals

Stephane is an 11-year old young chap that just arrived to the hospital. He complained repeatedly to his mother that his tummy aches. He has a very special way to express it, as he is quite introverted. Stephane suffers from autism.

Stephane and her mother meet the doctor whom they got an emergency appointment with. He's accompanied with a smiling guy. "Hello Stephane", says the doctor, "I'm Walter, this is Cédric. He's gonna help you feel more comfortable before we can remove your tummy ache". They move to another room. Stephane is impressed, it looks almost like his bedroom, with cosy lights and the stars shining over the ceiling as he likes in his own. Walter invites him to sit on some cloud-like sofa. It is comfier than the car seat! Stephane almost forgets his body. Cédric shows him some kind of table. There is a screen on it. It looks like monsters are displayed on it. How funny!

"Would you like to show us your favorite monster?" asks Walter. Stephane nods frantically, happy, feels like an invitation to play. Stephane notices a blank canvas in the center and some piles on the sides of the screen. Feels like his bedroom floor when he's playing with his toys. On the top-right corner there are already some monsters. He points one of them, and notices that when he inadvertently drags his finger that the creature pops up on the central canvas. On

Fig. 2. *Star Trails Over Solar Panels* taken by Mike Lewinski on April 13, 2013 in Embudo, New Mexico, US with a Sony NEX-5N camera. Available under a Creative Commons Attribution License (http://creativecommons.org/licenses/by/2.0/deed.en) on Flickr (http://www.flickr.com/photos/ikewinski/8646527486/).

the bottom-right corner, there are objects. He touches one of them: it produces a funny sound. He touches another one next to it: the sound is quite similar. He likes it. He drags it to the canvas, accustomed to the interface. By hovering the creature, it seems that the limbs get emphasized. He drops the sound onto a joint. The creature is now animated, every time Stephane touches the joint it produces this same sound. On the top-left corner, there are nice backgrounds and sceneries. One looks as if there was a bright blue sky on a dark house and garden. Is it really possible? It doesn't feel natural. Walter notices that Stephane understood it. Stephane opts for a background with nice yellow and green lights. He drops it onto the canvas. He's satisfied with the result. It seems they've just created the first frame of an interactive comic strip.

Walter invites Stephane to lay down on the cloud-like sofa: "Would you want to see your monster fly in the sky?". Stephane is amused, he accepts. As Stephane lays down, Cédric manipulates the screen. The ceiling now displays the monster over the shining stars. Stephane points it: the joint moves and it produces the sound. There is also a nice ambient music playing back. And a nice smell. Stephane falls into sleep.

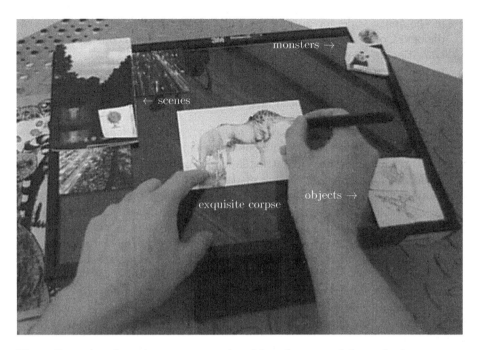

Fig. 3. Paper-based mockup on an actual multitouch screen of the authoring system

3   New Technologies Extending Comic Strips and Allowing Immersive and Participatory Environments

3.1   Intelligent Comic Strip Analysis

If comic strips compose visuals in space, cinema composes these as well in space but first and foremost in time. In [22], Lev Manovich establishes rules of actualized cinema which he coins "soft cinema" [23] :

- "algorithmic cinema" automates the layout on the screen, the number of windows and the content;
- "macro-cinema" borrows the paradigms of computer science: windows of variable ratios and sizes;
- "multimedia cinema" considers video as one type of representation among others: 2D animation 2D, motion animation, 3D scenes, diagrams...;
- "database cinema" sources media fragments from a larger database.

Plenty of research works have been studying "reverse-storyboarding", automated movie summaries styled like comic strips. Their graphical rendering can be affected in different steps:

- at the image scale, for instance by using phylacteries replacing subtitles of dialogues, or by replacing backgrounds by stylistic effects connoting the fast transitioning of events, or confusion [16];
- at the image scale, non-realistic photo rendering reviewed in [20] , also used in movies such as *A Scanner Darkly* (Richard Linklater, 2006);
- at the movie scale, the spatial layout of keyframes into panels [5].

Other works start from 3D models of video games and logs from the user behavior at game play and generate comics-styled summaries [30]. Quite fewer works go the opposite way: comic strips as source media. An Asian team proposed a workflow of automated extraction of panel through blob detection and text recognition, for the archival in an XML file format and the abstraction of the mobile device display type [2]. Chu et al proposed a system for transforming animations into comics to make their browsing more efficient and pleasurable on portable devices [11]. The authors of [8] propose an automated panel layout technique in manga style, weighting the importance of key-panels by their size in the rendering.

An inverse approach would be opted in our project. To do so we plan to integrate such image-based and audio-based feature extraction and segmentation into the MediaCycle framework[3].

3.2   Hypercomics and Web Content Copylefts

Exquisite corpses may be connected to ontologies of species, or even crypto-zoology, but this isn't our prime objective. All the exquisite corpses might be released on a web museum under carefully-chosen Creative Commons licenses.

3.3   Interfaces and Presentation Engines

Solutions for Media Authoring. Here follows a summary of functionalities and realizations that Scott Mc Cloud wishes would happen on tomorrow's interactive comic strips, as listed in his page about "infinite canvas"[4]:

- an alternative to vertical seeking bars that "hypercomics" inherited from HTML pages, an evolution motivated by the current omnipresence of tactile devices.
- new creative environments
- a "reader-centered" approach, with intuitive interaction cues, the ability to save the reading status and to go back to it later, download of the whole comic strip for offline reading...

Ruben R. Puentedura cites the Tarquin Engine et Prezi as potential tools for the authoring of comic books borrowing the concept of "infinite canvas" from Scott McCloud, in [27]. Celtx is an opensource framework for the management of

---

[3] http://www.mediacycle.org
[4] http://scottmccloud.com/4-inventions/canvas/

scripts and storyboards, for theater, cinema... and comic books [6][5]. Simplified version of Its desktop applications based on the Mozilla engine are free, mobile applications are paid. A workflow for reading and authoring of comic strips proposed in [1] lays out sheets of paper containing "high-level" elements of the story (characters, context) over a multitouch table associated with a projector that overlays "low-level" elements (objects, text) on the sheets. One of the authors tested the Virage platform [3] in a testbed of fuzzy temporal sequencing in a digital arts project. The Virage platform, now forked as i-score[6], allows to sequence multimedia events while assigning their cueing in to variable signals, from sensors or programming. We are considering using this solution for semi-prepared exquisite corpses.

Interaction Cues for the Authoring Interface. iOS application *The Three Little Pigs* is a game targeted to children, interactively tells the famous story, and drifts away from comic strips: a phylactery appears next to characters once touched thus animating. This game gets inspired from a concept popularized by the game Angry Birds born on mobile phones: direct manipulation of characters or objects of the scene so as to alter their trajectories then animated with physical effects. For instance, in the video[7], the young player touches the wolf which is so much hungry that he drives his lorry too fast, then the youngster swipes her finger imprinting a throwing trajectory, leading the wolf to end up in a car crash. Direct manipulation of objects in videos, complementary to temporal navigation, has been a research track for a decade. Christian Brockly's master thesis [7] provides a comprehensive state-of-the-art. Objects in the scene are "clicked" then moved and thus follow the gestural trajectory as long as such a trajectory is possible from the whole video content. implements this concept. SensorComix [29] proposes a visualization in a comic strip fashion of short text messages, augmented by emoticons generated by gestures of types *knock*, *shape*, *rub*, *tap*; through an inertial, magnetic field and capacitive sensor, placed on the backside of the mobile phone.An experiment with navigation in a game-book through gestures of a greater bounding box sensed by a Kinect camera has been proposed in [18]. Such an interaction technique is impossible with tablet devices when considering full-body gestures, but likely to work with heads and hands, depending on the aperture of the embedded camera, awaiting for the availability of future 3D sensors. Through other modalities of interaction than gestures and touch, Ozge Samanci proposes interactive comic strips using global positioning entitled "GPS comics" [28].

We plan to use a multitouch tabletop interface for augmented comic strip authoring. Additional pen-based interaction might be considered, especially for expert sketching, while recent usages of smartphones and tablets and their dedicated drawing applications shift the craft from pen to fingers. In public-based settings, styluses have the disadvantages to be hazardous for visitors, breakable or damaging the screen, lost.

---

[5] http://www.celtx.com

[6] http://www.i-score.org

[7] http://www.youtube.com/watch?v=WcIy9O344bI

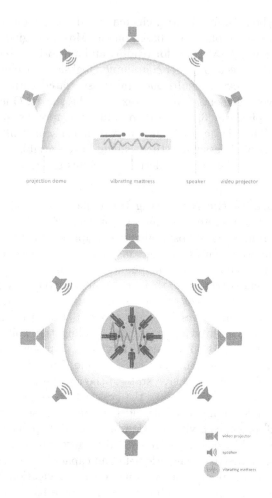

Fig. 4. Side/top views of the dome, vertorial sketch version

We will try Alexandre Quessy's Toonloop[8] and Michal Seta's Marionnect[9] for prototyping, both opensource software developed for the Metalab living lab by Société des Arts Technologiques (SAT) in Montreal, Quebec, Canada.

Experiencing an Immersive Audiovisual Rendering. An hemispherical projection technique using spherical mirrors described in a report from some of the current authors [14] might be used for the dome setting of this project. Figure 4 illustrate the required architecture for the dome.

---

[8] http://toonloop.com
[9] https://github.com/sat-metalab/marionnect

3.4   Contextualisation and Personalisation

In the dome setting, we plan to use a ceiling camera with wide lens so as to track the overall behavior of the participating audience using computer vision-based attention algorithms. The quantity of activity of all individuals would be mapped to the slow rotation of the audiovisual rendering around the vertical radius coming from the zenith of the dome. The quantity of activity of each individual would be mapped to the volume of the corresponding sound and to a magnification of its glyph-based representation.

An US team conducted user tests with gaze following to validate the efficiency of the spatial layout of comic strips [17]. Such a setup, for instance making use of the cheaper and less invasive Tobii Eye Trackers[10] available at some of the authors' lab, would allow to evaluate the authoring interface tailored for our scenarios.

Possible limited editions and sustainable amounts of souvenirs may be generated, such as t-shirts and posters.

4   Conclusion and Future Work

In his page about "infinite canvas"[11], Scott McCloud wrote: *"Comics strips work fine online and have been one of our biggest success stories of the last decade. They don't need any help from mad scientists."* Be them healthcare practitioners or user interface designers, we hope to provide a solution that uncovers new territories while preserving and respecting current practices. We aim that our solution supports collaborative creativity, for life improvement in living labs between patients and caretakers, and for social happiness between art creators and simply curious people.

Société des Arts Technologiques (SAT) and Hospital Sainte-Justine from Québec, Canada have already formed the Metalab living lab sharing close interests with our project. Similarly, hospital Ambroise Paré in Mons has shown interest for collaborations. We hope to collaborate altogether, as already discussed, so as to pursue the Mons-Montreal partnership already established by *L'Art-Chétype*.

Acknowledgements. Cédric Sabato, Aurélien Giraudet and Virginie Delattre's project *Art-Act-Need* has been selected to be featured for Mons 2015 EU Capital of Culture[12].

Christian Frisson, Guy Vandem Bemden and Thierry Dutoit will be partly funded by the SonixTrip GreenICT project from the Walloon Region of Belgium from autumn 2013.

---

[10] http://www.tobii.com

[11] http://scottmccloud.com/4-inventions/canvas/

[12] http://www.mons2015.eu

References

1. Andrews, D., Baber, C., Efremov, S., Komarov, M.: Creating and using interactive narratives: reading and writing branching comics. In: Proceedings of the 2012 ACM Annual Conference on Human Factors in Computing Systems, CHI 2012, pp. 1703–1712. ACM, New York (2012)
2. Arai, K., Tolle, H.: Automatic e-comic content adaptation. International Journal of Ubiquitous Computing (IJUC) 1(1) (2010)
3. Baltazar, P., Allombert, A., Couturier, R.M.J., Roy, M., Sédès, A., Desainte-Catherine, M.: Virage: une reflexion pluridisciplinaire autour du temps dans la création numérique. Actes des Journées d'Informatique Musicale, JIM (2009)
4. Blesser, B., Salter, L.-R.: Spaces Speak, Are You Listening? - Experiencing Aural Architecture. The MIT Press (2007)
5. Boreczky, J., Girgensohn, A., Golovchinsky, G., Uchihashi, S.: An interactive comic book presentation for exploring video. In: Proceedings of the SIGCHI Conference on Human Factors in Computing Systems. ACM (2000)
6. Borst, T.: Mastering Celtx. Course Technology PTR (2012)
7. Brockly, C.: Evaluation of direct manipulation techniques for in-scene video navigation. Master's thesis, Media Computing Group, Computer Science Department, RWTH Aachen University (2009)
8. Cao, Y., Chan, A.B., Lau, R.: Automatic stylistic manga layout. ACM Transactions on Graphics (Proc. of SIGGRAPH Asia 2012) 31 (2012)
9. Cavazza, M., Charles, F., Gersende, G.: How virtual narratives can improve patient-doctor communication. In: Proceedings of the ACM CHI 2013 Workshop, Patient-Clinician Communication: The Roadmap for HCI (2013)
10. Chion, M.: Audio-Vision: Sound on Screen. Columbia University Press (1994)
11. Chu, W.-T., Wang, H.-H.: Enabling portable animation browsing by transforming animations into comics. In: Proceedings of the 2nd ACM International Workshop on Interactive Multimedia on Mobile and Portable devices, IMMPD 2012, pp. 3–8. ACM, New York (2012)
12. der Borght, M.V.: Jouer avec Dada, jouer avec les surréalistes. Editions Aden (2010)
13. Desnos, Y., Segond, H., Maris, S.: Sensory integration devices as a new modality of therapeutic care for autistic people with mental retardation. In: Innovative Research In Autism (2009)
14. Filatriau, J.-J., Frisson, C., Reboursière, L., Siebert, X., Todoroff, T.: Behavioral installations: Emergent audiovisual installations influenced by visitors' behaviours. In: Dutoit, T., Macq, B. (eds.) QPSR of the Numediart Research Program, numediart Research Program on Digital Art Technologies, vol. 2, pp. 9–17 (March 2009)
15. Grau, O.: Virtual Art: From Illusion to Immersion. Leonardo. The MIT Press (2003)
16. Hwang, W.-I., Lee, P.-J., Chun, B.-K., Ryu, D.-S., Cho, H.-G.: Cinema comics: Cartoon generation from video stream. In: Proceedings of GRAPP 2006 - Computer Graphics Theory and Applications (2006)
17. Jain, E., Sheikh, Y., Hodgins, J.: Inferring artistic intention in comic art through viewer gaze. In: Proceedings of the ACM Symposium on Applied Perception, SAP 2012, pp. 55–62. ACM, New York (2012)
18. Kistler, F., Sollfrank, D., Bee, N., André, E.: Full Body Gestures Enhancing a Game Book for Interactive Story Telling. In: Si, M., Thue, D., André, E., Lester, J.C., Tanenbaum, J., Zammitto, V. (eds.) ICIDS 2011. LNCS, vol. 7069, pp. 207–218. Springer, Heidelberg (2011)

19. Kronenburg, R.: Flexible: Architecture that Responds to Change. Laurence King Publishers (2007)
20. Kyprianidis, J., Collomosse, J., Wang, T., Isenberg, T.: State of the 'art': A taxonomy of artistic stylization techniques for images and videos. IEEE Trans. on Visualization and Computer Graphics PP(99), 1 (2012)
21. Lefèvre, P., Salvia, M.D., Nakano, H.: Bande dessinée et illustration en Belgique: état des lieux et situation socio-économique du secteur. SMartBe Association Professionnelle des Métiers de la Création ASBL (2010)
22. Manovich, L.: The Language of New Media. The MIT Press (2001)
23. Manovich, L., Kratky, A.: Soft Cinema: navigating the database. MIT Press (2005)
24. McCloud, S.: Reinventing Comics. HarperCollins (2000)
25. Ponsard, C., Ramdoyal, R., Dziamski, D.: An OCR-enabled digital comic books viewer. In: Miesenberger, K., Karshmer, A., Penaz, P., Zagler, W. (eds.) ICCHP 2012, Part I. LNCS, vol. 7382, pp. 471–478. Springer, Heidelberg (2012)
26. Porteous, J., Charles, F., Cavazza, M.: A social network interface to an interactive narrative. In: Proceedings of the 2013 International Conference on Autonomous Agents and Multi-agent Systems, AAMAS 2013, pp. 1399–1400. International Foundation for Autonomous Agents and Multiagent Systems, Richland (2013)
27. Puentedura, R.R.: The infinite canvas reloaded: Digital storytelling, webcomics, and web 2.0. In: 2009 NMC Summer Conference Proceedings (2009)
28. Samanci, O., Tewari, A.: Expanding the comics canvas: Gps comics. In: Proceedings of the 4th International Conference on Fun and Games, pp. 27–34. FnG 2012. ACM (2012)
29. Setlur, V., Battestini, A., Sohn, T., Horii, H.: Using gestures on mobile phones to create sms comics. In: Proceedings of the Fourth International Conference on Tangible, Embedded, and Embodied Interaction, TEI 2010, pp. 217–220. ACM, New York (2010)
30. Shamir, A., Rubinstein, M., Levinboim, T.: Generating comics from 3d interactive computer graphics. IEEE Comput. Graph. Appl. 26(3), 53–61 (2006)

An Interactive Device
for Exploring Thematically Sorted Artworks

Aurélie Baltazar[1], Pascal Baltazar[1], and Christian Frisson[2]

[1] Les Baltazars, l'Arboretum SCOP, 99A Bd Descat, F-59200 Tourcoing
[2] University of Mons (UMONS, numediart Institute, Bd Dolez 31 B-7000 Mons
les@baltazars.org, christian.frisson@umons.ac.be

Abstract. This Mediadrom artful post-TV scenario consists in sketching the user interface of an interactive media content browsing system for exploring thematically sorted artworks, from the art field of plastic theater, merging art pieces at the intersection of the visual and the performing arts. Combining a touchscreen and an hypermedia browser of image and video content with expert annotations, this system can be installed in venues such as museum and media libraries, and performance spaces as satellite installation to plastic theater performances.

Keywords: hypermedia browser, multimedia annotation, interactive installation, plastic theater.

1   Introduction, Artistic Intention

This paper focuses on the work of the Baltazars[1], a duo of artists producing art forms between the performing and the visual arts. This artform, coined "plastic theater" [14], after Tennesse Williams' first usage of the term [7], uses the materials of the visual arts (named "arts plastiques" in French) in the context in performing arts. In concrete terms, their work is often presented in theaters, while no human figure is present on stage, nor any text visible or audible. Thus the spectators perception can be focused on the movements and variations of sensory materials such as light, sound, smoke, air movements, etc. All these materials are carefully arranged in time, in order to create a dramaturgy of matter, that brings the nature of this work, from the realm of the visual arts, towards those of the performing arts.

Our aim in this project is to design a collection management system that could be browsed both through a touch screen in venues such as festivals, media libraries and museums; but as well on a website, with standard keyboards and mice, or other peripherals left to the discretion of the remote visitors.

As any other artwork, the experience of this work benefits from being mediated to the audience, after or before the actual experience of the work itself.

---

[1] Image and video samples are available on
https://www.facebook.com/TheBaltazars.

C. Gurrin et al. (Eds.): MMM 2014, Part II, LNCS 8326, pp. 34–43, 2014.
© Springer International Publishing Switzerland 2014

Furthermore, the fact that this work drifts apart from the mainstream artistic production makes this need for mediation even stronger. In addition to implying human mediating agents, the current proposal would allow a more action-oriented and proactive mediation, based on the spectators needs, desires and curiosity.

The device proposed here could then, on the one hand prepare the spectator to the experience of the actual work, e.g. by clearing some a-priori about such a non-narrative/non-figurative type of work, by showing that similar works already existed for quite a long time, for some of them. On the other hand, the device would also allow him to go further the actual experience, by leading his own curiosity along the exploration, and hopefully helping him discovering some works and artists he had not heard of...

The content of the exploration would then be composed of such elements as a set of works by other artists, references to techniques and artistic materials, natural phenomena, etc. All of these would be selected for their connection to the work and statement of the Baltazars, as the purpose of this device is not a general introduction to art, but is rather intended to focus and relate to this specific work. In other words, the corpus of works presented here are done so along a certain artistic perspective.

2   Context, Similar Works

Paul Otlet (1868-1944), the founder of Mundaneum[2], a paper-based ancestry of Google, among other notable achievements, wrote two treaties on how to classify knowledge, *Traité de Documentation* (1934) and *Monde: Essai duniversalisme* (1935), long before nowadays online search engines. He theorized a whole workflow on documentation, including: the Universal Decimal Classification, and 35-inch cards to label items in catalogs. What could be written on such cards? Here follow examples of criteria used for the organization of media data:

- metadata usually provides factual data on media items, which can be generic such as: author, title, album or collection, location or geographical origin, often the date of creation or publication; or specific to the media collection;
- semantic data, such as tags, add subjectivity to media elements, and can often be organized into ontologies, which provide a relational structure to data that cant be classified into mutually-exclusive categories;
- by means of computer analysis and signal processing, content-based criteria are extracted from the data and stand as objective numeric data; from low-level criteria close to the signal properties as evocative as the algorithm designed to output these can be, to higher level including perceptual criteria adapted to the human perception: mean color or shape for images, motion orientation for videos, energy or loudness in audio.

---

[2] http://www.mundaneum.be

In [4], some of the authors evaluate how scientific works on multimedia information retrieval, particularly content-based similarity, can foster new practices in digital or new media arts. One artist and scientist that has been using such technologies as a theme for his artistic works is George Legrady[3], notably since *Pockets Full of Memories* (2001-2007), an installation where participants can scan everyday objects which are then virtually organized in a projected Kohonen self-organizing map, based on the textual description inputted by the visitors. This installation is described in the book *Database Aesthetics*, edited by Victoria Vesna, among other similar works [15]. In an other artistic field, American choreographer Bud Blumenthals project *Dancers!* proposes an online interactive website for browsing and comparing improvised recordings of dancers [4], developed in collaboration with scientists from the field of multimedia information retrieval [13]. The recent works of Lev Manovich combining information retrieval and media arts are labeled "cultural analytics" [8,9].

3   Description of the Scenario

3.1   Context and Setup

At venues such as festivals, theatres or media libraries, where some works of the Baltazars are possibly presented, a room is dedicated to the device described in this article. The room is darkened, there is a couch in the middle of it, facing a coffee table on which a touch screen is laid down, and a projection screen on the wall, with a pair of loudspeakers.

3.2   Interactive Content Exploration

When a visitor sits down and first looks at the touch screen, he sees the interface presented in Figure 1. At this first stage, the visitor is able to read a general presentation of the Baltazars' work, in which several keywords are marked in color, and will act as hyperlinks when tapped, as we will explain below. Clouds of thumbnails are floating around the central text-based block. Each of the thumbnails will reconfigure the screen of the application as described in the next figures. The bigger thumbnails are links to works from the Baltazars.

The visitor can then tap on:

- the Baltazars' works (bringing him to the screen described in Figure 2),
- any other items (leading to other artists' works as shown in Figure 3),
- the colored words, linking to the related themes (as presented in Figure 4)

Lets assume that the visitor taps on the *Nocturnes* thumbnail on the left side. The main block would then get smaller, go to the bottom of the page, and fill with some new content, as described in Figure 2. The floating thumbnails would then also smoothly re-arrange, disappearing for some of them, and letting other come on the front stage.

---

[3] http://www.georgelegrady.com
[4] http://www.dancersproject.com/interactive/

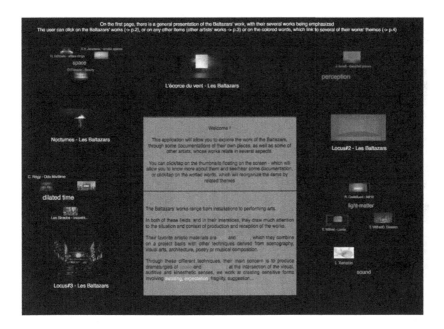

Fig. 1. The welcome screen of the application

The visitor would then be presented a short explanation of the main features of the work, and a more specific description of the particular sequence that is available by tapping on the central image (with the play button). This action would then trigger the projection (on the projection screen) of this sequence: a video documentation or a diorama of the presented work. In the body of the text - just like in the "welcome screen" - some words are colored. These are the main themes of this work and specific sequence. The elements related to these themes are displayed as clouds, arranged by theme and color around the central block. On the bottom of the main block, a space is devoted to a collection of specific hyperlinks that may allow the user to go further the short presentations by consulting online articles, websites, etc. Once the visitor is done with consulting this particular sequence or work, he can tap on another thumbnail, which will bring this very element to the center. Lets say that they tap on the top-right thumbnail, linking to Heiner Göbbels piece called *Stifters Dinge*. Thats exactly what Figure 3 describes thereafter.

The content of the main block has now been changed to fit Heiner Göbbels work, by following the same principles and design than in Figure 2. New thumbnails are now displayed and re-arranged in relation to the theme of the newly selected element (some of which being common with the previous element). If a lot of elements relate to the chosen, the most related ones will be displayed in a bigger size (with their titles), then less related being displayed as a cloud of micro-thumbnails... If the user wants to check something in the previous work, that brought him there, he can always go back to its previous choice by clicking on the arrow on the top left corner. If he is particularly interested in one of the

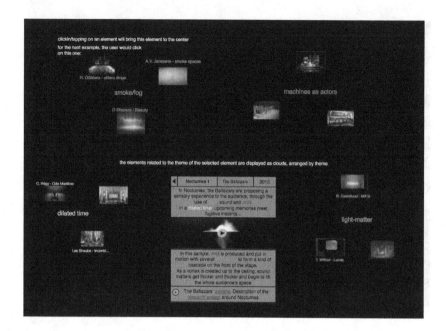

Fig. 2. Focusing on a specific work from the Baltazars (*Nocturnes*)

themes of the selected work, he could learn more about it, either by clicking on the colored word in the text, or in the middle of one of the clouds. Lets consider our visitor is interested by the concept of using machines as actors , this would bring him to the state described in Figure 4.

On the side of the main block, a kind of drawer will then swipe out, and display some specific information about the chosen theme. The visitor can now read on this new block a description of what this theme is, of which other artists are exploring it, and so on... Just like in the left part of the block, some words are colored and allow the visitor to further refine his exploration of this theme. In order to help focusing on this specific theme, only the elements related to this theme (here machines as actors) remain displayed on the screen, leaving more space for other works that were hidden away when considering all themes related to the work. Further refinement might be possible, e.g. by tapping on the date field, all related elements could be sorted by date, as in a kind of timeline, as shown in Figure 5. A line (or a triangle in the case of works spanned on a long period of time) would then help positioning this specific in comparison to the other presented works.

Other types of sorting could be devised, as the research progresses, in order to enrich even more the users experience and the possible interconnections between works and artists statements... Once the visitor goes away, the system goes back to the welcome screen described in Figure 1 and the whole process starts over again.

If possible, an online version and or a tablet/TV version of the system could be made available.

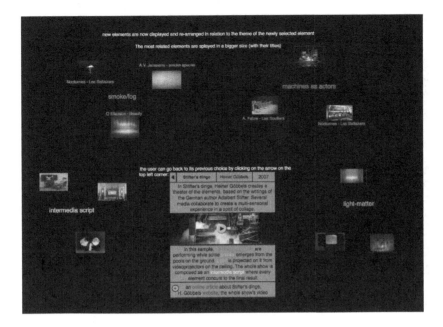

Fig. 3. Getting to know more about Heiner Göbble's work

4 Analysis of Technologies That Would Help to Realize the Scenario

4.1 Interfaces and Presentation Engines

Grafton, A., and Rosenberg, D. provide an in-depth overview of visualizations and maps of events with a strong emphasis on time in the graphical representation [5].

The works from La Médiathèque de la Communauté Francaise de Belgique are recent local examples from the two last years: two projects taking form of an online website and booth installed at media libraries. *Archipel*[5] proposes pathways to understand "unclassifiable" music from the last century, and *Beat Bang*[6] sorts representative albums of electronic music from 1988 to 2012 on a timeline against their average beat-per-minute (BPM). These illustrate that time is one parameter among others to organize and present data: the following section proposes an overview of other means.

4.2 Intelligent Hypervideo Analysis

In our case, analysis and tagging would be done manually by a team of scholars from the domains of art history, performing arts or cinema studies. Considering

---

[5] http://www.archipels.be
[6] http://www.beatbang.be

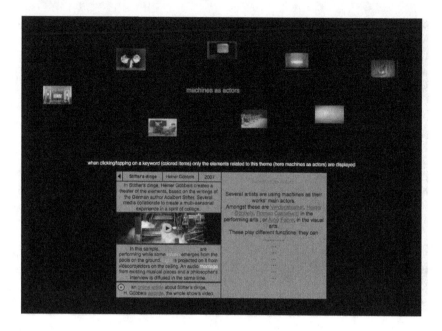

Fig. 4. Exploring a specific theme

the complexity of the classification proposed there is currently no automatic system to our knowledge that fulfils this task and that is available for download open-endedly.

There are promising works undertaken that could be combined so as to bene-fit from machine-based media analysis and human-contributed annotations. For instance Diogo Cabrals doctorate studies including the *TKB/Creation-Tool*[7] for the realtime annotation of dancers recordings [12] could be coupled with content-based analysis of such recordings [13] so as to benefit from both sides, human and machine, of semi-automated analysis. Alternative tools for intra-media au-diovisual annotation that focus on the user experience are *ChronoViz*[8] [2] and *VCode/VData*[9] [6]. Some authors of this paper have been using these for side projects after a summer school workshop project aiming at comparing and eval-uating such tools [3]. Both require Mac OSX, we might base our annotation tool upon these.

While plastic theater may involve scenes with danced movements, as it com-bines performances and plastic art forms, these arent its primary trademark. More major features are behaving similarly to natural phenomena, such as sceneries and entities involving matters and textures put to life by nature: the sea, tornadoes, flows, and so on. Vezzani et al propose a solution for smoke de-tection in surveillance videos by combining user-driven annotations, a built-in

---

[7] http://tkb.fcsh.unl.pt

[8] http://chronoviz.com

[9] https://code.google.com/p/vcode/

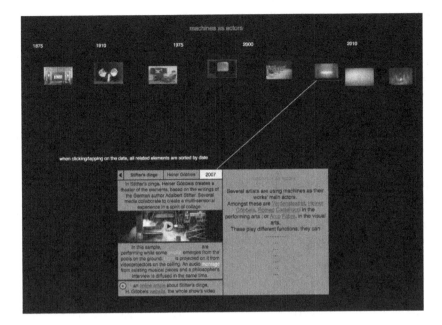

Fig. 5. Exploring a specific theme as a timeline

ontology and a content-based system, MediaMill, also allowing to detect objects from the scene and people [16]. Another paper from the same authors, Calderara et al. [1], describe the algorithm more into-depth.

4.3 Linking Hypervideo to Web Content

For each item of the media collection, a space in the bottom of the presentation will be dedicated to link online web content, as illustrated previously in the scenarios. Our goal is to gather experts from the various fields of performing arts and have them generate a corpus of analyses, concepts.

Richard Rinehart proposed an approach for the conceptualization of digital and media arts in [11], including a declarative model, a metadata framework, a notation system and ontology for these practices. To our knowledge, two ongoing international projects aim at preserving cultural heritage of non-mainstream art genres: e-clap[10] and i-Treasures[11].

4.4 Contextualisation and Personalisation

During the evaluation phases and cycles of the interface, we may use attention-based algorithms so as to determine the points of interest of the visual user interface so as to improve its layout, as usability experts do by analyzing heatmaps

---

[10] http://www.eclap.eu
[11] http://www.i-treasures.eu

computed from the pointer displacements. For that purpose we may use Nicolas Riche and Matei Mancas realtime computer vision algorithms for the automatic detection of salient events, inspired by human cognition and perception [10]. For inter-body-sized evaluation, a ceiling camera with wide-angle or fisheye lens may be used. For intra-body sized evaluation, a front depth-sensing camera such as the Microsoft Kinect or Asus Xtion may be used.

5   Sources of Web and Media Content and Their Related Copyright/Copyleft Issues

We plan two versions with different media collections, one being a subset of the other:

- a lighter copyright-friendly, mostly presenting pictures and images
- a denser version with video extracts, presented in media libraries, who are content providers

The video extracts of the denser version would be used for offline content-based analysis, and would still enhance the organization of media content of the lighter version through features extracted from these videos. We will investigate whether there is content appropriate to our scope in the e-clap database, mentioned in the previous subsection on hypermedia linking.

6   Conclusion

This paper presents a scenario and sketches of the desired user interface that will allow to create an interactive browser of art works related to the plastic theater genre. We plan to adapt an existing opensource multimodal annotation tool to allow experts from this genre to provide their insight on a collection of media fragments related to this genre. If we can get support from multimedia information retrieval and knowledge management scientists, it would be of help to improve the data organization of our browser. We plan to have this browser installed at venues such as media libraries (for instance La Médiathèque from Belgium), theaters where other works from Les Baltazars will be performed. We hope that such a project will be featured and presented during events related to Mons 2015 EU Capital of Culture.

Acknowledgements. The Baltazars received a grant from Experiences Interactives[12] to fund research activity around another of their plastic theater works: *Les Nocturnes.*

Christian Frisson has been funded by the NUMEDIART research program since its beginning in 2007 and up to summer 2013, then morphed into the numediart Institute [13].

---

[12] http://www.experiences-interactives.com
[13] http://numediart.org

References

1. Calderara, S., Piccinini, P., Cucchiara, R.: Smoke detection in video surveillance: A MoG model in the wavelet domain. In: Gasteratos, A., Vincze, M., Tsotsos, J.K. (eds.) ICVS 2008. LNCS, vol. 5008, pp. 119–128. Springer, Heidelberg (2008)
2. Fouse, A., Weibel, N., Hutchins, E., Hollan, J.D.: Chronoviz: a system for supporting navigation of time-coded data. In: CHI 2011 Extended Abstracts on Human Factors in Computing Systems, CHI EA 2011, pp. 299–304. ACM, New York (2011)
3. Frisson, C., Alaçam, S., Coşkun, E., Ertl, D., Kayalar, C., Lawson, L., Lingenfelser, F., Wagner, J.: CoMediAnnotate: towards more usable multimedia content annotation by adapting the user interface. In: Proceedings of the eNTERFACE 2010 Summer Workshop on Multimodal Interfaces, Amsterdam, Netherlands, July 12-August 6 (2010)
4. Frisson, C., Dupont, S., Siebert, X., Dutoit, T.: Similarity in media content: digital art perspectives. In: Proceedings of the 17th International Symposium on Electronic Art (ISEA 2011), Istanbul, Turkey, September 14-21 (2011)
5. Grafton, A., Rosenberg, D.: Cartographies of Time: A History of the Timeline. Princeton Architectural Press (2010)
6. Hagedorn, J., Hailpern, J., Karahalios, K.G.: VCode and VData: illustrating a new framework for supporting the video annotation workflow. In: Proceedings of the Working Conference on Advanced Visual Interfaces, AVI 2008, pp. 317–321. ACM, New York (2008)
7. Kramer, R.E.: The sculptural drama: Tennessee williams's plastic theatre. The Tennesee Williams Annual Review 5, 10 (2002)
8. Manovich, L.: Media Visualization: Visual Techniques for Exploring Large Media Collections. In: Media Studies Futures. Blackwell (2012)
9. Manovich, L.: Museum Without Walls, Art History without Names: Visualization Methods for Humanities and Media Studies. In: Oxford Handbook of Sound and Image in Digital Media. Oxford University Press (2012)
10. Riche, N., Mancas, M., Duvinage, M., Mibulumukini, M., Gosselin, B., Dutoit, T.: Rare2012: A multi-scale rarity-based saliency detection with its comparative statistical analysis. In: Signal Processing: Image Communication (2013)
11. Rinehart, R.: The media art notation system: Documenting and preserving digital/media art. Leonardo 40(2), 181–187 (2007)
12. Silva, J.A., Cabral, D., Fernandes, C., Correia, N.: Real-time annotation of video objects on tablet computers. In: Proceedings of the 11th International Conference on Mobile and Ubiquitous Multimedia, MUM 2012, pp. 19:1–19:9. ACM, New York (2012)
13. Tardieu, D., Siebert, X., Mazzarino, B., Chessini, R., Dubois, J., Dupont, S., Varni, G., Visentin, A.: Browsing a dance video collection: dance analysis and interface design. Journal on Multimodal User Interfaces 4(1), 37–46 (2010); Special Issue: eNTERFACE 2009
14. Tible-Cadiot, S.: Staging the abstract effects of light: Portraits of a few of the pioneers of plastic theatre. In: Les Cahiers de l'Atelier Arts-Sciences, vol. (6), pp. 19–20. Hexagone Scéne nationale/Atelier Arts-Sciences (2011)
15. Vesna, V. (ed.): Database Aesthetics: Art in the Age of Information Overflow. Electronic Mediations, vol. 20. University of Minnesota Press (2007)
16. Vezzani, R., Calderara, S., Piccinini, P., Cucchiara, R.: Smoke detection in video surveillance: the use of visor (video surveillance on-line repository). In: Proceedings of the 2008 International Conference on Content-based Image and Video Retrieval, CIVR 2008, pp. 289–298. ACM, New York (2008)

Hierarchical Audio-Visual Surveillance for Passenger Elevators

Teck Wee Chua, Karianto Leman, and Feng Gao

Institute for Infocomm Research,
A*STAR (Agency for Science, Technology and Research), Singapore
{tewchua,karianto,fgao}@i2r.a-star.edu.sg

Abstract. Modern elevators are equipped with closed-circuit television (CCTV) cameras to record videos for post-incident investigation rather than providing proactive event monitoring. While there are some attempts at automated video surveillance, events such as urinating, vandalism, and crimes that involved vulnerable targets may not exhibit significant visual cues. On contrary, such events are more discerning from audio cues. In this work, we propose a hierarchical audio-visual surveillance framework for elevators. Audio analytic module acts as the front line detector to monitor for such events. This means audio cue is the main determining source to infer the event occurrence. The secondary inference process involves queries to visual analytic module to build-up the evidences leading to event detection. We validate the performance of our system at a residential trial site and the initial results are promising.

1 Introduction

The use of CCTV cameras in the elevator for prevention of vandalism and crime has gained growing popularity in many countries. This has resulted more elevators being retrofitted with cameras. A major disadvantage for CCTV cameras is that they only perform passive recording and post-event video retrieval rather than active event monitoring. In the context of passenger elevator, law-enforcement units are concerned with anti-social behaviors such as vandalism, aggression, robbery, and urinating. Detecting such events using either audio or visual cue poses a challenging problem. Events such as urinating involve human actions that are very subtle thus video analytic system alone may not be able to detect them. Some of the vandalism acts (e.g. breaking, scribbling) are easier to be detected by the sounds they produce (especially when sharp object is used to carve on elevator structure). It would be unreliable if detection is simply based on image processing. In the crimes against vulnerable targets such as elderly, children, and women in the elevators, the incidents would have much stronger audio signals (such as crying, shouting, and screaming) compared to vigorous actions (as the result of fighting back). However, auditory cues can only tell 'what' types of events based on the audio signal received. The audio source could be originated from the elevator surrounding rather than the subjects in the elevator. On the other hand, visual cues are able to identify 'who' is involved and 'where'

C. Gurrin et al. (Eds.): MMM 2014, Part II, LNCS 8326, pp. 44–55, 2014.

the event takes place. In a confined space like elevator, although it is trivial to identify where the event happens visual information can be used to 'confirm' the existence of a person or a group of people after the suspicion arises from audio analytic. Such inference mechanism allows us to detect complex anti-social events stated above that are best detected from multimodal cues.

In this paper, we propose an audio-visual analytic framework that fuses computations of signals from three sensor types: acoustic audio signal from air-microphone vibration signal from contact microphones and video signal from CCTV camera. The proposed detection framework will be primarily driven by audio analytics. Upon a suspicious detection from audio source, computational result from video feeds will be used to verify the validity of the detection from audio signal. The contributions of this work are summarized in the following. As far as we know, we propose the first 1) audio-visual surveillance framework for elevators, 2) urination detection through processing of vibration signal collected from piezo contact microphone and detection of liquid patch from video.

2 Related Work

Since the whole research topic of automated surveillance research topic is rather wide, it is impossible to cover every aspect of it. This section focuses on some relevant work specific to elevator surveillance either by visual/audio analytic or both.

Automated video surveillance usually requires segmentation of foreground objects through background subtraction for further analysis. In the context of video surveillance for elevators, sudden illumination changes due to door opening/closing or transparent glass pose great challenge to background subtraction. Song et al. [1] presented a robust background subtraction technique which can handle such problem by using motion and texture information. Shao et al. [2] proposed a vision-based approach for elevator surveillance to detect abnormal events such as aggression, loitering, stain on the elevator wall. Foreground objects are segmented through background subtraction based on intensity and texture cues. Spatial-temporal features are extracted to perform motion analysis. Similarly, violence detection through motion analysis is proposed in [3, 4] whereby violent actions are characterized by unusual and chaotic motion flows computed from optical flow. Understanding the occupancy status or even the number of people in the elevators is useful to inference the types of events. For example, urinating is most likely to be associated with a person. However, due to the proximity of the people to the camera the extracted blobs are highly deformed. Schofield et al. [5] proposed a passenger counting method for elevators. The method uses neural network to separate foreground objects from background scenes. Serial search is performed on the output of neural network to determine the number of people. In [6], segmented objects are modeled as 3D parallelepiped models to estimate their dimensions, followed by classification and blobs merging to estimate the number of people in a confined space.

While video analytic remains as the mainstream, audio analytic receives less attention as surveillance tool. In fact, audio signals contain rich information about the ongoing events. The work presented in [7–9] extend the idea of background subtraction in video processing to audio domain. Unsupervised learning models are used to model audio background process and outliers are detected as foreground events. The foreground events will be further classified into different classes through supervised framework.

Combination of audio-visual analytic can greatly improve detection accuracy and increase the types of events that can be detected. Kim *et al.* [10] proposed an audio-video hardware and software architecture for elevator surveillance. However, they did not propose any analytic algorithm to detect events in the elevator.

In the context of urine detection in the elevator, it is worth to mention that there exists commercial chemical sensor, known as Urine Detection Device (UDD); however, the performance is far from acceptable while maintenance and installation are relatively cumbersome.

3   System Configuration

Fig. 1(a) shows the schematic drawing illustrating the configuration of audio and video sensors in an elevator. The elevator is installed with a CCTV camera and an audio microphone on the ceiling, and 14 contact microphones (3 on each side of the wall and 5 on the floor). The audio microphone aims to capture acoustic signal such as screaming while the contact microphones aims to capture vibration signals on the wall or floor surface possibly caused by water dripping, scribbling, and banging. The audio and video signals are transmitted to a computer located at the elevator control room for processing. Upon detection of a suspicious event, a short video segment that depicts the event is binded with the corresponding triggering audio source (either from acoustic or contact microphone) is sent to the server over internet. The authorities will receive an alarm SMS. Verification can be done by login to the admin webpage to playback the video clip.

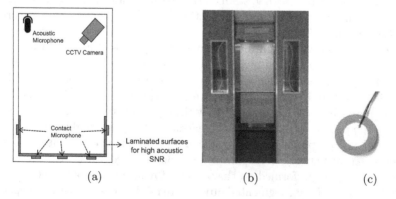

(a)   (b)   (c)

Fig. 1. (a) Hardware configuration, (b) actual photo of the elevator, (c) contact microphones

4   The Proposed Framework

In our proposed framework, audio classifier is designated as the master classifier. It shall trigger the system when a non-silent/abnormal audio event occurs, then it shall query to video classifier for low-level and mid-level visual evidences. The low-level query includes the number of estimated people in the elevator while mid-level query includes the aggressiveness of the motion vectors and whether two persons were in contact. Our framework aims to detect three types of anti-social events in the elevator: 1) urinating, 2) vandalism, and 3) crimes against vulnerable targets/predator-prey pattern (molestation and robbery). Fig. 2 illustrates the flow charts of the event detection mechanism. Different audio signals are used to detect different types of events. Urinating (liquid dripping sound), scratching/scribbling are expected to produce vibration on the elevator wall or floor surface. Likewise, suspicious acoustic events are expected to be triggered from screaming, shouting and crying of the victims. The detailed description of each event detection is as follows:

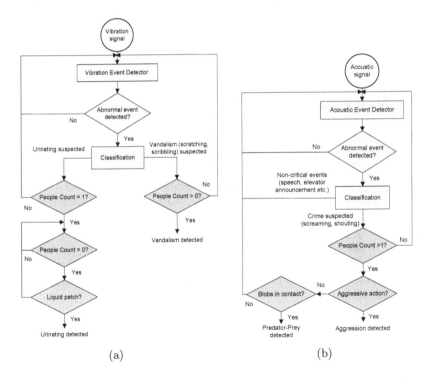

Fig. 2. (a) Vibration signal processing flow, (b) acoustic signal processing flow

Urinating. Vibration signal is continuously monitored. The act of urinating can be detected from the vibration created by liquid dripping on the surface. A query to video analytic is made to confirm that there is only one passenger in the lift. This is based on the observation that perpetrator urinates when there is no one else in the elevator. The next query is to check whether the perpetrator left the elevator before conducting liquid patch verification. If all the conditions are satisfied, the event is concluded as urinating.

Vandalism. The acts of banging, scratching and scribbling on the elevator structure produce more impact (thus vibration is stronger) compared to urinating. Upon the detection of such event, a query to video analytic is made to check the occupancy of the elevator. If the elevator is empty, the event will be discarded otherwise it is flagged as vandalism.

Crimes against vulnerable targets/Predator-prey. Crimes in the elevators are usually accompanied by screaming and shouting. These high pitch and narrow spectral signals are differentiated from other sound sources such as speeches, elevator beeping sound or announcement etc. Upon triggering, a video query is made to verify there exists more than one passenger in the elevator. Aggression is confirmed if chaotic and high energy motion vectors are observed. Otherwise a predator-prey event or crime against vulnerable targets (e.g. molestation and robbery) is concluded if the passengers were once in direct contact. Please note that in our framework, although it is possible to detect aggression, we only focus on detecting crimes against vulnerable targets which do not involve vigorous fight back hence less visual cues.

5 Audio Analysis

The audio analysis is divided into two separate components (see Fig. 3): event detector and event classifier. Event detector is responsible for listening to the input signal to determine if an event occurs. Detected events are then segmented out and transformed into Mel-Frequency Cepstrum Coefficients (MFCCs) feature before being passed to the classifier for identification. Event classifier identifies the event by selecting the most likely candidate from a pre-defined pool of events. If the incoming event is not among the pre-defined events (i.e., never before encountered), the classifier will output what it deems to be the closest candidate (winner-takes-all). Note that there are two layers of Gaussian Mixture Model (GMM) classifiers. The second classifier aims to suppress the false detection of urinating signal.

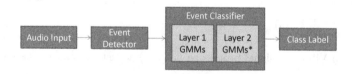

Fig. 3. The audio analytic module

5.1 Event Detector

The event detector functions by comparing the log-power of the signal to a given threshold value. The power value is computed from a sliding window with 400ms width and 200ms frame shift (i.e., 50% overlap between subsequent windows) as shown in Fig. 4(a). If the log-power of a window is greater than the set threshold, it is flagged as an event frame. Consecutive event frames are concatenated to form a single event. Fig. 4(b) shows how the detector works in practice.

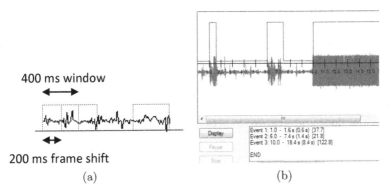

Fig. 4. (a) The sliding window of the audio event detector, (b) example of detector output, the value in the square brackets is the log-power of the given event

5.2 Event Classifier

The event classifier is based on GMMs which computes the log-likelihood of each GMM generating the given event. Each GMM is essentially two Gaussian distributions with a weighing factor that determines the relative importance of each Gaussian component in the GMM. In most cases, these two components describe the low and high magnitude elements of an event. The GMMs are trained using the Expectation-Maximization (EM) algorithm. As mentioned earlier, there are two layers of GMM classifiers. Training data are collected from an actual elevator of a residence building to train the first layer classifier separately with respect to the types of inputs:

Acoustic signal: scream, speech, elevator beep sound, announcement, junk (unknown random sounds)
Vibration signal: urination, impact, scratch, junk

The second layer is only activated for ambiguous vibration signal produced by urination. The false detections of urination samples from the first classifier are progressively added to the training pool of second classifier to better differentiate vibration caused by urinating from other sources such as rubbing between slipper and floor surface, and between the body and the wall surface when someone is leaning against the wall.

During event classification, each time frame is treated as an independent vector by the GMM classifier thus an event is essentially processed as a sequence of MFCC vectors. The GMM calculates the log-likelihood for each frame and the sum of these values is the log-likelihood for the entire sequence to be generated by that particular GMM. To improve the robustness, we also set the minimum duration for each type of event, events that are shorter than the minimum duration are discarded.

6   Visual Analysis

As the secondary classifier, visual analysis aims to provide more evidences to conclude the presence of an event after the initial triggering by audio analysis. In this work, one of the most fundamental task is to estimate the number of passengers in the elevator. In addition, it is also crucial to know whether a passenger is in direct to other passenger. Other tasks include detection of liquid patch on the floor and aggressive actions.

6.1   Passenger Counting

We made two assumptions that 1) when a person is inside the lift, the body is usually non-stationary over certain duration 2) human is of certain minimum size (this can be estimate from camera calibration). Based on these two assumptions, we devise an algorithm to estimate the number of passenger. Due to door opening/closing, conventional background subtraction is unable to handle sudden light changes. Instead of performing background subtraction based on the pixel intensity, we propose to use texture feature which is robust against light changes. During system initialisation, we capture an empty scene of the lift and extract its textured regions (Fig. 5(b)). Textured regions are defined as regions where the squared horizontal intensity gradient averaged over a square window of a given size is above a given threshold [11]. For new incoming frame, we also extract the textured regions (Fig. 5(c)). Next, we perform subtraction of both textured regions images and execute some postprocessing steps to remove noises (Fig. 5(d)). The remaining regions show the foreground objects, we retain only those objects whose areas are above a threshold which is the minimum total pixel size of a child. Fig. 5(a)shows the human candidate bounded by the green bounding box. If the bounding box is non-stationary for a certain period (typically 10-20 seconds), then the blob is considered as human. Next the binary pixels are projected along two axes (parallel and perpendicular to the elevator door) to form two histograms. For each histogram, we extract its skewness f_s, kurtosis f_k, and number of bins that exceed minimum height f_b as features. In addition, we also extract the percentage of occupancy f_o as the feature. It is the ratio of foreground pixels count over the total pixels count of the possible occupancy area. The denominator can be estimated from the predefined region-of-interest (ROI). The other two features are the height f_h and width f_w of the bounding box. Each bounding box is described by 9-tuple feature vector and it

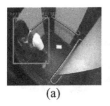 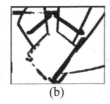 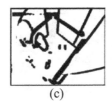 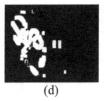

(a)   (b)   (c)   (d)

Fig. 5. Texture-based foreground segmentation: (a) detected passenger, (b) textured regions of the background (black), (c) textured regions of the current scene, and (d) foreground regions resulted from subtraction of (b) from (c) with postprocessing steps

is fed into Support Vector Machine (SVM) to classify whether there exists one or more than one passenger.

6.2 Passengers Interaction

Each blob is continuously tracked, two persons are considered in contact if their bounding boxes merged and split within a time frame before and after a screaming/shouting audio event triggered.

6.3 Liquid Patch Detection

Liquid patch detection is carried out when there is no passenger in the lift. This is to ensure that the liquid patch is not blocked by the passenger. Since liquid patch can only exist on the floor, we set the floor area as the ROI. The image is converted from color to grayscale for further processing. The algorithm is based on the fact that the floor area under the liquid patch appears darker than its surrounding area. This is due to that liquid increases the average scattering angle of light particles. The more the light penetrates the substance and the more it gets absorbed, the less it reaches the eye, and the darker the object appears. In addition, we made an assumption that liquid patches and floor area are almost textureless compared to non-liquid objects.

There are several image processing steps involved:

1. Perform Canny edge detection to detect the edge of an object on the floor.
2. Retain only exterior boundary of each blob and discard small blobs with less than minimum detection size.
3. Calculate the mean pixel values of the detected object and the remaining floor area respectively. If the mean value of the object is higher (hence the object is brighter) than the floors mean value, then the object is classified as non-liquid object. Otherwise proceed to the next step for texture verification.
4. Compute the mean Local-Binary-Pattern (LBP) [12] texture values of the object and the remaining floor area. If the objects mean texture value is higher than a threshold, the object is verified as non-liquid object, otherwise it is classified as a liquid patch.

(a)	(b)

Fig. 6. Examples of floor object detected as (a) liquid patch, (b) non-liquid object

Fig. 6 shows the examples of liquid detection algorithm.

6.4   Aggression Detection

Fast computation of motion vectors is realized by using block-matching algorithm [13]. If the mean magnitude and standard deviation of the motion vectors is high, there could possibly a fight occurring. Spatio-temporal (S-T) interest points [14] are computed to extract S-T motion patches. Histogram-of-oriented gradient and histogram-of-motion-flow are extracted at the S-T patches. A codebook is constructed by clustering the patches from the training data. Each video segment is represented as a codeword histogram. Finally, the histogram is classified by a trained SVM classifier with χ^2 kernel into aggressive action or normal action.

7   Experimental Results

We have performed extensive experiments of our framework at two elevators at a residential building. The reason why residential building is chosen over commercial building is because more crimes occur in residential buildings. Traffic flow in residential elevators is more sporadic and less often monitored by a professional security service. Commercial elevators on the other hand often see peak traffic during daylight hours and infrequent use at night. In our experiments, all audio signals are captured at sampling rate of 8KHz at 16 bits/sample and video resolution of 352×288 pixels at 10fps using a multi-channel audio-video capture card. It is almost impossible to collect sufficient training data from actual events especially for events such as urinating molestation and robbery. Therefore, professional actors and actresses are hired to perform realistic mockup sessions. The data collected are used to train and evaluate the system. To ensure the diversity of the data, the mockups are carried out on four different days by two actors and two actresses in different apparels. We use a small spray device to simulate urinating. The actors and actresses are required to adjust the force applied to the device such that the water pressure that reaches the wall (for actor only) and floor (for both actors and actresses) is as close as actual urinating. They are also required not to reveal the water spray device in front of the CCTV camera. The total number of mockup events we collected for urinating, vandalism (scratching/banging), and crimes against vulnerable targets (molestation and robbery)

Table 1. Performance of event detection

Event	Detection	
	Rate (%)	Accuracy (%)
Urinating	81.2	98.6
Vandalism	100.0	100.0
Vulnerable target crimes	95.0	100.0

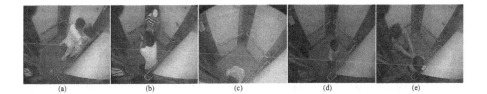

| (a) | (b) | (c) | (d) | (e) |

Fig. 7. Examples of mockup events: (a) molestation, (b) robbery, (c) urinating (female) and actual detected events at the trial sites: (d) scratching, (e) violence against woman

are 170, 32, 40 respectively. Half of the data are used for training and the other half are for testing. Table 1 shows the detection accuracy of each event. The performance metrics are defined as:

$$\text{Detection Rate} = \frac{\text{Number of Detections}}{\text{Total Mockup Events}} \tag{1}$$

$$\text{Detection Accuracy} = \frac{\text{Number of True Alarms}}{\text{Number of Detections}} \tag{2}$$

18.8% of urinating cases are missed due to failure of audio detection engine to pick up the weak vibration signal when the perpetrator urinates at the spots which are distance from the contact microphones. 5% of the vulnerable target crimes failed to be detected because the screams of the victims are shorter than minimum duration of 0.5s. The detection of scratching, scribbling, and banging the elevator surface is very reliable as those anti-social acts produce strong vibration characteristic. The events would not have been detected if the detection is purely based on visual information.

In order to evaluate the effectiveness of visual analytic module in assisting the event inference, we analyze the experiment results which consists of 104.5 hours of real-time monitoring of non-mockup events for two elevators. Out of 18 alarms triggered by audio analytic part, 8 of them are actually false alarms. The number of false alarms drops significantly to 2 with video verification step. Interestingly, none of the true event is correctly identified by audio analytic but rejected by video analytic. This shows that the proposed hierarchical audio-visual event inference framework is effective.

The prototype system has been on trial for 3 months. So far, the system managed to detect six cases of vandalism events (five scratching and one scribbling on the elevator structure), and one case of violence against women. Fig. 7 shows some examples of the mockup events and the actual detected events at the trial sites.

8   Conclusion

In this work, we propose a hierarchical audio-visual surveillance framework for elevators. Many anti-social acts and crimes targeting vulnerable groups in the elevators do not involve vigorous actions or distinct visual cues but most of them can be triggered from audio cues due to the confined space of the elevators. We demonstrated that the detection of such events can be triggered by specific audio signatures. Visual queries pertaining to specific event are then used to verify the validity of the event. The proposed framework can increase safety protection to the residences especially elderly, children and women against the growing trend of crimes in elevators. Moreover, by lowering the number vandalism and urinating cases the authorities can potentially save operation costs resulted from less elevator parts replacement and cleaning services. Due to different sizes and configurations of elevators, the variations in quality of the signals captured may be different. Currently, the audio signals (acoustic and vibration) are manually calibrated. In future, we aim to introduce auto-calibration step to facilitate mass deployment of the system.

References

1. Song, T., Han, D., Ko, H.: Robust background subtraction using data fusion for real elevator scene. In: Proc. of IEEE International Conference on Advanced Video and Signal-Based Surveillance (AVSS), pp. 392–397 (2011)
2. Shao, H., Li, L., Xiao, P., Leung, M.K.H.: ELEVIEW: an active elevator video surveillance system. In: Proc. of Workshop on Human Motion (HUMO), pp. 67–72 (2000)
3. Kage, H., Seki, M., Sumi, K., Tanaka, K., Kyuma, K.: Pattern recognition for video surveillance and physical security. In: Proc. of International Conference on Instrumentation, Control, Information Technology and System Integration (SICE), pp. 1823–1828 (2007)
4. Lee, Y., Song, T., Kim, H., Han, D.K., Ko, H.: Hostile intent and behaviour detection in elevators. In: 4th International Conference on Imaging for Crime Detection and Prevention, ICDP (2011)
5. Schofield, A.J., Stonham, T.J., Mehta, P.A.: Automated people counting to aid lift control. Automation in Construction 6, 437–445 (1997)
6. Zuniga, M., Bremond, F., Thonnat, M.: Fast and reliable object classification in video based on a 3d generic model. In: Proc. of IET International Conference on Visual Information Engineering (VIE), pp. 433–440 (2006)
7. Radhakrishnan, R., Divakaran, A.: Systematic acquisition of audio classes for elevator surveillance. In: Proc. of SPIE, pp. 64–71 (2005)
8. Radhakrishnan, R., Divakaran, A., Smaragdis, P.: Audio analysis for surveillance applications. In: Proc. of IEEE Workshop on Applications of Signal Processing to Audio and Acoustics (ASPAA), pp. 158–161 (2005)

9. Radhakrishnan, R., Divakaran, A.: Generative process tracking for audio analysis. In: Proc. of International Conference on Acoustics, Speech, and Signal Processing (ICASSP), pp. v1–v4 (2006)
10. Kim, W.-Y., Park, S.-G., Lim, M.-C.: The intelligent video and audio recognition black-box system of the elevator for the disaster and crime prevention. In: Kim, T.-H., Adeli, H., Robles, R.J., Balitanas, M. (eds.) ACN 2011. CCIS, vol. 199, pp. 245–252. Springer, Heidelberg (2011)
11. Scharstein, D., Szeliski, R.: A taxonomy and evaluation of dense two-frame stereo correspondence algorithms. International Journal of Computer Vision 47, 7–42 (2002)
12. Heikkilä, M., Pietikäinen, M.: A texture-based method for modeling the background and detecting moving objects. IEEE Trans. Pattern Anal. Machine Intell. 28, 657–662 (2006)
13. Nie, Y., Ma, K.K.: Adaptive rood pattern search for fast block-matching motion estimation. IEEE Trans. Image Processing 11(12), 1442–1448 (2002)
14. Laptev, I., Lindeberg, T.: Space-time interest points. In: Proc. of International Conference on Computer Vision (ICCV), pp. 432–439 (2003)

An Evaluation of Local Action Descriptors for Human Action Classification in the Presence of Occlusion

Iveel Jargalsaikhan, Cem Direkoglu, Suzanne Little, and Noel E. O'Connor

INSIGHT Centre for Data Analytics,
Dublin City University, Ireland
iveel.jargalsaikhan2@mail.dcu.ie

Abstract. This paper examines the impact that the choice of local descriptor has on human action classifier performance in the presence of static occlusion. This question is important when applying human action classification to surveillance video that is noisy, crowded, complex and incomplete. In real-world scenarios, it is natural that a human can be occluded by an object while carrying out different actions. However, it is unclear how the performance of the proposed action descriptors are affected by the associated loss of information. In this paper, we evaluate and compare the classification performance of the state-of-art human local action descriptors in the presence of varying degrees of static occlusion. We consider four different local action descriptors: Trajectory (TRAJ), Histogram of Orientation Gradient (HOG), Histogram of Orientation Flow (HOF) and Motion Boundary Histogram (MBH). These descriptors are combined with a standard bag-of-features representation and a Support Vector Machine classifier for action recognition. We investigate the performance of these descriptors and their possible combinations with respect to varying amounts of artificial occlusion in the KTH action dataset. This preliminary investigation shows that MBH in combination with TRAJ has the best performance in the case of partial occlusion while TRAJ in combination with MBH achieves the best results in the presence of heavy occlusion.

1 Introduction

Analyzing complex and dynamic video scenes for the purpose of human action recognition is an important task in computer vision. Therefore, extensive research efforts have been devoted to develop novel approaches for action-based video analysis. Action oriented event detection is an important component for many video management applications especially in surveillance and security [1], sports video [2], and video archive search and indexing domains.

In security applications CCTV footage can be analysed in order to index actions of interest and enable queries relating to actions such as anti-social or criminal behaviour or to monitor crowd volume or agression. This is an especially challenging example of human action recognition due to the volume and quantity

C. Gurrin et al. (Eds.): MMM 2014, Part II, LNCS 8326, pp. 56–67, 2014.

of video and the potentially low level of visual distinctiveness between the actions of interest. This can be seen in the performance of systems used in the TRECVid surveillance event detection (SED) task that has been operating for the last 6 years using the iLIDS dataset from the UK Home Office to annotate video segments with actions such as CellToEar, Embrace, ObjectPut, PeopleMeet, PeopleSplitUp, PersonRuns and Pointing [3]. Some of the unique challenges of this dataset are discussed in [4].

Fig. 1. Sample shots from TRECVid SED dataset show occlusion

Figure 1 shows some example frames from the TRECVid SED dataset illustrating occlusion of the main actor by other objects. Temporal occlusion by other actors (e.g., walking in front of someone who is using a cell phone) is also common. It is difficult to judge the extent of occlusion or the impact of the missing or mis-leading feature descriptors on the performance of human action classifiers trained on example data. Given the size of the TRECVid-SED dataset and the low accuracy levels thus far achieved, we have chosen to use the KTH action dataset to conduct preliminary investigations into the impact of occlusion of human action classification using local descriptors. Although a relatively simple dataset, KTH provides a "level playing field" for testing descriptors.

Despite the fact that existing action description methods have been tested on both artificial and real world datasets, there is no significant study that is directly focused on the problem of occlusion. Occlusion is a challenging problem in real-world scenarios where there are usually many people located at different positions and moving in different individual directions making it difficult to find effective descriptors for higher level analysis.

There are two main classes of human action description methods: global and local. The global methods [5,6] represent the actions based on holistic information about the action and scene. These methods often require the localization of the human body through alignment, background subtraction or tracking. These methods perform well in controlled environments, however exhibit poorer performance in the presence of occlusion, clutter in the background, variance in illumination, and view point changes. Local methods exist [7,8,9] that are less sensitive to these conditions. The local descriptors capture shape and motion information in the neighbourhoods of selected points using image measurements such as spatial or spatio-temporal image gradients and optical flow.

In this paper, we investigate the performance of state-of-art local descriptors for human action recognition in the presence of varying amounts of occlusion. Our objective is to understand how missing action features, i.e. because of static occlusion, affect action classification performance. In order to model static occlusion, we mask human action regions with a rectangular shaped, uniform colour object, so that the local descriptors are not extracted within that region.

We evaluate and compare four different local descriptors TRAJ [10], HOG [9], HOF [8], MBH [11] and their possible combinations. These descriptors are combined with a standard bag-of-features representation and a Support Vector Machine (SVM) classifier for action recognition. Our experiments are conducted on the KTH action dataset, and results show that the MBH in combination with TRAJ performs the best in the presence of partial occlusion while TRAJ in combination with MBH achieves the best results in the case of heavy occlusion (greater than 50% of the actor).

To our knowledge, evaluation and comparison of classification performance of local action description methods, in the presence of occlusion, has not been done in the past. However, several authors have evaluated the impact of occlusion on their own work. Weinland et al. [12] showed the robustness of his proposed work under occlusion and view-point changes using artificially imposed occlusions on the KTH and Weizmann datasets. Dollar et al. [13] evaluated the impact of occlusion in terms of pedestrian detection. Additionally, a number of key survey papers in human action recognition [14,15,16] stated the necessity of occlusion tolerant action recognition methods. In particular, Poppe [14] wrote "the question [of] how to deal with more severe occlusions has been largely ignored".

The rest of the paper is organized as follows: Section 2 explains the local action descriptors included in our evaluation. Section 3 presents the experimental setup describing how synthetic occlusion is applied to the KTH dataset and evaluation framework. Finally, Section 4 presents and discusses our results prior to the conclusion.

2   Local Action Descriptors

2.1   Trajectory Descriptor (TRAJ)

The Trajectory descriptor is proposed in the work of Wang et al. [10]. The descriptor encodes the shape characteristic of a given motion trajectory. Since motion is an important cue in action recognition, this representation allows motion characteristics to be exploited. The descriptor is straight-forward to compute using the points sampled on the trajectory in the image domain. Given a trajectory of length L, the shape is described by a descriptor vector S :

$$S = \frac{\Delta P_t, ..., \Delta P_{t+L-1}}{\sum_{t+L-1}^{j=t} |\Delta P_j|} \qquad (1)$$

where $\Delta P_t = (P_{t+1} - P_t) = (x_{t+1} - x_t, y_{t+1} - y_t)$. In our experiment, the trajectory length was chosen to be $L = 15$ video frames as recommended in [10].

2.2   The HOG/HOF Descriptor

The HOG/HOF descriptors were introduced by Laptev et al. in [8]. To characterize local motion and appearance, the authors compute histograms of spatial gradient and optical flow accumulated in space-time neighbourhoods of the selected points. The points can be detected using any interest point detectors [17] [7]. In our experiment, these points are selected along the motion trajectory as in [10]. For the combination of HOG/HOF descriptors with interest point detectors, the descriptor size is defined by $\Delta x(\sigma) = \Delta y(\sigma) = 18\sigma$, $\Delta t(\tau) = 8\tau$. Each volume is subdivided into a $n_x \times n_y \times n_t$ grid of cells; for each cell, 4-bin histograms of gradient orientations (HOG) and 5-bin histograms of optical flow (HOF) are computed. Normalized histograms are concatenated into HOG and HOF as well as HOG/HOF descriptor vectors and are similar in spirit to the well-known SIFT descriptor. In our evaluation we used the grid parameters $n_x = n_y = 3, n_t = 2$ as suggested by the authors [8].

2.3   The Motion Boundary Histogram (MBH) Descriptor

Dalal et al. [11] proposed the Motion Boundary Histogram (MBH) descriptor for human detection, where derivatives are computed separately for the horizontal and vertical components of the optical flow. The descriptor encodes the relative motion between pixels. The MBH descriptor separates the optical flow field $I_\omega = (I_x, I_y)$ into its x and y component. Spatial derivatives are computed for each of them and orientation information is quantized into histograms, similarly to the HOG descriptor. We obtain an 8-bin histogram for each component, and normalize them separately with the L_2 norm. Since MBH represents the gradient of the optical flow, constant motion information is suppressed and only information about changes in the flow field (i.e., motion boundaries) is kept. In our evaluation, we used the MBH parameters used in the work of Wang et al. [10].

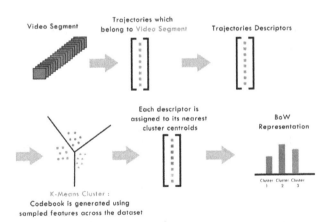

Fig. 2. The extracted trajectory features are represented by each descriptor: TRAJ, HOG, HOF and MBH. Then samples from training videos are used to generate the visual dictionary for respective descriptors. The test video is represented by the Bag-of-Features (BOF) approach using the built visual dictionary. For the case of descriptor combination such as TRAJ+MBH or HOG+HOF, the respective BOF histograms are concatenated together in order to train a SVM classifier.

3   Experimental Setup

3.1   Dataset

The KTH actions dataset [18] consists of six human action classes: walking, jogging, running, boxing, waving, and clapping. Each action class is performed several times by 25 subjects. The sequences were recorded in four different scenarios: outdoors, outdoors with scale variation, outdoors with different clothes and indoors. The background is homogeneous and static in most of the sequences.

In total, the data consists of 2391 video samples. We follow the original experimental setup of the dataset publishers [18]. Samples are divided into test set (9 subjects: 2, 3, 5, 6, 7, 8, 9, 10, and 22) and training set (the remaining 16 subjects). We train and evaluate a multi-class classifier and report average accuracy over all classes as the performance measure. The average accuracy is a commonly reported performance measurement when using the KTH dataset [18].

Fig. 3. The sample shots where the different degree of random occlusion is applied into KTH video sequence. The red boundary is manually drawn in order to set an action boundary for each action performer. The green rectangles are occlusion regions randomly selected with 4 different occlusion sizes: 10%, 25%, 50% and 75% of the active region.

3.2   Synthetic Occlusion

Occlusion may occur due to static and dynamic occluding objects. For example: If an action performer is occluded by a moving object like a moving car or a person, it is considered as dynamic occlusion. On the other hand, the occluding object may be static like a building or a table then in which case an occlusion represents static occlusion.

In our experiment, we focus our attention on *static occlusion*. Our objective is to understand how the missing action features, i.e. because of the static occlusion, affect the action classification performance. In order to model static occlusion, we mask human action regions with rectangular shaped uniform colour objects, so that the action descriptors are not extracted within that regions. The uniform colour ensures no interest points are detected.

Since the KTH action dataset does not contain any occlusion, we have integrated random static occlusion only for the test set sequences. First, action boundaries are manually selected in each test sequence as a bounding box as shown in red boundary in Figure 3 . The action boundary (AB) should be selected with a specific height H_{AB}, width W_{AB}, position (x_{AB}, y_{AB}), in order to accommodate the region of video where the action is performed. Once we label the action boundaries for all test video sequences, occlusion bounding box (OB) is automatically generated within the action boundary region specified by H_{AB},

W_{AB}, x_{AB}, y_{AB} with varying sizes of occlusion area $A(OB)$. The occlusion position is randomly generated and remained static for each test sequence. In our experiment, we have chosen the occlusion areas $A(OB)$ to be 10%, 25%, 50% and 75% of the action boundary area $A(AB)$ as shown in Figure 3. In given action boundary AB and occlusion percentage $Occ\%$, the parameters H_{OB}, W_{OB}, x_{OB}, y_{OB} of the occlusion bounding box OB are randomly selected as follows:

$$\forall H_{OB} \in [H_{AB} - (1 - Occ\%) \times H_{AB}, H_{AB}] \tag{2}$$

$$\forall W_{OB} \in [W_{AB} - (1 - Occ\%) \times W_{AB}, W_{AB}] \tag{3}$$

$$\forall x_{OB} \in [x_{AB}, x_{AB} + (W_{AB} - W_{OB})] \tag{4}$$

$$\forall y_{OB} \in [y_{AB}, y_{AB} + (H_{AB} - H_{OB})] \tag{5}$$

where $Occ\% = \frac{A(OB)}{A(AB)}$ and H_{AB}, W_{AB}, (x_{AB}, y_{AB}) is height, width and top-left corner coordinate of action the boundary box, AB, whereas H_{OB}, W_{OB}, (x_{OB}, y_{OB}) is height, width and top-left corner coordinate of the occlusion boundary box, OB, and $H_{OB}, W_{OB}, x_{OB}, y_{OB} \in \mathbb{N}$.

Table 1. The precision and recall rate for different combination of our evaluating descriptors. Here, the precision is defined as $P\% = (\frac{TP}{TP+FP}) \times 100$ and the recall (i.e. detection rate) is defined as $R\% = (\frac{TP}{TP+FN}) \times 100$, where TP is true positive, FP is false positive and FN is false negative. In this table, all of the measures must be high for a method to show that it can provide sufficient discrimination and classification.

					Recall					Precision			
Descriptor Combination				No Occ.	Partial Occ		Heavy Occ		No Occ.	Partial Occ		Heavy Occ	
TRAJ	HOG	HOF	MBH		10%	25%	50%	75%		10%	25%	50%	75%
			✓	91.2%	89.1%	87.3%	71.8%	49.1%	91.6%	89.8%	88.2%	77.5%	68.3%
		✓		87.0%	87.2%	79.7%	68.5%	45.8%	88.6%	88.0%	81.4%	73.9%	63.6%
		✓	✓	91.2%	89.1%	87.7%	76.9%	50.0%	91.8%	89.8%	88.5%	81.7%	67.6%
	✓			74.5%	69.4%	62.5%	46.8%	26.9%	82.0%	80.8%	74.2%	65.3%	60.4%
	✓		✓	89.8%	88.7%	84.0%	70.4%	46.8%	90.5%	89.6%	85.8%	76.7%	73.9%
	✓	✓		88.4%	87.2%	81.1%	72.2%	44.4%	89.8%	88.7%	83.2%	77.3%	69.1%
	✓	✓	✓	89.8%	89.6%	84.9%	74.1%	49.1%	90.7%	90.6%	86.4%	79.6%	74.4%
✓				87.4%	84.9%	81.5%	79.6%	57.0%	88.7%	86.6%	84.1%	84.3%	73.1%
✓			✓	92.1%	93.4%	86.7%	76.9%	56.5%	92.5%	93.7%	88.1%	82.3%	75.0%
✓		✓		91.2%	88.2%	82.9%	75.5%	52.8%	91.8%	89.0%	84.8%	80.5%	72.0%
✓		✓	✓	92.6%	91.5%	86.2%	76.9%	52.3%	92.9%	92.0%	87.6%	81.2%	70.9%
✓	✓			89.8%	87.8%	81.5%	74.1%	51.9%	90.9%	89.7%	84.9%	80.8%	72.3%
✓	✓		✓	91.6%	90.1%	84.8%	73.6%	51.9%	92.1%	90.9%	86.7%	80.2%	75.9%

Table 2. The ranking is computed on the F-Score measure. The F-score is a measure of accuracy that considers precision and recall rates to compute the score as follows: $F\% = 2 \times \frac{Precision \times Recall}{Precision + Recall}$. This table shows the ordered list of descriptor combination in terms their F-Score measure in partial occlusion case.

Rank	TRAJ	HOG	HOF	MBH	No Occ.	10%	25%	Avg.
1	✓			✓	92.0%	93.4%	86.7%	**90.1%**
2	✓		✓	✓	92.5%	91.5%	86.2%	**88.9%**
3			✓	✓	91.1%	89.1%	87.7%	**88.4%**
4				✓	91.1%	89.0%	87.2%	**88.1%**
5	✓	✓	✓	✓	91.5%	90.5%	84.8%	**87.7%**
6	✓	✓		✓	91.6%	90.1%	84.9%	**87.5%**
7		✓	✓	✓	89.6%	89.5%	84.9%	**87.2%**
8		✓		✓	89.6%	88.5%	84.0%	**86.2%**
9	✓		✓		91.1%	88.4%	83.1%	**85.7%**
10	✓	✓	✓		90.7%	88.8%	82.2%	**85.5%**
11	✓	✓			89.8%	87.8%	81.6%	**84.7%**
12		✓	✓		88.3%	87.3%	81.1%	**84.2%**
13			✓		86.9%	87.3%	79.8%	**83.5%**
14	✓				87.3%	85.0%	81.7%	**83.3%**
15		✓			74.0%	68.6%	59.4%	**64.0%**

3.3 Evaluation Framework

We adopted the approach of Wang et al. [10] as a video processing pipeline to evaluate spatio-temporal features under different occlusion settings. This approach extracts motion trajectories from the video and generates a set of trajectory with length of $L = 15$ frames.

We compute TRAJ, HOG, HOF and MBH descriptors for each motion trajectory. For volumetric features , HOG, HOF and MBH , we construct 3D volumes along the trajectory. The size of the volume is $N \times N$ pixels and L frames, with $N = 32$ and $L = 15$ in our experiments. The feature vector dimensions of HOG, HOF, MBH and TRAJ are respectively 96, 108, 192 and 30.

In order to represent human actions, we build a Bag-of-Features (BoF) model based on the four different types of descriptors. The Bag-of-Feature representation for each type of descriptor (i.e. HOG, HOF, MBH and TRAJ) is obtained as follows: First, we cluster a subset of 250,000 descriptors sampled from the training video with the mini batch K-Means algorithm proposed by Sculley [19]. In our experiments, the number of clusters is set to $k = 4,000$, the mini path size is 10,000 and the number of iterations for clustering is 500. These parameter values are selected empirically to obtain good results and avoid extensive computations. Then each descriptor type is assigned to its nearest cluster centroid using the Euclidean distance to form a co-occurrence histogram.

For combining descriptors, we concatenate the co-occurrence histogram of respective descriptors to generate a feature vector to train a SVM classifier.

In our evaluation, we train 15 different classifiers for each combination of our four descriptors.

3.4   Classification

A multi-class support vector machine (SVM) with a Gaussian radial basis function (RBF) kernel is used for classification. We apply a grid searching algorithm to learn the optimal values of the penalty parameter (C) in SVM and the scaling factor (γ) in Gaussian RBF kernel with the KTH dataset training set (without any occlusion). The grid searching is performed using 10 fold cross-validation. The optimal parameter values are : $C = 1$ and $\gamma = 32 \times 10^{-2}$. These parameters are fixed throughout our evaluation of the local descriptors and their possible combinations.

Table 3. Here shows the F-Score based ranking in heavy occlusion case for local action descriptors and their possible combinations

	Descriptor Combination				No	Heavy Occlusion		
Rank	TRAJ	HOG	HOF	MBH	Occ	50%	75%	Avg.
1	✓				87.3%	79.2%	56.1%	**67.7%**
2	✓			✓	92.0%	76.7%	57.2%	**66.9%**
3	✓		✓	✓	92.5%	76.6%	52.7%	**64.7%**
4	✓			✓	91.1%	74.9%	52.8%	**63.8%**
5	✓	✓		✓	91.6%	73.7%	53.0%	**63.3%**
6			✓	✓	91.1%	77.0%	49.7%	**63.3%**
7	✓	✓			89.8%	74.0%	51.5%	**62.8%**
8		✓	✓	✓	89.6%	74.3%	50.8%	**62.5%**
9	✓	✓	✓	✓	91.6%	73.8%	50.2%	**62.0%**
10				✓	91.1%	72.0%	50.1%	**61.1%**
11	✓	✓	✓		90.7%	74.3%	47.8%	**61.1%**
12		✓		✓	89.6%	70.8%	48.5%	**59.7%**
13		✓	✓		88.3%	72.3%	45.3%	**58.8%**
14		✓			86.9%	68.3%	45.8%	**57.0%**
15		✓			74.0%	42.2%	22.5%	**32.3%**

4   Experimental Results

Table 2 shows the ranking of different combinations of descriptors in the partial occlusion case based on F-Score. The best three combinations are TRAJ+MBH (90.1 %), TRAJ+HOF+MBH (88.9%) and HOF+MBH (88.4%). The worst performance is with HOG and HOF features. HOG descriptor obtained 64.3% and HOF descriptor obtained 83.3% and their combination is 83.5%.

The heavy occlusion ranking is presented in Table 3. TRAJ (67.7%), TRAJ+MBH (66.9%), TRAJ+HOF+MBH (64.7%) combinations perform best. The HOG, HOF and their combination perform poorly. Generally, the best descriptors are TRAJ, MBH and their combination. They consistently outperform any other combination for different scales of occlusion area in our experiments.

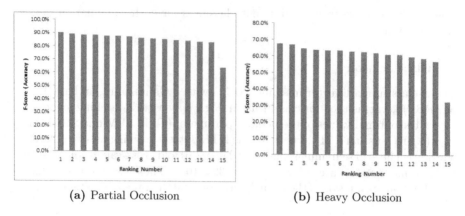

(a) Partial Occlusion   (b) Heavy Occlusion

Fig. 4. The graphical illustration of accuracy for partial and heavy occlusion cases. (a) The partial occlusion case. The ranking number corresponds to Table 2 (b) The heavy occlusion case. The ranking number corresponds to Table 3.

We now present experimental results for various descriptor combinations. We use multi-class classification where we apply the one-against-rest approach and compare the performance based on precision, recall and F-score. The scores are reported as an average of the 6 action classes. In order to measure the occlusion impact, we compute the above mentioned scores at four different cases of occlusion: 10%, 25%, 50% and 75% occlusion of the action area. We also group the cases into partial occlusion (10% 25% occluded) and heavy occlusion (50% 75% occluded). The classifier is trained with non-occluded training data. Therefore all occlusion cases are classified with the same trained classifier.

Table 1 shows the recall and precision scores for all combinations of the descriptors we evaluated. The recall is calculated for partial and heavy occlusion scenarios. In partial occlusion, MBH and its combination with other descriptors performed significantly better than other combinations. Especially the combination of TRAJ + MBH outperforms the without-occlusion case by 2%. This can be explained by the fact that occlusion also acts like a noise filtering. It increases the discriminative power of the representation. Regarding the heavy occlusion, the best performance is shown with all four combinations of trajectory descriptor. It makes the trajectory descriptor particularly suitable for scenarios with large occlusions. For example, with 75% occluded area, TRAJ individually obtained 57% recall rate which is the highest score compared to any other combination where most of them barely reached 50%.

In terms of precision, the same trend is observed in both occlusion scenarios. The partial occlusion is predominantly handled significantly better than others when there is combination of MBH descriptors. For heavy occlusion, TRAJ + MBH descriptors topped the precision rank.

The poorest performance is exhibited by HOG and its combination with other descriptors. In both partial and heavy occlusion cases, the HOG descriptor obtained the worst precision and recall rate. Therefore it is unsuitable to use HOG

	boxing	handclapping	handwaving	jogging	running	walking
boxing	100	0	0	0	0	0
handclapping	14	86	0	0	0	0
handwaving	8	11	81	0	0	0
jogging	0	0	0	94	6	0
running	0	0	0	11	89	0
walking	0	0	0	0	0	100

Fig. 5. Confusion matrix for the un-occluded KTH dataset

even with other occlusion tolerant features like MBH or TRAJ as it significantly decreases the performance.

5 Discussion

The experimental results confirm that the motion based descriptors (TRAJ, HOF and MBH) are more discriminative when recognizing human actions in an occluded scene. Among the motion based descriptors, MBH and TRAJ descriptors significantly outperform other descriptors. In the partial occlusion case, MBH is the best choice, whereas the TRAJ descriptor is good for heavy occlusion. Texture or appearance based descriptors (HOG) performed poorly in the presence of occlusion because the objects shape undergoes significant changes due to the occlusion artefact. We observed that combining MBH and TRAJ descriptors outperforms other possible combinations in both partial and heavy occlusion.

The performance under very heavy occlusion in particular is surprising. While showing a significant decrease in performance compared with no occlusion, average precision over the six actions of greater than 60% is still achieved. We speculate that this is due to the extremely simplified nature of the KTH dataset, a facet noted in a recent review of datasets for human action recognition [20] that described the unrealistic nature of KTH. The differentiation between classes is high (see the confusion matrix for the baseline classification with no occlusion, Figure 5) and the area of the action boundary is relatively large. Therefore actions can be successfully differentiated by the multi-class classifier using only a small number of local descriptors.

Performance with heavy occlusion in real-world surveillance datasets is predicted to be very poor. However the strong performance of the MBH descriptor either alone or combined with TRAJ is likely to transfer to the more complex scenes.

6   Conclusion and Future Work

We have presented an evaluation and comparison framework for the state-of-art human local action descriptors. We evaluated four different local action descriptors which are Trajectory (TRAJ), Histogram of Orientation Gradient (HOG), Histogram of Orientation Flow (HOF) and Motion Boundary Histogram (MBH). These descriptors are experimented with a standard bag-of-features representation and a Support Vector Machine classifier. We investigate the performance of these descriptors and their possible combinations with respect to varying amount of artificial occlusion in the KTH action dataset. Results show that the MBH and its combination with TRAJ achieve the best performance in partial occlusion. TRAJ and its combination with MBH perform the best results in the presence of heavy occlusion.

Indictations regarding the relative importance and robustness of local action descriptors will assist in designing systems that are more resilliant to the frequent occurences of occlusion. Particularly in developing classifiers for the more complex actions and scenes found in surveillance and security applications. We hope that weighting local action descriptors in scenarios where higher levels of occlusion are likely (such as the scenes shown in figure 1) will improve the overall accuracy of the classifier.

This work demonstrated that the choice of local descriptor has an impact on the classifier performance in the presence of occlusion. Further work will explore how this will transfer to real-world applications. Particularly we will expand our evaluation with more descriptors, as well as real-world datasets like TRECVid-SED [3] and Hollywood [21] with examples of realistic occlusion.

Acknowledgements. The research leading to these results has received funding from the European Union Seventh Framework Programme (FP7/2007-2013) under grant agreement number 285621, project titled SAVASA.

References

1. Liao, M.Y., Chen, D.Y., Sua, C.W., Tyan, H.R.: Real-time event detection and its application to surveillance systems. In: International Symposium on Circuits and Systems. IEEE (2006)
2. Direkoğlu, C., O'Connor, N.E.: Team activity recognition in sports. In: Fitzgibbon, A., Lazebnik, S., Perona, P., Sato, Y., Schmid, C. (eds.) ECCV 2012, Part VII. LNCS, vol. 7578, pp. 69–83. Springer, Heidelberg (2012)
3. Over, P., Awad, G., Fiscus, J., Antonishek, B., Michel, M., Smeaton, A.F., Kraaij, W., Quéenot, G.: An overview of the goals, tasks, data, evaluation mechanisms and metrics. In: TRECVID 2011-TREC Video Retrieval Evaluation Online (2011)
4. Little, S., Jargalsaikhan, I., Clawson, K., Nieto, M., Li, H., Direkoglu, C., O'Connor, N.E., Smeaton, A.F., Scotney, B., Wang, H., Liu, J.: An information retrieval approach to identifying infrequent events in surveillance video. In: Proceedings of the 3rd ACM International Conference on Multimedia Retrieval. ACM (2013)

5. Bobick, A.F., Davis, J.W.: The recognition of human movement using temporal templates. IEEE Transactions on Pattern Analysis and Machine Intelligence (2001)
6. Yilmaz, A., Shah, M.: A differential geometric approach to representing the human actions. Computer Vision and Image Understanding (2008)
7. Dollár, P., Rabaud, V., Cottrell, G., Belongie, S.: Behavior recognition via sparse spatio-temporal features. In: 2nd Joint IEEE International Workshop on Visual Surveillance and Performance Evaluation of Tracking and Surveillance. IEEE (2005)
8. Laptev, I., Marszalek, M., Schmid, C., Rozenfeld, B.: Learning realistic human actions from movies. In: Computer Vision and Pattern Recognition. IEEE (2008)
9. Dalal, N., Triggs, B.: Histograms of oriented gradients for human detection. In: Computer Vision and Pattern Recognition. IEEE (2005)
10. Wang, H., Klaser, A., Schmid, C., Liu, C.: Action recognition by dense trajectories. In: IEEE CVPR (2011)
11. Dalal, N., Triggs, B., Schmid, C.: Human detection using oriented histograms of flow and appearance. In: Leonardis, A., Bischof, H., Pinz, A. (eds.) ECCV 2006. LNCS, vol. 3952, pp. 428–441. Springer, Heidelberg (2006)
12. Weinland, D., Özuysal, M., Fua, P.: Making action recognition robust to occlusions and viewpoint changes. In: Daniilidis, K., Maragos, P., Paragios, N. (eds.) ECCV 2010, Part III. LNCS, vol. 6313, pp. 635–648. Springer, Heidelberg (2010)
13. Dollár, P., Wojek, C., Schiele, B., Perona, P.: Pedestrian detection: A benchmark. In: Conference on Computer Vision and Pattern Recognition. IEEE (2009)
14. Poppe, R.: A survey on vision-based human action recognition. Image and Vision Computing (2010)
15. Ballan, L., Bertini, M., Del Bimbo, A., Seidenari, L., Serra, G.: Event detection and recognition for semantic annotation of video. Multimedia Tools and Applications (2011)
16. Aggarwal, J.K., Cai, Q.: Human motion analysis: A review. In: Proceedings of the Nonrigid and Articulated Motion Workshop. IEEE (1997)
17. Laptev, I.: On space-time interest points. International Journal of Computer Vision (2005)
18. Blank, M., Gorelick, L., Shechtman, E., Irani, M., Basri, R.: Actions as space-time shapes. In: Tenth IEEE International Conference on Computer Vision. IEEE (2005)
19. Sculley, D.: Web-scale k-means clustering. In: Proceedings of the 19th International Conference on World Wide Web. ACM (2010)
20. Chaquet, J.M., Carmona, E.J., Fernández-Caballero, A.: A survey of video datasets for human action and activity recognition. Computer Vision and Image Understanding (2013)
21. Laptev, I., Marszałek, M., Schmid, C., Rozenfeld, B.: Learning Realistic Human Actions from Movies. In: IEEE Conference on Computer Vision & Pattern Recognition (2008)

Online Identification of Primary Social Groups

Dimitra Matsiki, Anastasios Dimou, and Petros Daras

Information Technologies Institute, Centre for Research and Technology Hellas,
6th km Charilaou-Thermi, 57001, Thessaloniki, Greece
{matsik,dimou,daras}@iti.gr
http://www.iti.gr

Abstract. Online group identification is a challenging task, due to the inherent dynamic nature of groups. In this paper, a novel framework is proposed that combines the individual trajectories produced by a tracker along with a prediction of their evolution, in order to identify existing groups. In addition to the widely known criteria used in the literature for group identification, we present a novel one, which exploits the motion pattern of the trajectories. The proposed framework utilizes the past, present and predicted states of groups within a scene, to provide robust online group identification. Experiments were conducted to provide evidence of the effectiveness of the proposed method with promising results.

Keywords: social groups, group identification, online, motion prediction.

1 Introduction

Surveillance video applications have attracted the interest of the research community throughout the years. The majority of the related literature focuses on single object activity. Given the significant improvement of such methodologies and the need for higher level semantic extraction, the interest of the research community is shifting towards more complex structures i.e. groups. A group is defined as a collection of people who interact with one another, share similar characteristics and collectively have a sense of unity. Groups in which individuals intimately interact and cooperate over a long period of time are also known as "Primary Social Groups" [1]. Additionally to the vagueness of the "Group" definition, the high variation of the recording conditions renders group identification a challenging task.

In the proposed work, a novel online methodology to identify groups is presented. The definition of a primary group requires that its members interact for a significant amount of time, requiring a critical amount of evidence to be accumulated. In an online scenario, this evidence accumulation introduces a delay before declaring a group. To alleviate this delay, the motion of the group members in subsequent frames is predicted, using a motion model that is created offline, using accumulated motion priors. Exploiting, collectively, the already identified trajectories and the prediction of the trajectory evolvement, robust, online group identification is made possible. Acknowledging the errors introduced by trackers

C. Gurrin et al. (Eds.): MMM 2014, Part II, LNCS 8326, pp. 68–79, 2014.

and challenges introduced by the scene characteristics, a probabilistic approach to identify a group is followed. Rather than taking hard decisions, the confidence that each individual belongs to a group is evaluated.

The main contribution of the proposed framework is that it exploits the predicted positions of individuals in subsequent frames, in addition to their present and past ones, to enable online group identification. The predictions made are based on prior accumulated trajectories from the scene, exploiting context awareness. A novel metric is introduced to assess trajectory similarity, which takes advantage of the motion pattern of the trajectories. It is observed that people forming a group follow a similar motion pattern through time. This pattern is captured, using the convex hull of the trajectories' points, and a similarity criterion is created, based on the area of convex hull.

The rest of the paper is organized as follows: in Section 2 related work on group analysis is presented. Section 3 describes the proposed methodology and in Section 4 the experimental results are drawn. Section 5 contains conclusions and discussion on the proposed method.

2 Related Work

Two main approaches have been proposed so far in the literature concerning group identification. The first approach considers groups as genuine atomic entities, without contemplating individual tracks, in an attempt to overcome people detection problems in highly cluttered scenes. The second approach detects and tracks individuals, and builds upon these findings towards group tracking. We briefly review literature based on these two main approaches.

Following the first approach, a tracking algorithm is developed in [2] that uses Correlated Topic Modeling (CTM) to capture different crowd behaviors in a scene. In [3], multiple-frame feature point detection and tracking is proposed, and crowd events are modeled for specific scenarios. In [4], Reisman et al. propose to use slices in the spatio-temporal domain to detect inward motion. Their system calculates a probability distribution function (PDF) for left and right inward motion and infers a decision for crowd detection, by thresholding left and right motion histograms. In [5], the authors create a crowd model using accumulated motion and foreground information. Occurrence PDF and orientation PDF are employed to find the most frequent path of the crowd.

Employing the latter approach, an agent-based behavioral model of pedestrians is proposed in [6]. An energy function is defined and its minimization leads to the estimation of pedestrian destination and social relationships (groups). In [7], the grouping between pedestrians is treated as a latent variable, which is estimated jointly together with the trajectory information. In [8], small groups of individuals travelling together are discovered by a bottom up hierarchical clustering, using a generalized, symmetric Hausdorff distance, defined with respect to pair-wise proximity and velocity. In [9], mobile objects in a scene are stored as moving region structures and the real groups are tracked by computing the moving regions' trajectories. An interpretation module recognizes the behavior of the tracked groups. In [10], a probabilistic grouping strategy is used. A path-based grouping scheme determines a soft segmentation of groups. Probabilistic models are derived to analyze individual track motion as well as group interactions.

3   Proposed Methodology

Due to the dynamic nature of groups, there are inherent difficulties in deciding the existence or evolution of a group, judging only from a single frame. In order to tackle the group identification problem, prior trajectories and sophisticated predictions of the trajectory evolvement are employed.

The proposed approach is examining a number of criteria related to the positions and trajectories of people present in the scene, to create a number of hypotheses regarding the possible existence of pairs, which are regarded as the building block of a group. A voting scheme is employed to decide upon the validity of the formed hypotheses. The future positions of the individuals are calculated using a prediction methodology [13], based on prior motion patterns in the specific scene. Previously validated relationships between people are propagated to the next frames and are tested again, using the future predicted positions of the respective people. Pairs with mutual individuals are merged to create larger groups. A general overview of the proposed framework is depicted in Figure 1.

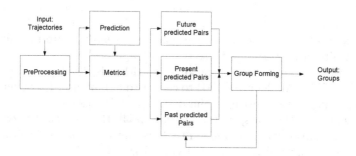

Fig. 1. System Overview

3.1   Group Identification Criteria

People appearing in the scene are detected in every frame and are tracked throughout time. The trajectory T of an individual i is defined as a set of locations w.r.t. time:

$$\mathbf{T}_i = \{x_t, y_t\}_{t=1}^K, \tag{1}$$

where x_t and y_t are the (x, y) coordinates, respectively, of person i at frame t, and K is the length of the trajectory in frames. The extracted trajectories are filtered with a median filter in order to remove noise and produce smoother paths.

Fig. 2. Example of trajectories of individuals belonging in the same group. The convex hull of each trajectory is drawn around it.

Similar to [7,8,10], it is assumed that the trajectories of group members share some common characteristics, namely spatiotemporal proximity and velocity similarity. An additional characteristic, overlooked until now, is the fact that the trajectories of the group members have a similar motion pattern. The properties of the motion pattern are captured, defining the convex hull of the trajectory as its shape descriptor. An example can be seen in Figure 2. All the above criteria are used in the proposed framework to assess the similarity of the generated trajectories.

The spatiotemporal proximity between group members is the most common similarity metric used. In the proposed work, the notion of the proximity area is utilized. For every individual, an area is set, within which, every other individual is regarded as being close. Due to the perspective effect introduced by the camera setup, the size of this area depends on the distance between the individual and the camera. The area size is defined relative to the width and height of its bounding box (w_{bbox}, h_{bbox}), in a naive attempt to rectify the effect of the distance to the camera. Thus, the proximity area of person i can be defined as $proxArea_i = \{x_i, y_i, th_w, th_h\}$, where (x_i, y_i) is the center of the bounding box of person i, and (th_w, th_h) are the width and height of the proximity area. Two individuals are considered a group when they are in each other's proximity area for more than a percentage λ of their trajectory length:

$$\{(x_j, y_j) \in proxArea_i\}_{j=1}^{K_j} > \lambda K_j, \tag{2}$$

where $\{(x_j, y_j) \in proxArea_i\}$ is the number of the trajectory points of person j in the proximity area of person i, and K_j is the length of the trajectory of j. The value of λ is experimentally set to 0.7, in all cases, to ensure that the candidate pair remains close most of the time, allowing, though, robustness to small diversions and/or tracking failures.

To identify successfully a group, though, the relative position of the individuals is equally important to their relative distance. Consider a case where two persons are moving close enough, but one is in front of the other. These two individuals do not form a pair, even though they are close. In order to filter such cases, the notion of the motion frontline is introduced. An axis, perpendicular to the motion orientation of a person, is defined as its frontline, depicted in Figure 3. The distance of a second person from this axis (d_{fl}) is defined in (3).

$$d_{fl} = d_{p_A, p_B} \sin|(\theta + sign(\tan\phi) \cdot \phi)|, \tag{3}$$

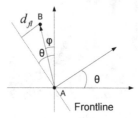

Fig. 3. Calculation of motion frontline

where p_A, p_B are the positions of the two persons, d_{p_A, p_B} is their distance, θ is the motion orientation translated to $\exists[0, \pi/2]$ set, and ϕ is the angle formed by the vector **AB** and the Y axis, as depicted in Figure 3.

A threshold (th_{fl}) concerning the acceptable distance of a person to the others frontline, relative to the size of the bounding box of the first one, is confronted as an additional criterion for pair validation.

$$d_{fl} < th_{fl} \cdot (h_{bbox} \sin \theta + w_{bbox} \cos \theta), \tag{4}$$

Another criterion to assess the similarity of two trajectories is their velocity (**V**) characteristics. Speed is an important similarity metric between trajectories. It is observed that individuals belonging to the same group have also similar speeds. In order to exploit this observation, we calculate the mean speed of every individual. Two individuals are assumed to have similar speeds if their speed ratio is below a threshold (th_{sp}):

$$\overline{v_x} = \frac{\sum_{l=2}^{K_i} x_l - x_{l-1}}{K_i - 1}, \quad \overline{v_y} = \frac{\sum_{l=2}^{K_i} y_l - y_{l-1}}{K_i - 1} \tag{5}$$

$$\frac{max(norm(\overline{v_{xi}}, \overline{v_{yi}}), norm(\overline{v_{xj}}, \overline{v_{yj}}))}{min(norm(\overline{v_{xi}}, \overline{v_{yi}}), norm(\overline{v_{xj}}, \overline{v_{yj}}))} < th_{sp}, \tag{6}$$

where $\overline{v_x}$ and $\overline{v_y}$ are the mean speed on X and Y axis, respectively, x_l and y_l are the coordinates on X and Y axis at the l-th frame of the trajectory, respectively, and K_i the length of trajectory. Moreover, $norm(\overline{v_{xi}}, \overline{v_{yi}})$ and $norm(\overline{v_{xj}}, \overline{v_{yj}})$ are the norm of the mean speed vector of person i and j, respectively.

The similarity in the orientation of the velocity is also a crucial criterion in group identification. It is defined as the angle formed between every point of the trajectory and its successive points, as depicted in Figure 4. In order to have an overall view of the orientation, the angle formed with every subsequent point of the trajectory is calculated and the mean value $\bar{a}$ is extracted. Since the arithmetic mean is not suitable for circular quantities [12], $\bar{a}$ is calculated using (8).

$$\bar{b}_p = arctan(\frac{1}{K - p} \cdot \sum_{q=p+1}^{K} \sin b_{q,p}, \frac{1}{K - p} \cdot \sum_{q=p+1}^{K} \cos b_{q,p}), \tag{7}$$

$$\bar{a} = arctan(\frac{1}{K-1} \cdot \sum_{l=1}^{K-1} \sin \bar{b}_p, \frac{1}{K-1} \cdot \sum_{l=1}^{K-1} \cos \bar{b}_p), \tag{8}$$

where $\bar{b}_p$ is the mean angle between point p and the subsequent trajectory points, K is the length of the trajectory, $b_{q,p}$ is the angle between point p and point q, and $\bar{a}$ is the mean angle. Two individuals are considered to have the same motion orientation if the difference between their mean angles is below a threshold ($th_{\bar{a}}$). The calculated mean angle differences $\exists [-\pi, \pi]$.

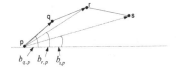

Fig. 4. Successive trajectory points and the formed angles

Another important characteristic of the trajectories that has not been properly addressed in literature is the motion pattern. A novel similarity criterion based on it, is introduced in this framework; the shape similarity of the trajectory. The shape is defined as the convex hull of the trajectory points (see Figure 2). It is argued that the convex hull contains valuable information about the evolvement of the trajectory, capturing properties of the motion pattern followed. As we can see in Figure 2, the trajectories of the two people walking together have almost the same pattern and this is depicted also on their convex hulls. The area of the convex hulls is utilized to measure the similarity between them. Two individuals are considered to have similar motion patterns if the ratio of their convex hull areas is within certain limits (th_{ch}), as described in (9).

$$\frac{max(CH_i, CH_j)}{min(CH_i, CH_j)} < th_{ch}, \tag{9}$$

where CH_i and CH_j are the areas of the convex hulls of person i and j, respectively.

The final decision on the validity of a candidate pair is taken using a voting scheme. All the criteria described are tested, and an elementary decision is taken for each criterion $c_m \exists \{0, 1\}$, where m is the identifier of the metric, based on the thresholds defined. The final decision is the average value of all elementary votes. The voting function is described in the following formula:

$$confidence = \begin{cases} 0 & if\ c_d = 0 \| c_{\bar{a}} = 0 \\ \frac{1}{C} \sum_{m=1}^{C} c_m & else \end{cases}, \tag{10}$$

where c_d is the vote of the distance criterion, $c_{\bar{a}}$ the vote of the orientation criterion, c_m the vote of criterion with label m, and C the number of criteria. The orientation similarity and the proximity between candidates are essential criteria of the group formation, and pairs that do not meet them are immediately excluded. For static people, only the proximity criterion is employed.

3.2   Prediction of Trajectory Evolution

In order to enable the online detection of groups, a prediction of the trajectory evolution is required to assist the validation of the group hypotheses created. The motion prediction methodology introduced in [13] is followed to provide the future positions of each individual in the current frame. It includes the offline, one-time creation of the motion models, and an online prediction module.

The first step in motion prediction is to create the motion model for the examined scene, based on prior motion patterns. The accumulated prior trajectories that are used as training material are divided into smaller tracklets with a fixed length $N_{tracklet}$. In order to use only the most informative tracklets, very small tracklets are removed and the remaining are filtered to produce smoother paths, producing a large set of tracklets that summarize the motion patterns observed in the scene. To reduce it, a grid of equally-spaced points is applied on the image. Mean shift clustering is performed on the local neighborhood of every grid point. The mean tracklet of every detected cluster is assigned on that point, and represents a local motion model. Thus, every grid point obtains multiple local motion models that reflect the underlying scene dynamics. Next, Gaussian Process (GP) [14] regression is used in order to model the dominant motion patterns.

The local motion models identified are exploited to create an online motion prediction module, extending the work of [13]. Given a person in the scene, a tracklet containing the $N_{tracklet}$ prior locations of the target is fed to the Motion Prediction module. This tracklet is assigned to a grid point of the scene, based on its localization. The motion models that correspond to this grid point are employed to estimate the next positions of the trajectory under investigation. Each model produces a predicted path that includes both the prior $N_{tracklet}$ locations and a set of subsequent predicted positions. This set of paths is filtered by removing non-fitting ones, choosing the most probable one.

3.3   Online Group Identification

A group is a dynamic structure that cannot be defined in a single frame. To identify groups, in an online fashion, an overlapping time window of N frames is defined, where the N_{th} frame is the current one. The remaining $N-1$ frames of the time window are the past frames. All frames after the N_{th} frame are considered as future ones.

The process of identifying groups is initiated from the current frame. The trajectories of all individuals in the defined time window are gathered, and their similarity is tested following the criteria described in Section 3.1. At the end of the voting procedure, a set of candidate pairs is identified. Then, for all individuals that have not been assigned a pair in the current frame, their pairing history in the last $N-1$ frames is examined. If they have constituted a pair with another individual for a significant amount of time and with high confidence, this pair is propagated to the current frame. Otherwise, pairs with short history and marginal confidence are discarded.

In order to boost the robustness of the proposed methodology, the pair hypotheses propagated, are tested using the evolution of the current trajectories, estimated using the motion prediction module described in Section 3.2. The similarity testing procedure described in Section 3.1 is applied for the predicted part of the trajectory and another set of candidate pairs is produced.

Finally, all the candidate pairs from the previous steps are combined to produce a final set. For every pair hypothesis based on history, it is examined whether it is propagated to the set of candidates based on prediction. If it does propagate, the pair's confidence value is updated to the mean of the history and prediction-based confidence. Otherwise, the pair's confidence is reduced to half of its initial value. This new set of candidate pairs is concatenated with the current frame's candidate pairs, and the final set of pairs is formed.

However, our goal is to identify groups and not just pairs of individuals. Therefore, all pairs with common individuals are merged so as to generate larger groups. The confidence value of the groups is calculated as the mean confidence of all included pairs.

4   Experimental Results

Our framework is implemented using MATLAB and tested on two datasets. All the similarity criteria thresholds employed, are presented in Table 1. Their value is set using a statistical analysis of the group characteristics in the same sequences used for training the motion models, and they are common for all datasets. The performance of the proposed group identification algorithm is evaluated, using Precision, Recall, and F-measure as metrics.

Table 1. Thresholds employed in the validation criteria for pair hypotheses

Metric	Threshold values
Proximity	$th_w = 4 \cdot w_{bbox}, \quad th_h = 0.8 \cdot h_{bbox}, \quad th_{fl} = 0.2$
Orientation	$th_{\bar{a}} = 30°$
Speed	$th_{sp} = 1.5$
Shape	$th_{ch} = 1.5$

4.1   Dataset

For the evaluation process, two publicly available datasets are used, namely the BEHAVE dataset [16] and the dataset from the European Community (EC) funded project CAVIAR [15]. These two datasets were chosen because they have ground truth annotation regarding the trajectories of the people present in the scene. Additionally, for the BEHAVE dataset there is also annotation regarding the groups. For CAVIAR, the respective annotation was done manually by the authors, and it will be made publicly available to enable comparisons with other methodologies.

The BEHAVE dataset is outdoors, and it comprises of various people interaction scenarios. The frame rate is 25 frames per second (fps), and the resolution is 640x480. A ground truth file of the annotated groups is also included. For our experiments Sequence 2 is used, which consists of 5700 frames.

The CAVIAR dataset is indoors (see Figure 1). It includes 26 video sequences, containing a varying number of individuals and groups. The average length of the video sequences is 1500 frames. The resolution of the frames is 384x288 pixels and the frame rate of each sequence is at 25 fps.

For the experiments, the publicly available ground truth trajectories of both video datasets are used, to prevent the tracking errors from affecting the results, rendering future comparison even more difficult. Since most trackers do not track individuals whose bounding box has a width less than 24 pixels, these individuals were excluded from the ground truth.

4.2   Group Identification Results

We evaluate the group identification algorithm output of our framework using the group-related ground truth. The accuracy of our results has been evaluated, using as metrics Precision, Recall and F-measure at multiple levels, namely group, frame and total.

$$P_{Gf} = \frac{|\{relevantGroupMembers\} \cap \{retrievedGroupMembers\}|}{\{retrievedGroupMembers\}} \qquad (11)$$

$$R_{Gf} = \frac{|\{relevantGroupMembers\} \cap \{retrievedGroupMembers\}|}{\{relevantGroupMembers\}} \qquad (12)$$

$$P_{frm} = \frac{\sum_{f=1}^{N_G} P_{G_f}}{N_G}, \ R_{frm} = \frac{\sum_{f=1}^{N_G} R_{G_f}}{N_G}, \qquad (13)$$

$$P = \frac{\sum_{l=1}^{N_{frm}} P_{frm}}{N_{frm}}, \ R = \frac{\sum_{l=1}^{N_{frm}} R_{frm}}{N_{frm}}, \qquad (14)$$

$$F = 2\frac{PR}{P + R}, \qquad (15)$$

where P_{G_f}, R_{G_f} are the precision and recall of the group f (G_f), N_G the total number of groups in frame frm, P_{frm}, R_{frm} are the overall precision and recall for frame frm, P and R are the overall precision and recall for the video sequence, respectively, N_{frm} is the number of frames of the video sequence that have at least two detected people, and F the F-measure, which combines precision and recall.

The video sequence of BEHAVE dataset contains the forming and deforming of groups of people. The video sequences of CAVIAR dataset contain different kind of group scenarios. From the 26 sequences 14 are chosen, since the rest do not contain groups.

The results of the video sequences, using the adopted metrics are presented in Table 2. Examples of correct group identification for the BEHAVE and CAVIAR dataset are depicted in Figures 6 and 7, respectively. The yellow bounding box in Figure 6c implies the change of the status of the group, since a new member has been added. As it can be seen, our algorithm produces accurate results in most cases. Group identification failures are usually due to sudden re-positioning of the group members within its limits. An example of such case is depicted in Figure 5, where a member of a group suddenly changes its intra group position (Figure 5b). Failures of this type are temporal and when the new ordering is finalized, the group is again correctly identified (Figure 5c).

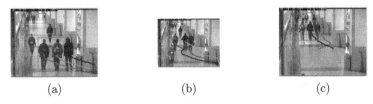

(a)   (b)   (c)

Fig. 5. A group of 3 people (a) is correctly identified, (b) until one changes position and the grouping fails to include him, (c) and he is again included when he establishes a new position in the group

Table 2. Precision, Recall, and F-measure results for the sequences tested

Dataset	Sequence	GT Group	Precision	Recall	F-measure
CAVIAR	c2es1	3	0.9757	0.9817	0.9788
	c2es2	3	0.9444	0.8342	0.8859
	c2es3	3	0.8953	0.8940	0.8946
	c2ls1	1	0.9957	0.9957	0.9957
	c2ls2	1	0.9815	0.9815	0.9815
	c3ps1	2	0.9923	0.9918	0.9920
	c3ps2	3	0.9501	0.8534	0.8992
	ceecp1	1	0.99	0.99	0.99
	cosow1	2	0.8880	0.8880	0.8880
	cosow2	5	0.9161	0.7748	0.8395
	csa1	1	0.8696	0.971	0.9175
	csa2	3	0.9459	0.9377	0.9418
	cwbs1	1	0.99	0.99	0.99
	cosme1	5	0.8691	0.9463	0.9060
BEHAVE	Seq. 2	11	0.9643	0.9310	0.9474

Comparison with other methodologies was not made possible, due to the different datasets which are not always available. Moreover, ground truth annotation is not provided and no standard evaluation metrics are employed. In this work, publicly available datasets and ground truth annotation (including the one that the authors will make public) are used, to encourage the research community to produce comparable results.

(a)   (b)   (c)

Fig. 6. Example of group identification from the BEHAVE dataset

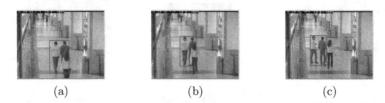

(a)   (b)   (c)

Fig. 7. Example of group identification from the CAVIAR dataset

5 Conclusions and Discussion

A novel approach for online primary social group identification is presented. The framework proposed combines tracking information for each individual in the present and recent past with a prediction of their trajectory in the near future, for robust group identification. The prediction is based on a model trained with trajectories that have been accumulated from the examined scene and used as training set. For the identification of the groups, a novel criterion based on the motion pattern is combined with established ones. The effectiveness of the proposed framework is demonstrated on two publicly available datasets. Ground truth annotation for groups will be made available by the authors. Further validation is necessary to examine the effectiveness of our framework in more complicated group scenarios and camera settings.

Acknowledgment. This work was supported by the European Unions funded project ADVISE (www.advise-project.eu) under Grant Agreement no. 285024.

References

1. CliffsNotes - Social Groups, http://www.cliffsnotes.com/sciences/sociology/social-groups-and-organizations/social-groups
2. Rodriguez, M., Ali, S., Kanade, T.: Tracking in unstructured crowded scenes. In: 2009 IEEE 12th International Conference on Computer Vision, pp. 1389–1396. IEEE (2009)

3. Saxena, S., Brémond, F., Thonnat, M., Ma, R.: Crowd behavior recognition for video surveillance. In: Blanc-Talon, J., Bourennane, S., Philips, W., Popescu, D., Scheunders, P. (eds.) ACIVS 2008. LNCS, vol. 5259, pp. 970–981. Springer, Heidelberg (2008)
4. Reisman, P., Mano, O., Avidan, S., Shashua, A.: Crowd detection in video sequences. In: 2004 IEEE Intelligent Vehicles Symposium, pp. 66–71. IEEE (2004)
5. Zhan, B., Remagnino, P., Velastin, S.A.: Mining paths of complex crowd scenes. In: Bebis, G., Boyle, R., Koracin, D., Parvin, B. (eds.) ISVC 2005. LNCS, vol. 3804, pp. 126–133. Springer, Heidelberg (2005)
6. Yamaguchi, K., Berg, A.C., Ortiz, L.E., Berg, T.L.: Who are you with and where are you going? In: 2011 IEEE Conference on Computer Vision and Pattern Recognition (CVPR), pp. 1345–1352. IEEE (2011)
7. Pellegrini, S., Ess, A., Van Gool, L.: Improving data association by joint modeling of pedestrian trajectories and groupings. In: Daniilidis, K., Maragos, P., Paragios, N. (eds.) ECCV 2010, Part I. LNCS, vol. 6311, pp. 452–465. Springer, Heidelberg (2010)
8. Ge, W., Collins, R.T., Ruback, R.B.: Vision-based analysis of small groups in pedestrian crowds. IEEE Transactions on Pattern Analysis and Machine Intelligence 34(5), 1003–1016 (2012)
9. Cupillard, F., Brémond, F., Thonnat, M.: Tracking groups of people for video surveillance. In: Video-Based Surveillance Systems, pp. 89–100. Springer (2002)
10. Chang, M.-C., Krahnstoever, N., Ge, W.: Probabilistic group-level motion analysis and scenario recognition. In: 2011 IEEE International Conference on Computer Vision (ICCV), pp. 747–754. IEEE (2011)
11. McPhail, C., Wohlstein, R.T.: Using film to analyze pedestrian behavior. Sociological Methods & Research 10(3), 347–375 (1982)
12. Bishop, C.M., Nasrabadi, N.M.: Pattern recognition and machine learning, vol. 1. Springer, New York (2006)
13. Lasdas, V., Timofte, R., Van Gool, L.: Non-parametric motion-priors for flow understanding. In: Proceedings of the 2012 IEEE Workshop on the Applications of Computer Vision, WACV 2012, pp. 417–424. IEEE Computer Society, Washington, DC (2012), http://dx.doi.org/10.1109/WACV.2012.6163049
14. Rasmussen, C.E., Williams, C.K.I.: Gaussian Processes for Machine Learning (Adaptive Computation and Machine Learning). The MIT Press (2005)
15. Caviar dataset, http://homepages.inf.ed.ac.uk/rbf/CAVIARDATA1
16. BEHAVE dataset, http://homepages.inf.ed.ac.uk/rbf/BEHAVE/

Gait Based Gender Recognition
Using Sparse Spatio Temporal Features

Matthew Collins, Paul Miller, and Jianguo Zhang

Centre for Secure Information Technologies
Queen's University Belfast,
Northern Ireland Science Park, Queen's Road, Queen's Island, Belfast, BT3 9DT
{m.collins,p.miller}@qub.ac.uk,
jgzhang@computing.dundee.ac.uk
http://www.csit.qub.ac.uk/

Abstract. A gender balanced dataset of 101 pedestrians on a treadmill is presented. Gait is analysed for gender classification using a modification of a framework which has previously proven effective when used in behaviour recognition experiments. Sparse spatio temporal features from the video clips are classified using Support Vector Machines. Tuning parameters are investigated to find an effective feature descriptor for gender separation and an accuracy of 87% is achieved.

Keywords: Gait Based Gender Classification, Video Dataset.

1   Introduction

In this work we utilise and further develop a framework previously used for characterising behaviour from video sequences in order to carry out the task of gender classification of subjects performing a specific behaviour, namely walking. In their paper, Behaviour Recognition via Sparse Spatio-Temporal Features, Dollar et al. [1] released a framework which used their own interest point detector to extract feature points within video clips. They extracted cuboids i.e. spatio-temporal windowed data surrounding each feature point and used established feature descriptors to characterize these cuboids. Cuboids are then clustered into a dictionary of cuboid prototypes for a bag of words model and a final cuboid type histogram is produced. This work showed promising results on a number of behaviour recognition datasets and prompted further investigation here in a new problem domain.

The focus of our own work is person profiling in a CCTV context. Specifically here we limit the scope to that of gender classification. Previous work has focused on the still image domain, extracting body shape information from single frame images and performing classification based on the resulting feature vectors. It has been shown that the feature extraction and classification techniques used in the realm of object recognition, specifically within pedestrian recognition can be further pushed into further categorisation of the subjects in the images as either Male Pedestrian or Female Pedestrian [2,3]. We now apply a similar hypothesis

C. Gurrin et al. (Eds.): MMM 2014, Part II, LNCS 8326, pp. 80–91, 2014.

to behaviour recognition, theorising that a walking behaviour descriptor should be possible to be further categorised into Male walking and Female walking.

Within psychology there are a number of papers that show that human visual systems are particularly well tuned to infer a number of cues including gender from observing biological motion, and that while structural information such as hip/shoulder ratio are useful in human gender perception, the visual system tends to rely more on dynamic information [4]. Pollick et al. showed that a point light display was sufficient for an observer to recognise the gender of a walking subject [5]. Troje et al. constructed a system for the analysis and synthesis of human gait patterns. Their framework transforms biological motion into a representation allowing for analysis using linear methods from statistics and pattern recognition. Their system also included a simple motion modeller which makes it possible to visualise and exaggerate the differences between male and female walking patterns [6].

This modeller confirms visually that the male gait is typified by lateral body sway in the shoulder region, whereas in females this lateral motion is more pronounced at the hips. Similar observations about what a human observer looks for when classifying gender from gait were noted by Barclay et al. in [7].

Similarly, computer vision researchers have recently begun to analyse gait as a bio-metric for use in a number of areas including subject identification and also gender classification specifically. Model based approaches such as that of Yoo et al. extract a stick figure model from the subjects body contour [8]. It is computationally expensive in extracting features robustly and gait sequences are required to be of a very high quality, with body parts requiring to be tracked in every frame [9]. Other techniques for gait analysis included appearance based approaches. Often a silhouette of a walking human is taken from the image of the walking subject and changes in this silhouette between frames are analysed, although Shan et al. have also shown that the Gait Energy Image (GEI), which represents human motion in a single image while preserving the temporal information, by averaging the silhouettes across frames into a greyscale image, can also be used as an effective representation for gender classification [10]. However GEI's are not always easy to extract in real scenarios, or very difficult without knowing the background images to perform background extraction such as in the popular CASIA database [11,12].

We investigate dynamic information (pure gait information) rather than the temporal shape information captured by GEI's. Our work tries to relate the interest points which are automatically extracted based on periodic motion within the video sequence, to the joints which are typically tracked in light point walker experiments and aims to capture similar lateral motion in areas such as hips and shoulders as were previously observed in the Troje et al. modeller.

2   System Description

2.1   Dataset

We recorded our own dataset of walking subjects for the purposes of this experiment. We recorded a gender balanced set of 101 subjects walking on a treadmill using a synchronised multi camera setup from multiple angles through 180 degrees of rotation around the subject. The aim was to record a video footage representation (without motion capture sensors as these are not available in a CCTV scenario) of the natural gait of the subject but in a controlled environment with consistent lighting, an uncluttered background and with fixed camera positioning.

Subjects were encouraged to walk on the treadmill as the speed was adjusted to one they were comfortable with and felt most closely matched their natural walking pace. They were allowed to walk for a period of time to acclimatise themselves with the sensation and to allow them to settle into a more natural gait before recording began and subjects were not told when recording had started. In as much as was possible, subjects were distracted with conversation during the recording process to prevent them over thinking about how they were walking and in this way their natural gait was able to be recorded. 60 seconds of footage at 25 frames per second was recorded for each subject.

Fig. 1. Sample subjects from Treadmill dataset

Cameras captured a 0° frontal view of each subject, a 180° view from behind, a 90° side angle view as well as both 45° and 135°. The cameras are synchronised so that corresponding timestamps from each of the angles correspond to the same frames. However, for the purposes of this experiment we focus only on the 0° frontal view videos as this most directly corresponds to point light walker gender recognition experiments seen previously. The feature extraction framework we use here is easily extendible to other viewing angles and might be interesting in future work to examine how classification is affected by a change in angle, or if a combination of cues from a multi angle setup could help improve results.

The dataset should also be of use in a number of other interesting areas of computer vision research such as multi camera subject reacquisition.

2.2   Feature Descriptors

We use a modified version of the Dollar cuboid feature extraction toolbox used in the paper Behaviour Recognition via Sparse Spatio-Temporal Features, with code available online. The histogram based descriptors used in their paper discarded positional knowledge of where the features were extracted from so we added an extra step in feature computation adding spatial constraints to the final computed feature vector. The following section gives an outline of how the features are computed and used in classification. A number of parameters can be modified during this computation and these are discussed in the experiments section.

Interest Point Detection: Firstly spatio temporal interest points are detected within the video which are then treated as the central point around which a cuboid of pixel values are extracted and processed to form the final descriptor. These interest points are the local maxima of a response function computed at every point. In the purely spatial domain, this typically takes the form of convolving the image $I(x, y)$ with a Gaussian smoothing kernel $g(x, y, \sigma)$ or similar. Here the parameter σ relates to spatial scale and the resulting response strength is an indication of the presence of interest points in the image such as corners and well defined edges.

Extending this into the temporal domain requires interest point detection to be applied to a stack of images $I(x, y, t)$. The 3D Harris corner detector is a popular and well-studied technique in which gradients are found along all 3 dimensions x,y and t and spatio temporal corners are regions where gradient directions are orthogonal over the 3 dimensions. In Dollars paper, they observed that sometimes, the Harris detector was not the most appropriate response function to use and proposed an alternative spatio temporal interest point detector for their framework [1]. It is tuned to fire in local image regions containing periodic motion and where there is complex motion in regions which are spatially distinguishable.

The response function takes the form of

$$R = (I * g * h_{ev})^2 + (I * g * h_{od})^2 \tag{1}$$

where I is an image cube $I(x, y, t)$, g is a 2D Gaussian smoothing kernel applied only to the spatial domain and h_{ev} and h_{od} are a quadrative pair of 1D Gabor filters applied along the temporal plane after the initial spatial smoothing has occurred.

They take the form:

$$h_{ev}(t; \tau, \omega) = -cos(2\pi t\omega)e^{-t^2/r^2} \tag{2}$$

and

$$h_{od}(t; \tau, \omega) = -sin(2\pi t\omega)e^{-t^2/r^2} \tag{3}$$

where $\omega = 4/\tau$, thus the response function effectively has two parameters σ and τ which correspond to the spatial and temporal scale of the detector respectively.

Fig. 2. Sample Interest point detection over a short 16 frame treadmill sequence Coloured points persist over multiple frames

Data Preparation and Feature Extraction: In [1] they used a number of short video clips of a small number of subjects performing a variety of behaviours which they wished to classify. Here each video of a subject in our treadmill database was 1500 frames in duration. We had 101 subjects and only a single behaviour, walking, which we aimed to sub classify into male walking and female walking.

One modification we made to the framework and experimental set up was that a parameter value was used to first divide each of our videos into a number of shorter clips to be processed individually and then, after feature extraction, data from clips corresponding to the same subject are recombined into a final feature vector preserving the temporal relationship between the clips.

For each clip, periodic interest point detection as described above is performed then a 25x25x25 cuboid of pixel values is extracted centred on the points which gave the highest responses. The size of this cuboid window is determined by a relationship between the spatial and temporal scale parameters σ and τ used in the interest point detector and is set to contain the majority of the data which contributed to the detector response. Padding is applied, should a cuboid window overlap the clip it is being extracted from. Then, for each extracted cuboid a descriptor is computed which can be compared using Euclidian distance. We investigated both Gradient and Optic Flow based features as in [1].

For Gradient features, the intensity gradient (G_x, G_y, G_t) is computed and the three channels are concatenated into one vector. For Optic Flow based features, Lucas Kanade optical flow is computed on each pair of consecutive frames, giving two channels (V_x, V_y), each one frame shorter than the original cuboid. Once again, these channels are flattened into a single feature vector for the cuboid. Regardless of the descriptor choice, PCA dimensionality reduction is performed reducing each vector to a dimensionality of 100.

Once all cuboids feature vectors have been computed for each clip in the dataset, Kmeans clustering is performed to produce a vocabulary of cuboid types. Once clusters are known, each cuboid is assigned a type using Euclidian distance from the computed clusters, then a histogram of cuboid types for each clip is constructed. Where in the original toolbox the histogram vectors would discard any positional information about the origin of the cuboid, we made a modification to maintain some of the spatial constraints. Using knowledge of previously provided per subject bounding boxes divided according to a parameter specified spatial gridding, each detected cuboid is assigned a spatial block location according to its spatial coordinates extracted during the interest point detection stage. Although the framework allows for variation and experimentation we chose a 4x8 spatial gridding as this intuitively corresponded with key regions of interest in the body and the general layout of the analogous Troje light point walker demo experiments.

Histograms of cuboid types are computed at each block location and concatenated preserving the spatial layout information. Finally each clip corresponding to the same subject is concatenated in order to bring in an element of the temporal relationship between the clips. Optionally at this temporal concatenation stage, there could also be a degree of histogram pooling to overlap clips, again parametrised.

Classification: The dataset consists of 101 videos of 101 subjects. For classification purposes these are divided into 4 gender balanced folds. Feature extraction was performed on all videos, then the final computed histograms were used in a 4 fold classification set up where 3 folds are used for training and the remaining fold used for testing. This is repeated four times holding a different fold back for testing each time. The final confusion matrix and classification error reported is a result of taking the mean confusion matrix across all four folds.

We allowed for the use of two classification methods at this stage. Initially, simple Nearest Neighbour classifier (using Chi Squared Distance) provided as part of the Dollar toolbox. This is a naïve classification method, but also ensured that observed changes in performance as parameters were adjusted, were only as a result of the features themselves, and not some quirk of a classifier.

We also modified the system to allow repeating the classification steps using an SVM using a Chi Squared Kernel Matrix (noted for its suitability for histogram based feature representations).

The SVM decision function takes the form:

$$g(x) = \sum_i a_i y_i K(x_i, x) - b \tag{4}$$

where the Chi squared kernel is defined as:

$$k(x_i, x_j) = \sum_{i=1}^{n} exp\left(-\gamma\left(\frac{(x_i - x_j)^2}{(x_i + x_j)}\right)\right) \tag{5}$$

This allowed use of the returned SVM confidence values to plot ROC curves, which was not possible with the NN classifier. It was noted that for some experiments classification performance using the two different classifiers could be markedly different. For example in experiments varying clip length keeping all other parameters constant the SVM appeared to be significantly poorer than the Nearest Neighbour approach, at least in the case where there was no temporal histogram pooling applied. However when a small amount of temporal pooling was applied the SVM classifier showed a marked improvement in performance (whereas the effect on the NN classifier versus no pooling was negligible) - see Experiment 2 section.

3   Experiments and Results

A number of key parameters have been identified in the feature extraction section. We established a framework to systematically vary a single parameter of interest keeping all other parameters constant so as to examine its effect.

3.1   Experiment 1   Clip Length

Initially we chose to investigate clip length i.e. the number of frames per clip to divide the 1500 frame subject videos into before performing feature extraction. 6 clip lengths were investigated keeping all other parameters constant: 25, 50, 75, 100, 125, 150 frames per clip and each video of 1500 frames was then divided into a number of non-overlapping clips depending on specified clip length. Thus the number of clips per video in each scenario were 60, 30, 20, 15, 12 and 10 respectively.

Other parameters which remained constant are detailed in Table 1

Table 1. Parameter setup for frame length experiments

Cuboids extracted per clip	50
Feature type computed on each cuboid	Optic flow
Cuboid dimensionality (controlled by other parameters)	25 x 25 frames x 25 pixels
Number of clusters for cuboid prototypes	350
Spatial gridding for per-subject bounding boxes	4x8
Temporal histogram pooling	None - straight concatenation

Features were extracted according to the process outlined earlier and classification performance was compared.

Table 2 presents accuracies and classification error results from Nearest Neighbour Classification as clip length is varied with all other parameters remaining constant. Note that the lowest classification error here, 100 frames does not actually seem to relate to the best overall separation. In fact there is a strong bias towards overclassifying as Female. For clip lengths of 125 and 150 frames, classification of male subjects is essentially random chance. Note also that as the clip length increases, the total number of clips decreases. Since the number of features sampled per clip remains constant, the total number of features included in the clustering step to determine the cuboid prototypes becomes much lower and this appears to have a negative effect on the discriminative power of the features.

Table 2. NN Classifier results (Experiment 1)

Clip Length (No. Frames)	Male Acc/Missclass	Female Acc/Missclass	Overall Error Rate
25	63% / 37%	76% / 24%	0.3069
50	61% / 39%	72% / 28%	0.3366
75	57% / 43%	78% / 22%	0.3267
100	57% / 43%	84% / 16%	0.2970
125	51% / 49%	78% / 22%	0.3564
150	51% / 49%	84% / 16%	0.3267

SVM: Next, in Table 3, we present results for the same underlying data as before but this time using a chi squared kernel SVM classifier. (libSVM using custom kernels). Using the SVM classifier provides a classifier confidence score which can be used to plot ROC curves etc.

The results presented though, are considerably worse than those of the nearest neighbour classifier. They are presented here purely so the ROC curves can be examined, though their value is questionable. Were it not for an interesting boost in performance, noted when other parameters are also changed (see Experiment 2), it would perhaps be best not to consider this classifier at all.

Table 3. SVM Classifier results (Experiment 1)

Clip Length (No. Frames)	Male Acc/Missclass	Female Acc/Missclass	Overall Error Rate
25	73% / 27%	24% / 76%	0.5149
50	75% / 25%	26% / 74%	0.4950
75	76% / 24%	42% / 58%	0.4059
100	75% / 25%	44% / 56%	0.4059
125	75% / 25%	48% / 52%	0.3861
150	76% / 24%	48% / 52%	0.3762

These results show little discriminative power at all, and a clear bias towards overclassification as male (as opposed to the relatively small bias towards female classification shown by the NN classifier).

We also present in Figure 3 a set of ROC curves generated from the full dataset, concatenating ground truth and svm scores from each of the 4 folds, in order to give a complete picture. (As accuracy between folds could vary greatly). Here the male class has been taken as the positive class.

The bias towards male noted in the accuracy table appears to level off a little as clip length increases, however overall it is clear that the change in classifier has led to a marked drop in performance. It appears that clip length is not really an appropriate parameter to vary in order to assess its effect on accuracy of the

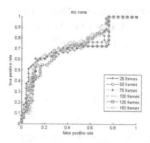

Fig. 3. ROC curves from SVM results for frame length experiment

system. The reason for this is likely that clip length doesn't have a direct effect, but rather a knock on effect on a number of other factors in the system. For example, increasing clip length reduces the number of clips, and as a result the total amount of features sampled, since there is another parameter for number of features sampled per clip.

It is important to have a clip length long enough to capture the relevant motion, as clip length will affect the spatial and temporal locations of the extracted cuboids. However it doesn't change the temporal length of the cuboids themselves as this would be controlled by other parameters, so after a certain size the effect of clip length on features, other than how many are extracted, should be negligible.

3.2　Experiment 2　Temporal Pooling and Number of Cuboids Extracted Per Clip

It was noted while testing the SVM classifier that whilst a direct repetition of the Clip Length experiments keeping all other parameters constant resulted in a drop in performance, a minor change to one parameter actually dramatically improved results, namely the degree of temporal histogram pooling. Setting this to 3 resulted in a marked improvement in performance, particularly using the Chi Sq kernel SVM (noted for its suitability for histogram based feature representations).

For temporal pooling, rather than simply concatenating the computed histograms for each clip in temporal order a degree of pooling in performed. For example if the parameter is set to 3, the histograms for clips 1-3 are added together then concatenated with the pooled sum of clips 2-4, then clips 3-5 and so on. In this way the temporal relationship between the clips is more strongly emphasised than straight forward one to one concatenation.

A sample of results for 50 frame clip length and 100 frame clip length with temporal histogram pooling set to 3 clips are shown below. In both cases the Chi sq kernel SVM has been used for classification. The results are significantly better than with either classifier (SVM or NN) with no temporal pooling, though there is still a strong bias towards one class, in this case male. In this case, unlike the earlier experiments, it is the SVM classifier which gives the better performance.

Using the NN classifier for in the 50 frame clip length, 3 clip temporal pooling scenario has a negligible effect against no pooling. But it is clear that in the SVM case performance is greatly improved.

It was felt that while clip length might not be the most appropriate parameter to vary, temporal pooling and the number of cuboids extracted per clip might play a more significant part and deserved fuller investigation. As the degree of temporal pooling is just a simple final step in the histogram construction process after all the intensive feature extraction, clustering etc. has been performed; these two parameters were investigated in tandem. The number of features sampled was varied in steps of 10 from 20 cuboids up to 100 and varied temporal pooling in each case. Clip length was set at 50 frames per clip and all other parameters remained constant.

Table 4. Parameter setup for temporal pooling and number of cuboid experiments

Feature type computed on each cuboid	Optic flow based
Cuboid dimensionality (controlled by other parameters)	25 x 25 frames x 25 pixels
Number of clusters for cuboid prototypes	350
Spatial gridding for per subject bounding boxes	4x8
Temporal histogram pooling	Varied 1-10 for each case

Figure 4a shows the effect on classification error as temporal pooling is varied. We show this for a sample of experimental setups each examining the number of features sampled per clip. All other parameters remain constant and these results are from the SVM classifier.

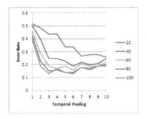 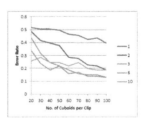

Fig. 4. Sample classification error change as (a) temporal pooling and (b) No. of Cuboids sampled is varied

Note the high classification error with no temporal pooling and how it consistently improves with even a minor level of pooling. This positive effect levels off though at a point between 4 and 6 frames being pooled and results begin to get worse again as indicated by the almost bowl shaped graphs. This implies that while an element of temporal pooling reinforces the temporal relationship between the clips and helps with the overall classification, too much most likely results in saturation of the histogram bins in the feature vectors, reducing their sparcity and the meaningfulness of the information held within, adversely affecting classification again.

Figure 4b shows the reverse. Here, in each case, the degree of temporal pooling remains constant and the number of features sampled per clip remains constant. Again we see trends in the direction of the classification error reported by the SVM classifier. Generally, as the number of features sampled per clip increases, the classification error falls. It becomes a little erratic in the cases where temporal pooling has got particularly high, reflecting similar trends in the previous table where classification error started to rise again as temporal pooling rose about 6 or 7.

Across both sets of experiments the lowest classification score achieved was 0.1287 which occurred when the number of features was set at 100, and with temporal pooling set at either 3 or 6. Figure 5 shows the corresponding confusion matrix and ROC curve for this best result (Cuboids:100, Temporal pooling: 6). There is a very high degree of separation here and results are very promising. Across the 4 folds, of the 101 videos there were only 13 misclassified examples. Of those examples, some can be visually examined and explanations could be offered as to why the might be considered outliers.

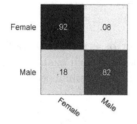 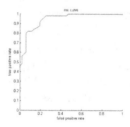

Fig. 5. Confusion Matrix & ROC Curve for experimental setup with 100 cuboids sampled and a temporal pooling of 6 (Err: 0.1287)

Equally it is important not to get too excited about this high degree of separation as it is a relatively small dataset and it could be argued that over tuning the parameters is training the classifier to recognise the data rather than the problem. However the consistent trends in the tables presented here seem to indicate a solid foundation for the result so it is encouraging.

4   Conclusion

We present a study of using spatio-temporal features for gait gender recognition in video sequences. We technically show that spatial temporal features are effective for gender recognition, and further propose an improved version of feature representations using spatial partition and temporal pooling, which is motived by the psychology study in [5]. Though this is similar to the popular spatial pyramid matching used in object recognition and action recognition, this is the first time its effectiveness has been shown in the context of gender recognition.

Parallels are drawn between the automatically extracted interest points at sites of high periodic motion, and the key points at subjects joints tracked in light point walker experiments which have been shown for example to capture the lateral motion in shoulders in male walkers, and in hip sway in females.

The dataset contains synchronised recordings of subjects walking captured from multiple angles, though experiments here only focus on the frontal view. In future work it would be good to see if accuracies can be matched or improved upon from other angles, or if data captured from multiple angles could be fused. It would also be worth exploring fusing static image full body gender recognition features such as those explored in [2] with the dynamic information presented here to see if it could further improve performance.

References

1. Dollár, P., Rabaud, V., Cottrell, G., Belongie, S.: Behavior recognition via sparse spatio-temporal features. In: VS-PETS (October 2005)
2. Collins, M., Zhang, J., Miller, P., Wang, H.: Full body image feature representations for gender profiling. In: IEEE Workshop on Visual Surveillance, ICCV (2009)
3. Collins, M., Zhang, J., Miller, P., Wang, H., Zhou, H.: Eigenbody: Analysis of body shape for gender from noisy images. In: IMVIP, University of Limerick, Cambridge Publications (2010)
4. Mather, G., Murdoch, L.: Gender discrimination in biological motion displays based on dynamic cues. Proc. R. Soc. B 258(1353), 273–279 (1994)
5. Pollick, F.E., Kay, J.W., Heim, K., Stringer, R.: Gender recognition from point-light walkers. J. Exp. Psychol. [Hum. Percept.] 31(6), 1247–1265 (2005)
6. Troje, N.F.: Decomposing biological motion: a framework for analysis and synthesis of human gait patterns. J. Vis. 2(5), 371–387 (2002)
7. Barclay, C.D., Cutting, J.E., Kozlowski, L.T.: Temporal and spatial factors in gait perception that influence gender recognition. Percept. Psychophys. 23(2), 145–152, 643509 (1978)
8. Yoo, J.-H., Hwang, D., Nixon, M.S.: Gender classification in human gait using support vector machine. In: Blanc-Talon, J., Philips, W., Popescu, D.C., Scheunders, P. (eds.) ACIVS 2005. LNCS, vol. 3708, pp. 138–145. Springer, Heidelberg (2005)
9. Ng, C.B., Tay, Y.H., Goi, B.M.: Vision-based human gender recognition: A survey. CoRR abs/1204.1611 (2012)
10. Shan, C., Gong, S., McOwan, P.W.: Fusing gait and face cues for human gender recognition. Neurocomput. 71(10-12), 1931–1938 (2008)
11. Yu, S., Tan, D., Tan, T.: A framework for evaluating the effect of view angle, clothing and carrying condition on gait recognition. In: ICPR, vol. 4, 441–444 (2006); ID: 1
12. Zheng, S., Zhang, J., Huang, K., He, R., Tan, T.: Robust view transformation model for gait recognition. In: ICIP, pp. 2073–2076 (2011); ID: 1

Perspective Multiscale Detection and Tracking of Persons

Marcos Nieto, Juan Diego Ortega, Andoni Cortes, and Seán Gaines

Vicomtech-IK4, Paseo Mikeletegi 57, San Sebastian, Spain
mnieto@vicomtech.org

Abstract. The efficient detection and tracking of persons in videos has widespread applications, specially in CCTV systems for surveillance or forensics applications. In this paper we present a new method for people detection and tracking based on the knowledge of the perspective information of the scene. It allows alleviating two main drawbacks of existing methods: (i) high or even excessive computational cost associated to multiscale detection-by-classification methods; and (ii) the inherent difficulty of the CCTV, in which predominate partial and full occlusions as well as very high intra-class variability. During the detection stage, we propose to use the homograhy of the dominant plane to compute the expected sizes of persons at different positions of the image and thus dramatically reduce the number of evaluation of the multiscale sliding window detection scheme. To achieve robustness against false positives and negatives, we have used a combination of full and upper-body detectors, as well as a Data Association Filter (DAF) inspired in the well-known Rao-Blackwellization-based particle filters (RBPF). Our experiments demonstrate the benefit of using the proposed perspective multiscale approach, compared to conventional sliding window approaches, and also that this perspective information can lead to useful mixes of full-body and upper-body detectors.

Keywords: Object Detection, Machine Learning, Person Detection, Person Tracking, Homography, Camera Calibration.

1 Introduction

People detection and tracking in video sequences using computer vision methods has become a hot topic in the related scientific community due to its potential in CCTV applications like surveillance or forensics. Significant progresses have been made, specially in the object detection-by-classification approaches [16], object tracking using appearance [8,2], and also to extract semantic information from the sequence [14,11].

Detection-by-classification is the most promising family of techniques, using the sliding window technique [15], which consists on exhaustively scanning the whole image searching for objects at different scales or levels. Although this methodology is adequate for general problems, it is too much exhaustive for CCTV applications. On the one hand they may require low computational cost

C. Gurrin et al. (Eds.): MMM 2014, Part II, LNCS 8326, pp. 92–103, 2014.
© Springer International Publishing Switzerland 2014

(to analyze many video files in large installations), but on the other hand are typically static enough to use useful prior information of the scene.

Using contextual information is a way enhance such approaches [4]. We propose to exploit the perspective information of the scene to determine the maximum and minimum expected size of persons at different locations in the images and use them to reduce the number of levels to be used. In the context of CCTV systems, it is broadly accepted an initial set-up or installation stage in which prior information can be retrieved using an appropriate GUI. We have observed that the generation of the perspective information with a GUI takes only about 1-2 minutes and might allow for significant speedups (in our experiments from 30% to 80% depending on the perspective), which can result on more video sequences processed with the same computer or less time to process a given video file, and also better results in terms of false positives using the same detectors.

Most related works focus on the detection of full-body [12], upper-body [16,10,13], or heads [2], according to the type of targer application. We propose to use both type of detectors and combine them using the perspective. On the one hand, full-body detections are really distinctives when the person is seen completely in the image. On the other hand, upper-body detections are useful in scenarios in which partial occlusions happen.

However, using two detectors imply more computational load, and also the necessity to handle more false positives. The use of the perspective information help us to control these two problems. Particularly, we combine these two type of detections so that (i) each upper-body detection generates a full-body estimation; and (ii) the location of detections is projected into the plane to filter out false positives by checking if its size is between the expected minimum and maximum sizes of persons.

To complete our contributions, we apply a tracking approach based on the Rao-Blackwellization Data Association Particle Filter (RBDAPF) [3] that provides the required temporal coherence to detections by linking detections through time and generating predictions according to object appearances.

The results presented at the end of the paper demonstrate the benefits of using the proposed approach, specially the usage of the perspective of the scene, plus the combination of detectors in surveillance sequences (for this purpose we have used the available dataset from Oxford Active Vision group [2]).

2 Approach Overview

Figure 1 illustrates the modular architecture of the proposed approach. Details of each module are given in the next sections. The first step is the generation of the perspective information, that can be done offline, and it is only done once. This information is encoded as the projection matrix P, which is composed by the camera calibration matrix K and the relative pose of the camera, R and $\mathbf{t}$ with respect to a coordinate system placed in the dominant plane of the scene.

The full-body and upper-body detectors load the respective SVM models, and the multiscale sliding window parameters are set according to the perspective. The detector then detects candidate regions of the images likely containing

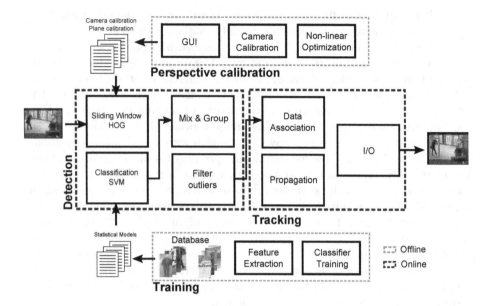

Fig. 1. Perspective multi-scale using detection-by-classification

full-bodies and upper-bodies. These regions are mixed (upper-bodies can be up-graded to full-bodies using approximate human dimensions), and filtered (many false positives are removed applying the perspective restriction which determines the expected sizes of human beings at different positions in the image). Also, for long sequences, conventional background substraction methods could be applied, and only those regions which contain a certain amount of foreground pixels are considered as valid detections (in this paper we have not included this module because we wanted to focus on the detectors alone).

As a result, a set of detections is obtained and fed to the tracker, which associates the detections with the tracks. Entering and exiting persons or tracks are handled by the tracker, which creates a new track when detections not associated to existing tracks show time coherence (e.g. appear consecutively during a number of frames), and deletes an existing track when it is not associated to detections during a certain amount of frames.

3   Perspective Multiscale Detection

The calibration of the camera and the computation of its relative pose with respect to the ground plane offers valuable information for the detection of persons in images under the hypothesis that there is a dominant ground plane in the scene and persons are on it. In this work we propose to formalize the exploitation of the perspective of the scene by means of computing the projection matrix and defining a multi-scale detection approach according to it.

3.1   Perspective Multiscale

Figure 2 illustrates the difference between the typical use of multiscale sliding window detectors and our proposed approach. The simplest way to proceed is to run a multi-scale scanning of the image evaluating each image patch with the classifier in order to determine the presence of objects in the image. Starting from the smallest size, which is determined by the window size parameter of the SVM model (e.g. 64×128 pixels for instance), L copies of the images are created, down-scaled by a factor that is typically 1.05 or 1.1 in the hope that this exhaustive scan will likely find small, medium and large objects; we have called this method *brute-force multiscale*. The total amount of evaluations of image patches against the SVM classifier is given by $N_e = \sum_{i=1}^{L} \left(\frac{W_i}{s} - \left(\frac{w}{s} - 1 \right) \right) \left(\frac{H_i}{s} - \left(\frac{h}{s} - 1 \right) \right)$, where $W_i \times H_i$ is the size of the image in pixels at each level, s is the window stride, and $w \times h$ is the size of the model. For instance, for a 1920×1080 image with a 64×128 model and $s = 16$, and $L = 10$, then $N_e = 144880$.

In our approach, since the projection matrix P is known, we can reproject a human model to any position in the scene (see Figure 3) and determine the smallest and largest sizes of it in a region of interest. Therefore we can know the exact scale we have to apply to the multilevel scan procedure to start from the smallest possible detections to the largest. In our experiments we have observed that we can reduce L to 3-5 levels to achieve similar results that the brute-force approach using $L = 10$. Note that the minimum number of scales we propose to use is 2, one corresponds to the smallest person size, and the other to the largest. Any additional scale is an intermediate scale between these two sizes.

The main difference between these alternatives is that the perspective analysis of the scene focuses significantly the effort of the classifier resulting in a much more efficient scan of the image.

3.2   Ground Plane Calibration

The calibration of the scene required to apply the proposed perspective multi-scale approach can be obtained in a single-step process. The user must introduce 4 points in the image that corresponds to a rectangle in the ground plane of the scene, plus the longitudinal and transversal distances between the points.

This information is enough to compute the homography H between the image plane (in pixels) and the ground plane (in metric units) using the DLT (Direct Linear Transform) algorithm [7].

The coordinate system in the ground plane can be selected such that it is defined by $Z = 0$. In such situation, the projection of a point $\mathbf{X} = (X, Y, Z, 1)^{\top}$ into a image point $\mathbf{x}$ yields:

$$\mathbf{x} = K(R|\mathbf{t})\mathbf{X} = K(\mathbf{r}_1 \, \mathbf{r}_2 \, \mathbf{r}_3 \, \mathbf{t})\left(X \, Y \, 0 \, 1 \right)^{\top} \tag{1}$$

and therefore:

$$\mathbf{x} = K(\mathbf{r}_1 \, \mathbf{r}_2 \, \mathbf{t})\left(X \, Y \, 1 \right)^{\top} \tag{2}$$

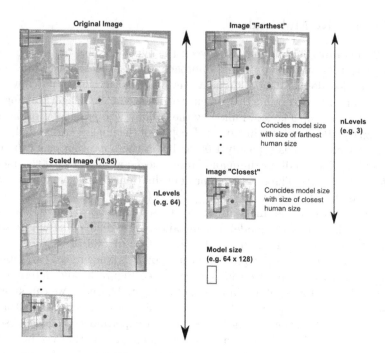

Fig. 2. Sliding window approaches: (left) Brute-force and (right) Perspective multi-scale

which is a 3×3 homography between the image and world plane points:

$$H = K(\mathbf{r}_1\ \mathbf{r}_2\ \mathbf{t})\qquad(3)$$

As expected, the homography matrix contains all the information about the intrinsics and extrinsics parameters of the projection process. We use the following procedure to estimate the values of K, R and $\mathbf{t}$:

First, the calibration matrix can be assumed to have 1-DoF with the principal point as the center of the image so the only unknown is the focal length that can be computed solving the following expression with SVD (Singular Value Decomposition):

$$\begin{pmatrix} h_{0,0}h_{0,1} + h_{1,0}h_{1,1} \\ h_{0,0}^2 - h_{0,1}^2 + h_{1,0}^2 - h_{1,1}^2 \end{pmatrix} \mathbf{x} = \begin{pmatrix} -h_{2,0}h_{2,1} \\ -h_{2,0}^2 + h_{2,1}^2 \end{pmatrix}\qquad(4)$$

as $f = \|x_0\|^{-\frac{1}{2}}$, where x_0 is the first eigenvalue of $\mathbf{x}$.

Second, given this initial value of K, we can calibrate the expression of the homography:

$$K^{-1}H = (\mathbf{p}_1\ \mathbf{p}_2\ \mathbf{p}_3)\qquad(5)$$

This way, once we have computed and calibrated the homography we can extract the rotation and traslation from the columns of the resulting matrix. Please

Fig. 3. Example use of the GUI for calibrating a typical CCTV scene. The rulers and the projected boxes help to fit the expected sizes of pedestrians.

note that since these are homogeneous matrices it is necessary to normalize the columns of the matrix in order to get the vectors: $r_1 = \frac{p_1}{\|p_1\|}$, $r_3 = \frac{p_2}{\|p_2\|}$ and $r_2 = r_1 \times r_3$.

Finally, we can use a refinement step that optimizes simultaneously the reprojection error over the set of parameters given by K, R and t. We propose to use the Levenberg-Marquardt non-linear optimization method for which many implementations can be found (e.g. lmfit-3.5, 2013, by Joachim Wuttke, http://apps.jcns.fz-juelich.de/lmfit).

3.3   Model Reprojection

The obtained projection matrix $P = K[R|t]$ can be used to project 3D points into the image. Therefore, we can roughly model a person (more specifically a bounding volume around a person) as a parallelepiped and project it at the closest and farthest point of the defined quadrilateral used for calibration. The projection of a parallelepiped in an image is a convex polygon whose bounding box can be easily computed and used to determine the sizes of the persons that will configure the perspective multiscale approach. Figure 3 shows the two projections of the box model in an image and the corresponding bounding boxes.

4   Detection

Discriminative learning methods have been used in the majority of works referred to object and person detection. Within this kind of methods, variations of SVM and Adaboost algorithms stand out in the literature. The main idea underlying the training of classifiers is to find a model which could map the input feature vector to a set of output labels. The training stage involves the application of supervised training algorithms to a set of feature vectors extracted from the image database. This database must have positive images (e.g. "person") and negative images (e.g. "non-person").

Fig. 4. Full-body and upper-body detectors can be combined to achieve better resutls: (green) upper-body detections, (blue) full-body detections

Fig. 5. The perspective can be used to filter out detections that do not represent coherent human sizes

In this work we have used two detectors, which correspond to full-body person and upper-body person, both using linear SVM and HOG features [5] due to its outstanding capabilities to detect persons in images. Specifically, we have used the Daimler full-body detector [12] available as an SVM file within OpenCV-2.4.5, which corresponds to a dataset of approximately 25k images. For the upper-body we have trained our own detector using a database of approximately 1k positive and negative images taken from TRECVID2013 training dataset, which contain images from 5 different cameras. Although this dataset is not large yet, it has been useful to comprobe that even a simple upper-body detector can help.

The full-body detections and extended upper-bodies as full-bodies (see Figure 4) are filtered to group detections which have significant overlap. Also, the detections are filtered according to perspective: a given detection is assumed to correspond to a person in the ground plane, so that it can be reprojected to that plane, and the approximate width and height can be computed. Figure 5 illustrates examples of detections filtered out.

5   Tracking

Tracking is the stage in which intra-frame detections are analyzed through time in order to group them in time and create an inter-frame entity called track. Also, tracking helps to alleviate the problems associated to detection-by-classification methods. Namely, the detections tend to generate noisy, incorrect, missing, and time sparse observations.

First, the tracking methods act as filters, so that noise can be reduced using appropriate models (such as bivariate Gaussian models). Second, tracking methods work on a two-steps fashion: observe and predict. The observation stage reads the observations coming from the detectors and associates them to the existing objects. Incorrect detections are mitigated using association schemes as the one described in the next sections. Also, missing detections can be handled using the prediction step of the filter, in which each object, given its estimated dynamics (e.g. velocity and acceleration) is projected onto the next frame. Therefore, the nature of tracking methods deal well with the drawbacks imposed by detectors and thus a combination of these two types of methods provide a good solution for object detection in video sequences.

5.1 Rao-Blackellized Data Association Particle Filter

The RBDAPF [6] has a special structure that allows to analytically compute the object magnitudes (positions, size, etc.) while the data associations between tracks and detections are approximated by a sampling approach [3]. This method defines a state vector $\mathbf{x}_t = \{\mathbf{x}_t^m, \mathbf{x}_t^a\}$ at time t, where $\mathbf{x}_t^m$ contains the 2D object magnitudes, and $\mathbf{x}_t^a = \{x_k^{a(j)}\}$ encodes the data associations between the tracks and the detections. The j-th data association component relates the j-th detection with a track $x_k^{a(j)} = id0$ or with clutter $x_k^{a(j)} = idC$, where $id0$ is a unique identifier of the track, and idC is the generic identifier for clutter. The mixed and filtered full-body detections are represented by the random variable $\mathbf{z}_t$.

As a Bayesian inference method, the RBDAPF aims to provide an estimate of the posterior density function. The idea of this technique is that solving analytically part of the state vector, and leaving only the non-linear part to the sampling approach gives a more accurate representation of the posterior probability. Intuitively, the variance is smaller because some variables are computed exactly and the non-linear dimensionality is lower that the dimension of the complete state-vector. The Rao-Blackwellization of the posterior density leads to the following expression:

$$p(\mathbf{x}_t|\mathbf{z}_{1:t}) = p(\mathbf{x}_t^m, \mathbf{x}_t^a|\mathbf{z}_{1:t}) = p(\mathbf{x}_t^m|\mathbf{z}_{1:t}, \mathbf{x}_t^a)p(\mathbf{x}_t^a|\mathbf{z}_{1:t}) \qquad (6)$$

where $p(\mathbf{x}_t^m|\mathbf{z}_{1:t}, \mathbf{x}_t^a)$ is assumed to be conditionally linear Gaussian, and therefore, with an analytical expression, given by the Kalman filter.

The data association posterior density $p(\mathbf{x}_t^a|\mathbf{z}_{1:t})$ can be expressed as:

$$p(\mathbf{x}_t^a|\mathbf{z}_{1:t}) = \frac{p(\mathbf{z}_t|\mathbf{z}_{1:t-1}, \mathbf{x}_t^a)p(\mathbf{x}_t^a)}{p(\mathbf{z}_t|\mathbf{z}_{1:t-1})} \qquad (7)$$

where $p(\mathbf{z}_t|\mathbf{z}_{1:t-1}, \mathbf{x}_t^a)$ is the data association likelihood, $p(\mathbf{x}_t^a)$ is the data association prior, and $p(\mathbf{z}_t|\mathbf{z}_{1:t-1})$ the normalization constant.

The data association prior determines the possible associations between tracks and detections, for which several criteria can be applied, such as: (i) each track can be associated only with one or none of the detections; (ii) each detection can be associated only to one track, although several detections can be associated to the clutter object. In this work we are using as likelihood function the Euclidean distance plus a constant clutter model. The data association posterior is approximated using importance sampling [1].

5.2   Data Association Filter

In practice, the use of the RBDAPF imposes handling very carefully entering and exiting objects. The reason is that the data association matrix is sampled using an importance sampling algorithm, and therefore, there are many association hypotheses, and the detections may be associated to different tracks for each sample or hypothesis. Even in the case that the posterior distribution this way defined shows unimodal behaviour (i.e. point-wise estimators can be applied to the set of samples), the different history of associations between observations and tracks at each sample is problematic at the time of considering input and output objects. A track is labeled as exiting the scene if it is not associated to new detections for a period of time (e.g. 5 frames). In that case, the object is considered to have left the scene and removed from the estimation (delete event). On the contrary, observations that are not associated to any existing objects are initially considered as clutter (i.e. erroneous measurements). In our work we create a new clutter object associated to that new observation just in case it receives new observations in the subsequent frames. If this is the case for a number of frames (e.g. 3 frames), the clutter object is upgraded to track, and added to the list of tracks (new event). These events are related to the history of associations and therefore each sample of the RBDAPF filter has its own association history that might lead to different new/delete events. Although filters like RBDAPF can be upgraded to consider samples with different numbers of objects inside it (by means of adding a dimension that spans the number of objects in the scene [9]), the complexity of the filter increases significantly, and the generation of a point-wise estimate from the posterior density function might become a tough task.

We define the Data Association Filter (DAF), which works exactly the same as the RBDAPF but selecting only the best hypothesis (Maximum A Posterior, MAP) during the data association step. Since this is a single sample, the point-wise estimate can be directly retrieved from its components.

6   System Test and Discussion

To evaluate the improvements derived from the use of the proposed perspective multiscale method we have used the TownCentre sequence made available by the Active Vision Group from Oxford [2] (1920×1080, 4500 frames, with 71460 persons labeled).

We wanted to evaluate the following hypotheses: (i) using the perspective multiscale method the optimum parameters for the detector are found automatically; (ii) our approach reduces significantly the number of SVM evaluations required to achieve similar results than brute force multiscale; (iii) using a combination of full-body and upper-body detectors provide better results; (iv) perspective allows also filtering out numerous false positives; (v) the DAF tracking stage helps to increase the performance of intra-frame detectors thanks to its prediction capabilities.

First, Table 1 shows the performance of the proposed perspective multiscale ($L = 3$ and $L = 5$) at different stages in terms of true positives (TP), false positives (FP), false negatives (FN), and the related Recall (R), Precision (P) and F-measure. In that sense, we have defined a detection to be a TP if the overlap it has with a ground-truth rectangle is larger than half their non-overlapping union area (i.e. overlap is above 50%).

Table 1. Comparison of the performance of the different combination of detectors (FB: full-body, UB: upper-body, FBUB: both mixed, FBUB*: mixed and filtered according to perspective) and tracker (DAF)

| | $L = 3$ | | | | | | $L = 5$ | | | | | |
	TP	FP	FN	R	P	F	TP	FP	FN	R	P	F
FB	21521	358	49926	0.301	0.984	0.461	27725	505	43722	0.388	0.982	0.556
UB	3395	23	68052	0.048	0.993	0.091	3395	23	68052	0.047	0.993	0.091
FBUB	23339	381	48108	0.327	0.984	0.490	29075	528	42372	0.407	0.982	0.575
FBUB*	20099	239	51348	0.281	0.988	0.438	25659	307	45788	0.359	0.988	0.527
DAF	27106	503	44341	0.379	0.982	0.547	32485	642	38962	0.455	0.981	0.621

As expected, the usage of UB joint with FB increases the performance of FB alone, even when the UB by itself does not reach good values. After filtering (FBUB*) we can see that many FP have been removed (although TP has slightly decreased, possibly due to the elimination of detections that did not fit exactly to the ground truth). The application of the tracker dramatically enhances these numbers, since a significant number of FN become TP thanks to the prediction capabilities of the filter. Better values are obtained for $L = 5$.

Table 2 compares the proposed scheme with the brute traditional force multiscale method with different values of levels L. We can see that the perspective multiscale gives good values with very few levels. Actually, we have found that the performance is stabilized at 3-5 levels, depending on the sequence, and adding more levels shows no significant improvement while increases the number of operations to carry out. Another remarkable advantage of our approach is that it automatically determines the optimum sizes of the images inside the multiscale pyramid. We deem this feature very practical because otherwise (with brute force multiscale), the optimum values for L, and *scale* must be found by try-and-error. For instance, Table 2 shows that for these large images, the multiscale approach

Table 2. Results of Perspective Multiscale and Brute-force multiscale

Perspective Multiscale					Brute-force Multiscale				
L	$W \times H$	N_e	R	P	F-measure	L	N_e	R	P	F-measure
3	1920×1080	75929	0.03	0.08	0.05	2	33051	0.15	0.96	0.27
6	1920×1080	117471	0.12	0.18	0.14	3	46226	0.30	0.98	0.46
10	1920×1080	144880	0.47	0.48	0.48	4	60105	0.29	0.96	0.45
20	1920×1080	162968	0.66	0.58	0.62	5	74141	95	263	0.55

can not find good performance given the small size of the model unless using a large number of scales. Therefore, the number of levels that best work can only be found launching and evaluating the detector spanning different values for L and scale which can take time and requires the existence of ground truth and evaluation tools.

In the TownCentre video, with the perspective we have computed, the farthest person is represented by a 101×184 bounding box, such that using a 64×128 SVM model, the largest image to be scanned is approximately 1805×1014. Therefore, the reduction of computational load is noteworthy: from 75929 to 46226 (a reduction of 39%). In the case of sequences with lowest perspective, where persons are not so small, (such as those in CAM1 or CAM3 of TRECVID dataset), the gain can reach much largest values, up to $80\% - 90\%$.

7   Conclusions

In this paper we have presented a methodology to apply contextual perspective information of the scene to traditional detection-by-classification schemes that use sliding window scanning. The multiscale procedure typically implies massive amounts of comparisons between windows of the image with a certain model, resulting in variable, a priori unknown, and possibly excessive computational load to achieve good results. Our scheme automatizes the sliding window technique, so that when the perspective information is injected into the solution, the optimum values of the parameters that govern the multiscale approach are found. The experiments carried out show that we can get results comparable to those of traditional (brute force) multiscale with only 3-5 levels, which can lead to computational loads reduction between 30% to 80% depending on the perspective of the scene (more reductions are achieved for images where persons are imaged larger). The addition of a tracking stage based on the Rao-Blacwellization concept helps as well to enhance the detection rates, since the nature of detection-by-classification is often sparse and noisy.

Acknowledgements. This work has been partially supported by the European project SAVASA (grant agreement number 285621) under the 7th Marco Framework, and by the program ETORGAI of the Basque Government with the BERRITRANS project.

References

1. Arulampalam, M.S., Maskell, S., Gordon, N.: A tutorial on particle filters for online nonlinear/non-gaussian bayesian tracking. IEEE Transactions on Signal Processing 50, 174–188 (2002)
2. Benfold, B., Reid, I.: Stable multi-target tracking in real-time surveillance video. In: Proc. Computer Vision and Pattern Recognition, pp. 3457–3464 (2011)
3. del Blanco, C.R., Jaureguizar, F., Garcia, N.: An advanced bayesian model for the visual tracking of multiple interacting objects. EURASIP Journal on Advances in Signal Processing 130 (2011)
4. Carbonetto, P., de Freitas, N., Barnard, K.: A statistical model for general contextual object recognition. In: Pajdla, T., Matas, J(G.) (eds.) ECCV 2004. LNCS, vol. 3021, pp. 350–362. Springer, Heidelberg (2004)
5. Dalal, N., Triggs, B.: Histograms of oriented gradients for human detection. In: CVPR, pp. 886–893 (2005)
6. Doucet, A., Gordon, N.J., Krishnamurthy, V.: Particle filters for state estimation of jump markov linear systems. IEEE Transactions on Signal Processing 49, 613–624 (1999)
7. Hartley, R.I., Zisserman, A.: Multiple View Geometry in Computer Vision, 2nd edn. Cambridge University Press (2004) ISBN: 0521540518
8. Kalal, Z., Mikolajczyk, K., Matas, J.: Tracking-learning-detection. Pattern Analysis and Machine Intelligence 6(1) (2010)
9. Khan, Z., Balch, T., Dellaert, F.: Mcmc-based particle filtering for tracking a variable number of interacting targets. IEEE Transactions on Pattern Analysis and Machine Intelligence 27, 2005 (2005)
10. Li, M., Zhang, X., Huang, K.Q., Tan, T.N.: Estimating the number of people in crowded scenes by mid based foreground segmentation and head-shoulder detection. In: Proc. International Conference on Pattern Recognition (2008)
11. Little, S., Jargalsaikhan, I., Clawson, K., Li, H., Nieto, M., Direkoglu, C., O'Connor, N., Smeaton, A., Scotney, B., Wang, H., Liu, J.: An information retrieval approach to identifying infrequent events in surveillance video. In: ACM International Conference on Multimedia Retrieval, pp. 223–230 (2013)
12. Munder, S., Gavrila, D.M.: An experimental study on pedestrian classification. IEEE Transactions on Pattern Analysis and Machine Intelligence 28(11), 1863–1868 (2006)
13. Park, L.J., Moon, J.H.: Exploiting global self similarity for head-shoulder detection. World Academy of Science Engineering and Technology 0076 (2013)
14. Thonnat, M.: Semantic activity recognition. In: 18th European Conference on Artificial Intelligence, pp. 3–7 (2008)
15. Viola, P., Jones, M.: Robust real-time face detection. International Journal of Computer Vision 57(2), 137–154 (2004)
16. Zeng, C., Ma, H.: Robust head-shoulder detection by pca-based multilevel hog-lbp detector for people counting. In: ICPR, pp. 2069–2072 (2010)

Human Action Recognition in Video via Fused Optical Flow and Moment Features – Towards a Hierarchical Approach to Complex Scenario Recognition

Kathy Clawson[1], Min Jing[2], Bryan Scotney[1], Hui Wang[2], and Jun Liu[2]

[1] School of Computing and Information Engineering, University of Ulster
Cromore Road, Coleraine, BT52 1SA, UK
[2] School of Computing and Mathematics, University of Ulster
Shore Road, Newtownabbey, BT37 0QB, UK
{k.clawson,m.jing,bw.scotney,h.wang,j.liu}@ulster.ac.uk

Abstract. This paper explores using motion features for human action recognition in video, as the first step towards hierarchical complex event detection for surveillance and security. We compensate for the low resolution and noise, characteristic of many CCTV modalities, by generating optical flow feature descriptors which view motion vectors as a global representation of the scene as opposed to a set of pixel-wise measurements. Specifically, we combine existing optical flow features with a set of moment-based features which not only capture the orientation of motion within each video scene, but incorporate spatial information regarding the relative locations of directed optical flow magnitudes. Our evaluation, using a benchmark dataset, considers their diagnostic capability when recognizing human actions under varying feature set parameterizations and signal-to-noise ratios. The results show that human actions can be recognized with mean accuracy across all actions of 93.3%. Furthermore, we illustrate that precision degrades less in low signal-to-noise images when our moments-based features are utilized.

Keywords: Video pattern recognition, optical flow, Zernike moments, Hidden Markov Model.

1 Introduction

Scenario recognition and event detection in video streams has gained increasing attention in the computer vision research community due to the needs of many applications, such as surveillance for security [1]. The ubiquity of CCTV capture systems combined with the sheer volume of generated data has amplified the necessity for retrospective, automated (or semi-automated) evaluation, indexing, annotation and retrieval of surveillance video. Intelligent analysis of surveillance data faces unique challenges. Current CCTV systems suffer limitations, including the existence of: diversified and non-interoperable video archiving systems and proprietary technologies; massive volumes of recorded data; low resolution video; scene occlusion, background clutter; noise; low inter-class variance between events of interest; infrequently

C. Gurrin et al. (Eds.): MMM 2014, Part II, LNCS 8326, pp. 104–115, 2014.
© Springer International Publishing Switzerland 2014

occurring events of interest; limited training data; conflicting legal frameworks across geographically disparate locations; and privacy and ethical issues relating to semantic video analysis.

Our research constitutes part of the SAVASA project [1], which seeks to develop a standards–based approach to surveillance video archive, search and analysis. A key component of SAVASA is video analytics for automatic detection of concepts and scenarios, to allow video indexing via an operational ontology and enable authorised users to perform semantic queries (reducing their search space and retrieving potentially pertinent data). Within the literature, computer vision techniques can identify complex semantic events with generally low precision and recall. SAVASA proposes to improve upon the state-of-the-art by decomposing complex events into their atomic components and regarding high level event detection as a reasoning process [2] which searches for both the existence of multiple sub-events and for spatio-temporal sub-event associations / properties (Figure 1). The perceived benefits of a hierarchical approach include the ability to analyse data across multiple levels of detail, the ability to model sparsely occurring events based on their more frequently occurring compositions, the ability to allow for inaccurate / noisy lower level event detections via uncertainty reasoning [2], and the reduction in dependency on high volumes of training data.

This paper describes lower-level components of hierarchical scenario recognition, and constitutes video feature extraction and single layer recognition, with focus on human action recognition. The components presented, although not suitable for detection of complex events in isolation, will act as the lower level in a complete system for hierarchical scenario recognition as illustrated in Figure 1. In order to facilitate complex event detection, it is desirable to utilize features which are robust given a variety of image conditions, including noise. The main contribution of this work is an evaluation of the applicability of motion features, specifically optical flow features, as the sole basis of feature generation. We propose a set of fused features which combine and build upon existing features in the literature, and consider the robustness of our methodology across varying signal-to-noise (SNR) ratios. We wish to reduce a known limitation of classification using optical flow features, specifically that degradation occurs in reliability of pixel-wise motion vectors given low SNR ratios or given motion patterns which defy underlying assumptions upon which flow has been calculated (for example that subsequent frames are in approximate registration). For this reason, we follow the methodology of Efros et al [3] and regard the complete set of motion vectors as a spatial pattern of noisy measurements which have been smoothed and aggregated, as opposed to (more commonly) viewing them as precise pixel displacements. We extend upon this idea, and generate moment-based features which are subsequently used for classification.

The remainder of this paper is structured as follows. We present a brief overview of related literature in Section 2. Our methodology for atomic classification is described in Section 3. Our experimental overview and results are given in Section 4, and conclusions are offered in Section 5.

2   Related Work

The detection of complex events within video has received much research focus in recent years [5-7]. Single layer approaches which model and recognize actions of a single actor (characterized by simple motion patterns and typically executed by a single entity), constitute the focus of this review. Due to their nature, single-layered approaches are suitable for the recognition of human actions and gestures with sequential characteristics. The fundamental problem is to learn reference scenarios from training samples, and to devise training and matching methods for coping effectively with small variations of the feature data within each scenario class [8]. Existing methods of single layer scenario recognition include space-time solutions and general sequential matching techniques [4].

Space-time solutions characterize scenarios based on features extracted from 3D volumes [9], space-time trajectories [10] or 3D local features. Standard features such as HOG, HOOF [10], SIFT and SURF have been explored. Such methods provide a straightforward solution, can be robust given illumination variations, but can have difficulties handling speed and motion variations. The major limitation of space-time methods is that they are not suitable for modeling more complex activities as the relations among non-periodic activity features are often lost.

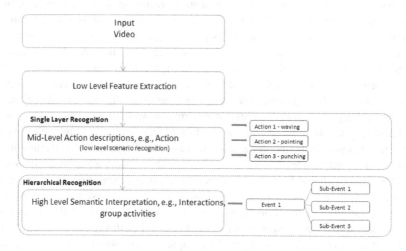

Fig. 1. Hierarchical Scenario Recognition Framework

Sequential matching for scenario recognition views an activity of interest as a sequence of observations and recognizes similar activities within test data by searching for similar sequences. Template-based or state-based classification may be performed, where descriptions of event classes are derived using training samples directly. Efros et al. [3] utilize optical flow features within a nearest neighbour framework for detection of human actions at a distance. State-based methods such as Hidden Markov

Models, 'Semi-Hidden' Markov models [11], Bayesian Networks, Dynamic Bayesian Networks [7] and Neural Networks [12] have also been considered. In [7] the eigen-shapes from silhouette images are extracted by PCA and used for human and animal behaviour classification. HMMs are well suited for facilitating recursive probability estimation given uncertainty. However, like space-time solutions, state-based methods have high overhead in terms of training requirements. It is difficult to have a training video archive covering all possible appearances of a scenario.

3   Methodology

For single layer scenario recognition we perform Hidden Markov Model classification, where input vectors describe Kanade Lucas Tomasi (KLT) optical flow vector properties [13]. The KLT approach, first published in 1981 as a method for image registration, is a point tracking algorithm which compares the spatial intensity between windowed regions of each video frame and utilizes the fact that, in many instances, subsequent time frames within video datasets are already approximately registered [14]. Optical flow calculation via KLT is equivalent to quantifying displacement across two images, $A(x)$ and $B(x) = A(x + v)$, where v is an n-dimensional vector. To achieve this, the behaviour of $A(x)$ is approximated across the neighborhood of x [13]. It is assumed that the following linear relationship exists:

$$A(x+v) \approx A(x) + vA'(x) \tag{1}$$

Based on this, we may then solve displacement by minimizing the energy function [15]:

$$E_x = [A(x+h) - B(x)]^2 \tag{2}$$

After per-frame flow vector calculation, we generate features capturing motion orientation, magnitude, and relative location. Subsequent frame-based feature vectors are regarded as time-varying sequences, represented using Principal Components decomposition, and used as inputs for Hidden Markov Model classification. Specifically, we generate histograms of oriented optical flow magnitude (HOOF), and 2D Zernike moments of Efros Optical Flow Channels [3], as described below.

3.1   HOOF Features

For each individual frame we compute a normalized histogram of oriented optical flow magnitude (n bins, of equal width) where the magnitude of each bin corresponds to the sum of magnitude of optical flow. The entire histogram is normalized to unit vector. The number of bins is set to 90.

3.2   2D Zernike Moments of Efros Optical Flow Channels

The KLT algorithm can reliably perform pixel–wise motion vector computation when noise levels are low and whenever frame-wise displacement is less than the chosen size of integration window. However, when SNR is low, velocity computation becomes erroneous and constitutes an unreliable basis for feature generation. Efros et al., [3] compensate for the limiting factor of noise by regarding the full set of motion vectors corresponding to a single frame as a spatial pattern of noisy measurements. After tracking and stabilizing each human figure, they characterize motion by computing optical flow, projecting it onto a number of motion channels, and convolving with a Gaussian filter. Action recognition is subsequently viewed as a template–matching exercise and conducted within a nearest neighbour framework. We draw upon elements of this methodology and offer extensions to generate moment-based features. Efros et al., [3] do not consider all motion information, due to the stabilization phase. However, given that within surveillance applications one may be faced with multiple scales of view (near, medium and far field cameras) and considering there may be both statio-nary and moving persons within the scene, we bypass this stage and work directly from the entire video frame.

Like Efros, we view temporal flow displacements as a global spatial pattern and generate new channels by decomposing smoothed optical vectors into a number of half-wave rectified, directed channels. For each frame, we calculate the optical flow vector field F and split F into two scalar fields, Fx and Fy, corresponding to the horizontal and vertical components of the flow. Fx and Fy are half-wave rectified to form 4 non-negative channels: $Fx+$ $Fx-$ $Fy+$ and $Fy-$ and blurred with a Gaussian (sigma = 0.5) to remove spurious motions. We extend upon this idea by viewing each directed channel as a new image (Figure 2) and generating feature vectors which represent properties of the new images. Specifically, we calculate 2D Zernike moments of each of the new channels and concatenate, resulting in a $4n \times 1$ feature vector per frame, where n is the number of moment features per channel. We then append this feature vector to the HOOF feature vector, resulting in a ($4n$ + *numberHOOFBins*) $\times$ 1 feature vector per frame.

Generally, moments characterize the numeric quantities of an image in relation to some reference point or axis. Moments and moment-based invariants have been used in image recognition for biometrics, such as fingerprint recognition, face recognition, and gait recognition, with promising results. Hu [16] first introduced 7 moment-based features invariant under translation, scaling and rotation. However, computation of higher order Hu moments is computationally expensive. Unlike Hu moments, Zernike moments [17] are produced using an orthogonal basis set, specifically, they project $f(x,y)$ onto a set of orthogonal polynomials defined on the unit disk. The advantages of such an approach include: rotational invariance; reduction in information redundancy due to the orthogonal property; and higher accuracy for more detailed shapes / signals. Additionally, Zernike moments have been shown to be less sensitive to noise than alternative methods [18]. It is for these reasons that we choose Zernike moments as our method for feature set generation.

Fig. 2. Efros Channel Images. From left to right: Original Frame; Positive X Efros Channel; and Positive Y Efros Channel.

4 Experimental Results

We apply our methodology to the KTH dataset [19]. KTH consists of 2391 video sequences across 6 event classes: boxing, hand waving, hand clapping, walking, running and jogging. Each action is performed multiple times by 25 individuals, across 4 different scenarios: outdoors (d1), indoors (d2), outdoors with scale variations (d4), and outdoors with different clothes (d3). The background in most cases is homogeneous. KTH has been frequently used as a performance benchmark within the literature, and current performance is within the range of 93 – 95% [10]. As in the original paper [19], we train and evaluate a multi-class classifier and report average accuracy across all classes as a measure of performance. Specifically, we perform hidden Markov model classification, where the number of hidden states = 6, and the number of Gaussians under each state = 3. Accuracy is calculated as $(TP + TN) / N$ where TP is the number of true positive classifications, TN is the number of true negative classifications, and N is the number of test cases under consideration. We adopt 10-fold cross validation as our training / test paradigm, for outdoors scenes (scenes d1 and d3).

Our initial investigation considers the effect of feature set parameterization on classification accuracy. Specifically we vary the order of Zernike moments calculated as features and train classifiers for each new feature set generated. Additionally, we evaluate classifier performance across multiple PCA dimensions, to determine the optimal feature space representation. Secondary evaluation considers robustness of features across varying SNRs, and evaluates whether our features can compensate for image quality degradation. Preliminary results are presented below.

4.1 Comparison of Feature Parameter Settings

Figure 3 illustrates mean classification accuracy across varying orders of Zernike moments, and varying feature set sizes (PCA dimensions). It can be seen from Figure 3 that accuracy is lowest when the PCA dimension is less than 10. Additionally, it is apparent that increasing the order of Zernike moments results in small increases in accuracy. Within Figure 3, there is no visually identifiable advantage in increasing the PCA dimension beyond 25. This observation is confirmed in Figure 4, where mean classification accuracy is calculated for each PCA dimension across all orders of Zernike moments. A general trend is identifiable: as the PCA dimension increases there is initially an increase in classification accuracy (the extended feature space provides greater discriminative capabilities). Once the PCA dimension is increased beyond 22, a slight degradation in accuracy occurs. The feature space representation at higher PCA dimensions does not necessarily constitute discriminative content. Across all parameterization experiments, maximum mean accuracy was 93.3 % (order of Zernike moments = 12 and number of principal components per feature vector = 14).

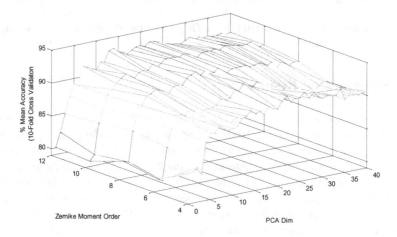

Fig. 3. Classification Accuracy for KTH using Optical Flow Features with PCA Feature Space Representation

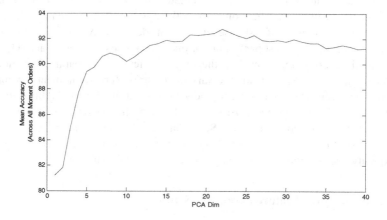

Fig. 4. Average Classification Accuracy across all Moment Orders

4.2 Comparison of HOOF and Fused Features

To illustrate the advantage of moment features against existing optical flow features, we evaluate the performance of both HOOF and fused feature sets (at order 12) under PCA dimension 14; results are illustrated in Fig. 5. The y-axis is the mean accuracy from 10-fold cross validation. The x-axis shows the types of features being evaluated. From the figure it is observed that the fused feature approach yeilds better classification accuracy than the HOOF features alone. In addition, we compare the associated average precision, computed across all cross validation trials. HOOF provides true positive (TP) classification across all events of 50.42%, but when fused with moment features, the TP increases to 79.86%. The mean confusion matrix for fused features is illustrated in Table 1. Results again demonstrate the advantage of the proposed fused feature approach.

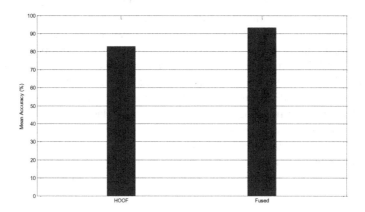

Fig. 5. Comparison of mean accuracy under PCA dimension 14

Table 1. Mean Confusion Matrix (10-fold cross validation) based on the fused feature (Zernike Order = 12, PCA Dimension = 14)

Event	Classifier Output (TP: 79.86%)					
	Boxing	Handclapping	Handwaving	Jogging	Running	Walking
Boxing	0.63	0.23	0.13	0.00	0.00	0.01
Handclapping	0.00	0.96	0.04	0.00	0.00	0.00
Handwaving	0.02	0.02	0.97	0.00	0.00	0.00
Jogging	0.17	0.00	0.00	0.73	0.01	0.09
Running	0.12	0.00	0.00	0.19	0.69	0.00
Walking	0.16	0.00	0.00	0.03	0.00	0.81

4.3   Effect of Additive Noise

A known limitation of optical flow analysis is degradation in the reliability of pixel-wise motion vectors given low SNRs. Theoretically, the moment-based features proposed in this paper compensate for the limiting factor of noise by regarding flow displacements as a global spatial pattern of smoothed, aggregated, pixel-wise vectors and capturing properties of those spatial patterns. To test this assumption, we introduce synthetic Gaussian noise of known properties into each video prior to feature extraction, and perform classification using 10-fold cross validation (as previously). The objective is to evaluate the extent to which classification performance degrades when data is erroneous / noisy, and to compare results with those achieved using only HOOF features. Illustrations of noisy KTH images, and their related Efros images (positive X channel), are given in Figure 6. Example feature data are illustrated in Figure 7.

Tables 2 and 3 summarize mean accuracy and mean precision for our fused features (moment order = 10), across varying SNRs and PCA dimensions, respectively. It can be seen from Table 2 and Table 3 that when SNR falls, mean accuracy and precision also degrade. However, the degradation in classifier performance as SNR drops from 4 to 2 is relatively small. Maximum precision was 69.7% for SNR = 2, 69.4% for SNR = 3, 70.6 % for SNR = 4 and 79% for zero noise. For comparison, we perform classification experiments using HOOF features only, where SNR = 2. Results are illustrated in Figure 8. It can be seen from Figure 8 that at SNR of 2, classifier performance using HOOF features only is significantly lower than that achieved when fused features are utilized. Mean precision using HOOF features is 38.7% using 5 PCA dimensions and 38.9% using 15 PCA dimensions. Mean precision using fused features is 57.8% and 66.3% for PCA = 5 and 15, respectively. The fused features are more robust when video data are noisy.

Table 2. Mean accuracy, all events, by SNR and PCA Dim (Fused Features with Moment Order 10)

SNR	5	10	15	20	25	30	35	40
						PCA Dim		
2	85.9	89.3	88.7	88.7	89.3	89.9	89.9	89.4
3	85.9	88.4	88.8	89.8	89.2	88.8	88.6	88.8
4	85.7	87.2	88.7	90.1	90.2	89.2	89.8	89.1
No Noise	89.9	91.5	92.6	93	92.4	92.8	92.8	92.1

Table 3. Mean precision, all events, by SNR and PCA Dim (Fused Features, Moment Order 10)

SNR	5	10	15	20	25	30	35	40
						PCA Dim		
2	57.8	67.9	66.3	66.1	67.9	69.7	69.6	68.2
3	57.9	65.4	66.4	69.4	67.8	66.5	66.3	66.4
4	57.2	61.4	66.3	70.4	70.6	70.6	69.3	67.4
No Noise	67.9	74.6	77.9	79	77.1	78.3	78.3	76.4

Fig. 6. Noisy Versus Noise Free KTH Frames: (a) KTH frame with no noise; (b) Positive X Efros Channel for (a); (c) KTH frame, SNR = 4; (d) Positive X Efros Channel for (c)

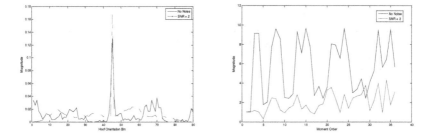

Fig. 7. Example HOOF Features and Zernike Moments for Zero Noise and SNR = 2

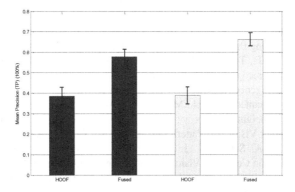

Fig. 8. Comparison of mean precision with std under PCA dimension: 5 (blue) and 15 (yellow)

5   Conclusions

In this paper, we propose that human actions may be recognized in both noisy and noise-free environments using optical flow properties only. To facilitate robust recognition given noise, we regard optical flow vectors as per-frame spatial patterns within directed channels. We extend upon the existing feature representation defined by Efros et al [3], and utilize 2D Zernike moments to generate feature sets from these representation. Through evaluation of feature set parameterizations, specifically moment order and Principal Component dimensionality, we achieve average accuracy of 93.3%. This performance is within range of the 93 – 95% classification accuracy currently cited within the literature. Finally, we illustrate an additional benefit of Zernike moment-based features, specifically that when incorporated with other commonly used features, they can significantly improve classifier performance when SNR is low.

Acknowledgements. The research is funded by the European Union Seventh Framework Programme (FP7/2007-2013) under grant agreement number 285621, project titled SAVASA.

References

[1] The SAVASA project, http://www.savasa.eu
[2] Yang, J.B., Liu, J., Sii, H.S., Wang, H.W.: Belief rule-base inference methodology using the evidential reasoning approach—RIMER. IEEE Trans. Syst. Man Cybern. 36(2), 266–284 (2006)
[3] Efros, A., Berg, A.C., Mori, G., Malik, J.: Recognizing action at a distance. In: Proceedings of the Ninth IEEE International Conference on Computer Vision, pp. 726–733 (2003)
[4] Aggarwal, J., Ryoo, M.S.: Human Activity Analysis: A Review. ACM Computing Surveys 43(3), 1–43 (2011)
[5] Regazzoni, C., Cavallaro, A., Wu, Y., Konrad, J., Hampapur, A.: Video analytics for surveillance: theory and practice. IEEE Signal Processing Magazine 5, 16–17 (2010)
[6] Ziani, A., Motamed, C.: Temporal Bayesian Networks for Scenario Recognition. In: Ersbøll, B.K., Pedersen, K.S. (eds.) SCIA 2007. LNCS, vol. 4522, pp. 689–698. Springer, Heidelberg (2007)
[7] Ziani, A., Motamed, C., Noyer, J.: Temporal reasoning for scenario recognition in video-surveillance using Bayesian networks. Computer Vision 2(2), 99–107 (2008)
[8] Vu, V., Bremond, F., Thonnat, M.: Automatic video interpretation: a novel algorithm for temporal scenario recognition. In: Proceedings of the 18th International Joint Conference on Artificial Intelligence (IJCAI 2003), Acapulco, Mexico, pp. 523–533 (2003)
[9] Rodriguez, M., Ahamed, J., Shah, M.: Action MACH: A spatio-temporal maximum average correlation height filter for action recognition. In: Proceedings of the IEEE Conference on Computer Vision and Pattern Recognition (CVPR). IEEE, Los Alamitos (2008)
[10] Wang, H., Klaser, A., Schmid, C., Liu, C.: Action recognition using dense trajectories. In: IEEE Conference on Computer Vision & Pattern Recognition, pp. 3169–3176 (2011)
[11] Hongeng, S., Nevatia, R.: Large scale event detection using semi-hidden Markov models. In: Proceedings of the International Conference on Computer Vision, vol. 2, pp. 1455–1462 (2003)

[12] Szarvas, M., Sakai, U., Ogata, J.: Real-time pedestrian detection using LIDAR and convolutional neural networks. In: IEEE Intelligent Vehicles Symposium, pp. 213–218 (2006)

[13] Lucas, B., Kanade, T.: An Iterative Image Registration Technique with an Application to Stereo Vision. In: Proceedings of 7th International Conference on Artificial Intelligence, pp. 121–130

[14] Wu, H., Sankaranarayanan, A., Chellappa, R.: Online Empirical Evaluation of Tracking Algorithms. IEEE Transactions on Pattern Analysis and Machine Intelligence 32(8), 1443–1458 (2009)

[15] Tomasi, C., Kanade, T.: Detection and tracking of point Features, Technical Report CMU-CS-91-132, Carnegie Mellon University (1991)

[16] Hu, M.: Visual pattern recognition by moment invariants. IRE Transaction Information Theory IT-8(2), 179–187 (1962)

[17] Mukundan, R., Rmakrishnan, K.: Moments functions in image analysis theory and applications. World Scientific Publishing, Singapore (1998)

[18] The, C., Chin, R.: On image analysis by the methods of moments. IEEE Transactions on Patten Analysis Machine Intelligence 10(4), 16–19 (2004)

[19] Schuldt, C., Laptev, I., Caputo, B.: Recognizing human actions: a local svm approach. In: Proceedings of the International Conference on Pattern Recognition, Cambridge (2004)

Sparse Patch Coding for 3D Model Retrieval

Zhenbao Liu, Shuhui Bu*, Junwei Han, and Jun Wu

Northwestern Polytechnical University, Xi'an, 710072, China
{liuzhenbao,bushuhui,jhan,junwu}@nwpu.edu.cn

Abstract. 3D shape retrieval is a fundamental task in many domains such as multimedia, graphics, CAD, and amusement. In this paper, we propose a 3D object retrieval approach by effectively utilizing low-level patches with initial semantics of 3D shapes, which are similar as super-pixels in images. These patches are first obtained by means of stably over-segmenting 3D shape, and we adopt five representative geometric features such as shape diameter function, average geodesic distance, and heat kernel signature, to characterize these low-level patches. A large number of patches collected from shapes in a dataset are encoded into visual words by virtue of sparse coding, and input query compares with 3D models in the dataset by probability distribution of visual words. Experiments show that the proposed method achieves comparable retrieval performance to state-of-the-art methods.

Keywords: 3D object retrieval, Patch, Sparse coding.

1 Introduction

3D model as an important media contains rich 3D information preserving real object surface, color, and texture, which has been extensively accepted in the domain of multimedia, graphics, virtual reality, amusement, design, and manufacturing. A huge number of publicly usable models such as in Google 3D Warehouse has been quickly spread, and many researchers attempt to provide content based retrieval techniques [1] for reusing these models and accurately searching desirable objects.

In order to deal with the problems of rigid transformation, non-rigid deformation, and scale, in contrast with early works characterizing gross feature, recent works mainly focus on local point description and organization, topological structure, and appearance utilization. Local point descriptors can encode enough local context while keeping it rotation or deformation invariant. Representative local descriptors include Laplace-Beltrami operator defined on local manifold [2], heat kernel signature and its variants [3], 3D Harris [4], 3D SURF [5], 3D intrinsic shape context [6], and 3D SIFT [7]. To avoid high computational burden from comparison between two sets of dense points, and efficiently organize these descriptors, bag of features has been borrowed from text and image processing to address correspondence and matching of 3D local descriptors [8] [9]. Retrieval

---

* Corresponding author.

C. Gurrin et al. (Eds.): MMM 2014, Part II, LNCS 8326, pp. 116–127, 2014.

based on topological structure assumes that a 3D shape is represented by means of a topologically connected graph consisting of nodes and edges, such as common undirected graph adopted in [10], Reeb graph used in [11], bipartite graph used in [12], skeleton graph adopted in [13] and binary tree used in [14]. Each segmented meaningful part is identified by a single node, and edges in the graph represent adjacency relations between these segments. Therefore, the problem of shape retrieval is easily solved resorting to detecting graph isomorphism and matching graphs of two shapes. Appearance based 3D object retrieval tends to address how to effectively utilize many views of a 3D object, for example, selecting query views [15] and weighted bipartite graph matching of views [16], camera constrained-free view generation [17], constructing multiple hypergraphs of views [18], and panoramic views [19].

We propose a 3D object retrieval approach by effectively utilizing low-level patches perceptually meaningful of 3D shapes, which are analogous to superpixels in images. In our framework, 3D shape is first over-segmented into many patches, and different types of geometric features are extracted from these patches. Then we encode a large number of patches over a 3D model dataset via sparse coding, and extract representative visual words. Input query compares with 3D models in the database by probability distribution of visual words. This method improves the retrieval performance of state-of-the-art methods.

The main contributions of the paper are described as follows.

1. Compared with point descriptors based retrieval, we introduce low-level patch to represent a 3D object, and each object only requires a small number of patches to discriminate other objects. Moreover, these patches are not randomly generated but according to initial semantics.
2. Compared with retrieval methods based on meaningful segments and graph structure, our method avoids directly generating several meaningful parts because the techniques of semantic segmentation are not mature which possibly leads to many mis-segmented parts. Moreover, we do not adopt graph structure to organize these parts because of many topologically variable objects such as vases with different number of handles. Patch based representation will make retrieval robust against topology variation.
3. The motivation of adopting sparse coding to represent 3D objects is based on our observation, that not only the same category of objects but also irrelevant objects have many similar patches. For example, human body has many similar patches as horse body. Therefore, we extract sparse visual words to approximate these shapes and model occurrence frequency of these common features using relatively sparser coefficients than previous bag of words retrieval algorithms.

2 Low-Level Semantic Patch

The segmentation based retrieval methods directly segment 3D shapes into high-level semantic parts, for example, human is represented by several components,

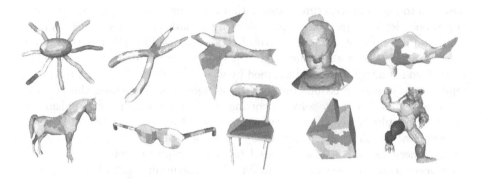

Fig. 1. Patch generation results for 3D objects from ten classes. Each object is partitioned into 50 patches, and we can see that these patches have already low-level semantics, and can be regarded as smallest semantical patch.

head, arms, legs, etc. A graph is used to organize these parts, and 3D model matching is simplified to a problem of graph isomorphism. However, this type of methods is sensitive to objects with different topology, for example, vases with different number of handles, because their graphs significantly change. It results in that objects with similar semantics are easily mis-recognized into different objects. In this paper, to our best knowledge, it is the first time to introduce low-level semantic patch into the domain of 3D shape retrieval. It is similar as the concept of superpixel in image processing, which has been successfully applied to recognition of complex scenes. However, we can not resort to spatially uniform patch generation on 3D model like [20] because it loses low-level semantic information.

It is regretful, however, that up to now there are seldom segmentation methods aiming at generation of low-level semantic patches because most segmentation methods are designated to generate high-level meaningful parts. We slightly modify a classical and fully automatic segmentation method based on randomized cuts [21], which is considered as highly discriminative and robust segmentation [22] because it generates a random set of segmentations and measure the frequencies that each edge of a mesh lies on a segmentation boundary in the randomized set. It is less sensitive to surface noise, tessellation, pose, and intra-class shape variations, and these properties will help to accurately match shapes with same semantics but large geometrical variations and non-rigid deformations. Nevertheless, it can not be directly applied to generate our desired low-level patches because it defines concave weights to control mesh boundaries, which will result in extremely larger patches and also smaller patches. To overcome this problem, we add a control item to prevent these extreme patches by considering patch size, and smaller patches will be automatically merged with its neighbors. In order to provide sufficient low-level patches for next high-level processing, we set patch number to 50 for each object and generate the same number of patches for each category of objects, as illustrated in Fig. 1.

3   Patch Description

After we obtained these low-level semantic patches for each 3D model, another problem we face is how to effectively describe these patches because it is different from point features. Each patch is composed of large points, edges, and faces, and describing the patch requires enough consideration on its geometrical characteristics, topological features of connecting its neighbors, and also its relative position or function in the global shape. We tend to adopt conformal geometry signature to describe its geometrical features, for example, easily differentiating sharp or smooth surface regions. Moreover, shape diameter function of each region is also adopted to describe its relative thickness, an important geometrical attribute which can recognize its part that it belongs to. For topological features, we consider its connection relationship with local neighbor regions, and adopt Laplace-Beltrami operator to describe its topological connection. In order to describe its global position or function, two descriptors, average geodesic distance and heat kernel descriptor, provide messages about the global relationship between this region and other regions. We first compute point descriptors and then generate patch features by estimating distributions of point descriptors. The detailed point descriptors are described as follows.

Conformal Geometry Signature. When a 3D manifold model is conformally transformed into plane, each conformal scaling factor is used to locally scale the neighborhood of a vertex in order to achieve the target curvature at the vertex [23]. High scaling factor corresponds with cone singularity so that its plane parameterization has low distortion. We use the scaling factor on each vertex as our point feature.

Shape Diameter Function. The shape diameter function (SDF) [24] is a volume-based scalar function measuring the diameters of different parts. The SDF value is computed by sending 30 rays inside a small cone with angle of $30°$ to intersect with the opposite side of the boundary, and averaging these weighted ray lengths. The values remain similar on the neighborhood of the same part, and oblivious to articulated deformation. The SDF values are presented in Fig. 2.

Fig. 2. Shape diameter functions of points in a class of vases with different topology or geometrical variant. The red discriminates higher diameters of parts from thin parts.

Laplace-Beltrami Descriptor. This descriptor extends Laplace operator in Euclidean domain onto manifold to achieve the divergence of vertex gradients on a mesh. Laplace-Beltrami operator of each vertex is commonly discretized into sum of the distances with cotangent and area weights from this vertex to its one-ring neighbors. Local surface feature is extracted based on the eigendecomposition of the Laplace-Beltrami operator defined on the local region [2]. The set of larger eigenvalues for the local region are adopted to form a local descriptor for describing this point, which is isometry-invariant, and consequently it is able to handle non-rigid transformation.

Average Geodesic Distance. We compute approximate geodesic distances [25] on 3D mesh between each point and other points, and average these distances for the point. The average geodesic distance is used for our point feature. We computed this distance for each vertex, and visualize these vertices in Fig. 3.

Fig. 3. Average geodesic distances of points in a class of chairs with different topology or geometrical variant. The red shows higher values.

Scale Invariant Heat Kernel Signature. Heat kernel signature is derived from a heat diffusion equation by using Laplace-Beltrami operator on surfaces. The fundamental solution of the heat equation is called the heat kernel. Sun et al. [26] proposed a heat kernel signature $h_t(x, x)$ composed of different diffusion times t_i, and set the time of heat diffusion to obtain local neighborhood information at a easily controlled scale. The heat kernel descriptor is intrinsic, insensitive to isometric deformation, and robust to surface noises. In order to make it scale invariant, Fourier transform of heat kernel signatures is proposed in [3], and we attempt to adopt this feature and then visualize point values of relevant objects in Fig. 4.

After we obtained five descriptors for each point on the low-level semantic patch, the description of the patch is defined by constructing a histogram of all the point values for each type of descriptors. Five histograms are naturally concatenated into a whole feature vector so as to form the final patch description. In this paper, we consider these five types of features are equally important for each patch and hence their weights are same.

Fig. 4. Scale invariant heat kernel signature of points in a class of teddy bears with different poses. The red shows higher heat kernel values.

4 High-Level Patch Coding

Sparse coding [27] has attracted many researchers from the domain of image and vision to solve tasks of image analysis, e.g., image retrieval [28] [29], classification, recognition, and segmentation. The advantages of sparse coding are two-fold. The first one is that it can capture higher-level features via learning basis functions from unlabeled data, and these features contain more semantic information and are adaptable to complex recognition tasks on variable 3D shapes. Moreover, sparse coding can learn over-complete basis sets, which can more adequately represent objects than limited orthogonal basis and then capture a large number of pattern in the input data. In this paper we introduce the concept of sparse coding into 3D shape retrieval to improve the state-of-the-art performance.

4.1 Sparse Coding

Given each feature point $\mathbf{x} \in R^k$, assume that it is sampled from the feature space representing all the types of patches from different 3D shapes. Sparse coding aims to interpret the point with an over-complete set of n bases $\{\mathbf{b}_1, ..., \mathbf{b}_n\} \in R^k$, which are linearly combined with a sparse weight vector $\boldsymbol{\omega} \in R^n$ such that

$$\mathbf{x} \approx \sum_{j=1}^{n} \omega_j \mathbf{b}_j, \tag{1}$$

where n satisfies $n > k$, which means that the basis are over-complete. These bases are also known as a dictionary $\mathbf{B} = [\mathbf{b}_1, ..., \mathbf{b}_n] \in R^{k \times n}$. The problem of coding is converted to discover these bases only dependent on unlabeled data, which are considered as a training set of m input vectors $\{\mathbf{x}_1, ..., \mathbf{x}_m\}$. These bases and their corresponding weights $\{\boldsymbol{\omega}_1, ..., \boldsymbol{\omega}_m\}$ are the solutions to the following optimization problem,

$$\underset{\mathbf{B}, \boldsymbol{\Omega}}{\text{minimize}} \sum_{i=1}^{m} \|\mathbf{x}_i - \mathbf{B}\boldsymbol{\omega}_i\|_2^2 + \lambda \|\boldsymbol{\omega}_i\|_1 , s.t. \|\mathbf{b}_j\|_2^2 \leq 1. \tag{2}$$

where $\boldsymbol{\Omega} = \{\boldsymbol{\omega}_i\}$. The above cost function is union of the reconstruction error approximating the input vector via linear combination of basis, and the sparsity penalty of basis. We use l_1 penalty as the sparsity function, and the optimization problem is convex while holding two set of parameters $\mathbf{B}$ and Ω alternatively fixed. In order to avoid trivial solutions, the l_2 norm of each basis $\mathbf{b}_j$ is limited to be less than or equal to 1. Although several approaches such as QCQP solver can be applied to solve the problem, we adopt an efficient algorithm proposed in [30], where a feature-sign search strategy converges to global optimum while keeping the bases fixed and then bases are learned by the Lagrange dual given these unchangeable coefficients. We finally obtain a set of bases and coefficients of each feature.

4.2 Algorithm

The retrieval algorithm based on sparse patch coding compares 3D models by three main steps, which are illustrated in Fig. 5.

Low-Level Patch Generation. Each object is partitioned into 50 different patches with initial semantics, and each low-level patch is characterized utilizing histograms of five types of descriptors, that is, conformal geometry signature, shape diameter function, Laplace-Beltrami descriptor, average geodesic distance, and scale invariant heat kernel signature.

Vocabulary Construction. After extracting features for each patch from all the models in a large data set, a set of bases also known as visual words in a vocabulary, is learned from large numbers of patch features. The data set we adopt is composed of 400 3D models subdivided to 20 classes. Since each model is described with 50 patch features, 20K patch features are obtained in the whole data set, which are used to construct visual words in the vocabulary. We solve the optimization mentioned above to get an optimum solution, which contains a set of visual words and a series of coefficients. We also studied the influence of the number of visual words while constructing vocabulary, and the size is set to 8, 16, 32, 64, 128, 256, respectively. The overall performance improved with the increase of the vocabulary size from 8, however, if the size exceeds 128, retrieval error rate has turned to rise and furthermore the run time becomes longer because of optimization of a large number of parameters.

High-Level Object Representation. Given a new object as the input shape, the problem of representing it with high-level visual words is converted to optimize its coefficients. In patch level, the coefficients which are linearly combined with visual words are considered as a distribution of occurrences Ω of these words. Dissimilarity between a pair of 3D objects P and Q is defined by comparing two groups of patch coefficients in the following equation.

$$Dis(P,Q) = \sum_{i=1}^{50} \left\| \boldsymbol{\omega}_i^P - \boldsymbol{\omega}_i^Q \right\| \tag{3}$$

where $\boldsymbol{\omega}_i^Q$ identifies the closet match of the sparse coefficient $\boldsymbol{\omega}_i^P$. The distance between two sparse coefficients can be l_1 norm, Kullback-Leibler Divergence

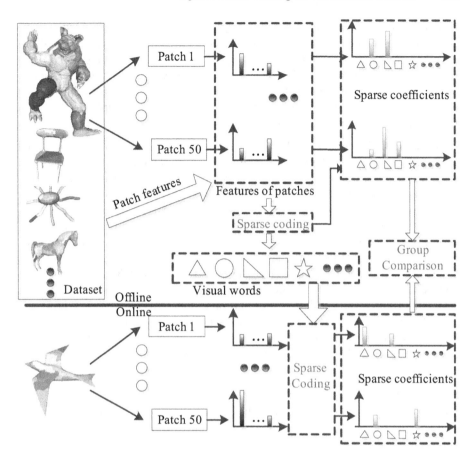

Fig. 5. Overview of the proposed algorithm based on sparse patch coding. The online and offline process of 3D shape retrieval is composed of three main steps, low-level patch generation, vocabulary construction, and high-level object representation.

(KLD), and Earth Mover's Distance. In this paper we adopt KLD to measure the difference among the distributions of coefficients by considering relative entropy.

5   Experiments

In this section we adopt a challenging data set and corresponding metrics to evaluate the proposed retrieval algorithm, and also compare the retrieval performance to that of representative methods.

5.1   Data Set and Measures

Data set. A common data collection, composed of SHREC 2007 watertight models, is adopted to test the retrieval performance of the proposed method.

This collection is made up of 400 watertight mesh models, subdivided into 20 classes, each of which contains 20 objects with different geometrical variations and also articulated deformations. The data set contains not only natural objects but also man-made objects. Moreover, some shapes with redundant parts exist, for example, sculpture head model with extra base, and there are also shapes with different geometric genus, for example, vases with different number of handles. It is considered as a challenging data set.

Evaluation Criteria. To evaluate the retrieval performance of the proposed method, we adopt recall precision values, two fundamental measures. Recall is the ratio of the number of retrieved relevant objects to the total number of relevant objects in the test database. Precision is the ratio of the number of retrieved relevant objects to the number of returned objects. Each object is selected as query, and compared against all the objects in the database. A retrieval list with length equal to database size is returned. For each query, the number of relevant objects in the retrieval list is same as the size of its class. Accordingly its value is always 20 in the collection of SHREC 2007 watertight models. Given a query, the desired retrieval result is that the relevant objects lies in the front of the list. The recall and precision values are averaged on each class of objects, and then the whole data set. For easily comparing with other methods, we report the recall and precision values of returned 20, 40, 60, 80 items, which are 1, 2, 3, and 4 times the size of class.

5.2   Comparative Methods

We will compare the Sparse Patch Coding descriptor against the following state-of-the-art methods, and these methods have also been evaluated on the same data set.

- **Augmented multi-resolution Reeb graph** [31] [32]. They defined Reeb graph to describe a contour relationship mapping vertices of a 3D shape to a geodesic space. Each contour level is represented as an edge of the Reeb graph, and each region between contour levels in a regular interval is coded into a node of Reeb graph, which is seen as a semantically segmented surface. The multi-resolution graph structure is then formed by hierarchically linking these nodes of connected regions. Topological, geometrical and visual information is attached to each graph node for enhancing graph matching and model comparison.
- **Spherical trace transform** [33] . This method first scales and places a 3D model into an unit sphere, and defines a set of planes tangential to defined concentric spheres. Each plane intersects with the object, and intersection area is analyzed via 2D Krawtchouk moments, 2D Zernike Moments, and Polar Fourier Transform. Spherical Fourier transform is applied on intersection functions in order to generate rotation invariant descriptors.
- **Depth line encoding** [34]. In their method, each 3D model is represented by a set of depth lines, extracted from depth buffer images projected onto the six faces of its bounding box after being normalized and scaled.

Table 1. The recall values of 20, 40, 60, and 80 return items

Method	20	40	60	80
Depth line encoding	0.55	0.66	0.72	0.76
Spherical trace transform	0.56	0.69	0.76	0.80
Augmented multi-resolution Reeb graph	0.71	0.83	0.87	0.90
Sparse patch coding	0.77	0.87	0.91	0.93

Table 2. The precision values of 20, 40, 60, and 80 return items

Method	20	40	60	80
Depth line encoding	0.55	0.33	0.24	0.19
Spherical trace transform	0.56	0.35	0.25	0.20
Augmented multi-resolution Reeb graph	0.71	0.41	0.29	0.23
Sparse patch coding	0.77	0.43	0.30	0.23

5.3 Numerical Results

Numerical values for the averaged recall and precision on all the models in the dataset are reported in Table 1 and 2, respectively. We list these values of returned 20, 40, 60, 80 items, which are 1, 2, 3, and 4 times the size of class. These results show that our method is comparable to the representative methods evaluated on the same dataset. The proposed algorithm based on sparse patch coding improved the retrieval accuracy.

6 Conclusion

In this paper, we introduced the concept of low-level patch with initial semantics to represent a 3D object, which reduces the number of point descriptors, and makes retrieval robust against topology variation. These patches were further encoded into visual words in a sparse form, and the distribution of visual words was used to improve retrieval performance.

In the future, we will investigate the performance influence of setting different patch number for each model, and also consider the structural relationship of visual words. Specifically we plan to combine l_1 regularized sparse coding with the spatial pyramid representation to enhance the retrieval accuracy.

Acknowledgments. The work was supported by NSFC (61003137, 61202185, 61005018, 91120005), NPU-FER(JC201202, JC201220, JC20120237), Shaanxi NSF(2012JQ8037), and Open Fund from State Key Lab of CAD&CG of Zhejiang University, and Program for New Century Excellent Talents in University under grant NCET-10-0079.

References

1. Liu, Z., Bu, S., Zhou, K., Gao, S., Han, J., Wu, J.: A survey on partial retrieval of 3D shapes. Journal of Computer Science and Technology 28(5), 836–851 (2013)
2. Wu, H., Zha, H., Luo, T., Wang, X., Ma, S.: Global and local isometry-invariant descriptor for 3D shape comparison and partial matching. In: Proceedings of IEEE Computer Vision and Pattern Recognition, pp. 438–445 (2010)
3. Bronstein, M.M., Kokkinos, I.: Scale-invariant heat kernel signatures for non-rigid shape recognition. In: Proceedings of IEEE Conference on Computer Vision and Pattern Recognition, pp. 1704–1711 (2010)
4. Sipiran, I.: Local features for partial shape matching and retrieval. In: Proceedings of ACM Multimedia, pp. 853–856 (2011)
5. Knopp, J., Prasad, M., Willems, G., Timofte, R., Van Gool, L.: Hough transform and 3D SURF for robust three dimensional classification. In: Daniilidis, K., Maragos, P., Paragios, N. (eds.) ECCV 2010, Part VI. LNCS, vol. 6316, pp. 589–602. Springer, Heidelberg (2010)
6. Kokkinos, I., Bronstein, M.M., Litman, R., Bronstein, A.M.: Intrinsic shape context descriptors for deformable shapes. In: Proceedings of IEEE Conference on Computer Vision and Pattern Recognition, pp. 159–166 (June 2012)
7. Berretti, S., Bimbo, A.D., Pala, P.: Partial match of 3D faces using facial curves between SIFT keypoints. In: Proceedings of Eurographics Workshop on 3D Object Retrieval, pp. 117–120 (2011)
8. Bronstein, A.M., Bronstein, M.M., Guibas, L.J., Ovsjanikov, M.: Shape google: geometric words and expressions for invariant shape retrieval. ACM Transactions on Graphics 30(1), 1–20 (2011)
9. Lavoué, G.: Combination of bag-of-words descriptors for robust partial shape retrieval. The Visual Computer 28(9), 931–942 (2012)
10. Mademlis, A., Daras, P., Axenopoulos, A., Tzovaras, D., Strintzis, M.G.: Combining topological and geometrical features for global and partial 3-D shape retrieval. IEEE Transactions on Multimedia 10(5), 819–831 (2008)
11. Biasotti, S., Marini, S., Spagnuolo, M., Falcidieno, B.: Sub-part correspondence by structural descriptors of 3D shapes. Computer-Aided Design 38(9), 1002–1019 (2006)
12. Shapira, L., Shalom, S., Shamir, A., Cohen-Or, D., Zhang, H.: Contextual part analogies in 3D objects. International Journal of Computer Vision 89(2-3), 309–326 (2010)
13. Cornea, N.D., Demirci, M.F., Silver, D.E., Shokoufandeh, A.C., Dickinson, S.J., Kantor, P.B.: 3D object retrieval using many-to-many matching of curve skeletons. Proceedings of Shape Modeling International, 366–371 (June 2005)
14. Liu, Z., Zhou, K., Bu, S., Sun, X.: Geometrically attributed binary tree for 3D shape matching. In: Computer Graphics International Conference (2011)
15. Gao, Y., Wang, M., Zha, Z., Tian, Q., Dai, Q., Zhang, N.: Less is more: efficient 3D object retrieval with query view selection. IEEE Transactions on Multimedia 11(5), 1007–1018 (2011)
16. Gao, Y., Dai, Q., Wang, M., Zhang, N.: 3d model retrieval using weighted bipartite graph matching. Signal Processing: Image Communication 26(1), 39–47 (2011)
17. Gao, Y., Tang, J., Hong, R., Yan, S., Dai, Q., Zhang, N., Chua, T.S.: Camera constraint-free view-based 3D object retrieval. IEEE Transactions on Image Processing 21(4), 2269–2281 (2012)

18. Gao, Y., Wang, M., Tao, D., Ji, R., Dai, Q.: 3-D object retrieval and recognition with hypergraph analysis. IEEE Transactions on Image Processing 21(9), 4290–4303 (2012)
19. Papadakis, P., Pratikakis, I., Theoharis, T., Perantonis, S.: PANORAMA-a 3D shape descriptor based on panoramic views for unsupervised 3D object retrieval. International Journal of Computer Vision 89(2-3), 177–192 (2010)
20. Shan, Y., Sawhney, H.S., Matei, B., Kumar, R.: Shapeme histogram projection and matching for partial object recognition. IEEE Transactions on Pattern Analysis and Machine Intelligence 28(4), 568–577 (2006)
21. Golovinskiy, A., Funkhouser, T.: Randomized cuts for 3D mesh analysis. ACM Transactions on Graphics 27(5) (2008)
22. Liu, Z., Tang, S., Bu, S., Zhang, H.: New evaluation metrics for mesh segmentation. Computers and Graphics (SMI) 37(6), 553–564 (2013)
23. Ben-Chen, M., Gotsman, C., Bunin, G.: Conformal flattening by curvature prescription and metric scaling. Computer Graphics Forum (Eurographics) 28(2), 449–458 (2008)
24. Shapira, L., Shamir, A., Cohen-Or, D.: Consistent mesh partitioning and skeletonisation using the shape diameter function. The Visual Computer 24(4), 249–259 (2008)
25. Surazhsky, V., Surazhsky, T., Kirsanov, D., Gortler, S.J., Hoppe, H.: Fast exact and approximate geodesics on meshes. ACM Transactions on Graphics 25(4), 553–560 (2005)
26. Sun, J., Ovsjanikov, M., Guibas, L.J.: A concise and provably informative multiscale signature based on heat diffusion. Computer Graphics Forum (SGP) 28(5), 1383–1392 (2009)
27. Bach, F., Mairal, J., Ponce, J., Sapiro, G.: Sparse coding and dictionary learning for image analysis. In: Proceedings of IEEE International Conference on Computer Vision and Pattern Recognition (2010)
28. Ji, R., Yao, H., Liu, W., Sun, X., Tian, Q.: Task-dependent visual-codebook compression. IEEE Transactions on Image Processing 21(4), 2282–2293 (2011)
29. Ji, R., Duan, L.Y., Chen, J., Xie, L., Yao, H., Gao, W.: Learning to distribute vocabulary indexing for scalable visual search. IEEE Transactions on Multimedia 15(1), 153–166 (2011)
30. Lee, H., Battle, A., Raina, R., Ng, A.Y.: Efficient sparse coding algorithms. In: Proceedings of Neural Information Processing Systems, pp. 801–808 (2007)
31. Tung, T., Schmitt, F.: The augmented multiresolution reeb graph approach for content-based retrieval of 3D shapes. International Journal of Shape Modeling 11(1), 91–120 (2005)
32. Tung, T., Schmitt, F., Matsuyama, T.: Topology matching for 3D video compression. In: Proceedings of IEEE International Conference on Computer Vision and Pattern Recognition, pp. 1–8 (2007)
33. Zarpalas, D., Daras, P., Axenopoulos, A., Tzovaras, D., Strintzis, M.G.: 3D model search and retrieval using the spherical trace transform. EURASIP Journal on Advances in Signal Processing, Article 23912 (2007)
34. Chaouch, M., Verroust-Blondet, A.: 3D model retrieval based on depth line descriptor. In: Proceedings of IEEE International Conference on Multimedia and Expo. 599–602 (2007)

3D Object Classification Using Deep Belief Networks

Biao Leng, Xiangyang Zhang, Ming Yao, and Zhang Xiong

School of Computer Science & Engineering,
Beihang University, Beijing, 100191, P.R. China
lengbiao@buaa.edu.cn

Abstract. Extracting features with strong expressive and discriminative ability is one of key factors for the effectiveness of 3D model classifier. Lots of research work has illustrated that deep belief networks (DBN) have enough power to represent the distributions of input data. In this paper, we apply DBN for extracting the features of 3D model. After implementing a contrastive divergence method, we obtain a trained-well DBN, which can powerfully represent the input data. Therefore, the feature from the output of last layer is acquired. This procedure is unsupervised. Due to the limit of labeled data, a semi-supervised method is utilized to recognize 3D objects using the feature obtained from the trained DBN. The experiments are conducted in the publicly available Princeton Shape Benchmark (PSB), and the experimental results demonstrate the effectiveness of our method.

Keywords: 3D model classification, Deep belief networks, Semi-supervised learning.

1 Introduction

Due to the explosion in the number of 3D objects over various digital archives, there is an increasing demand for effectively 3D model understanding. Therefore, high-effective technologies for dealing with 3D models are becoming more and more important.

In the last decades, a great number of model descriptors have been proposed for 3D model retrieval[14,23,20,22,25,28,38,11]. Among these algorithms, different model features are introduced, including statistics-based feature descriptors [32], topological representation [33], geometric-based feature descriptors [7,8] and visual representation [1,10]. All these methods just capture only one side of object's features, and the retrieval performance is not satisfied. Then, hybrid features are introduced to combine several different feature descriptors [6]. These models just describe the low-level features of 3D objects, and they are not able to acquire the semantic information of objects. In order to improve the retrieval effectiveness and capture the user's semantic information for 3D models, the technique of relevance feedback [26,24,13,27], narrowing the gap between high-level semantic knowledge and low-level object representation, is introduced to 3D model understanding.

C. Gurrin et al. (Eds.): MMM 2014, Part II, LNCS 8326, pp. 128–139, 2014.

There are few related work about the 3D model classification in the field of 3D object understanding. Classification needs test data set, however, getting test data set for classification needs lots of human labor to label models. As a result, the desire of acquiring semantic information from a few labeled models has attracted extensive attentions. Obtaining labeled data for machine learning is often difficulty, but unlabeled data is usually easily accessed.

In this paper, we propose a novel 3D object classification model based on deep belief networks (DBN). First of all, the DBN is applied to extract the features from the projected images of 3D model. Then, a contrastive divergence method is implemented, and we obtain a trained-well DBN, representing the input data. Thus, the feature from the output of last layer is acquired. Due to the limit of labeled data, a semi-supervised method is utilized to train the labeled and unlabeled data using the feature obtained from the trained DBN. The experimental results show the effectiveness of the proposed method in the publicly available 3D model databases.

2   Related Work

2.1   Model Descriptors

In the framework of 3D model retrieval, a feature descriptor is an essential and crucial part. Due to various characteristics of 3D objects, diverse descriptors are invented to capture different features.

The statistics-based feature descriptors sample the 3D models and utilize the histogram to compare characteristics. Park et al. [32] developed a sliced image histogram to represent PCA normalized 3D objects. For the topological representation, Patane et al. [33] presented a minimal contouring algorithm to rapidly compute the Reeb graph.

The geometric-based feature descriptors [7,30] extract surface characteristics and spherical harmonics to present the objects. Daras et al. [7] applied voxelization for 3D models before extracting radial and spherical information via a radial integration transform (RIT) and a spherical integration transform (SIT). Spherical-based feature descriptors are frequently applied in the area of 3D model retrieval because of their robustness, rotational invariance and computational efficiency. Papadakis et al. [30] proposed the novel spherical functions, representing not only the intersection points of the model's surface with rays emanating from the origin but also the whole points in the direction of each ray that are closer to the origin than the furthest intersection points.

The visual representations [1,31,10,9] in 3D model firstly convert the 3D object into 2D projection images, and then extract various features. In [1], adaptive views clustering (AVC) selects the best characteristic views from more than 320 projected views. The K-means derived method and a Bayesian information criteria are employed to evaluate how likely the characteristic views fit the data. Papadakis et al [31] proposed a 3D shape descriptor called PANORAMA, standing for PANormic object representation for accurate model attribution. Gao et al. [10] adopted a probabilistic graph model, in which they divided the captured camera views into several sets, modeling them as a first order Markov Chain. The query comparison is then converted to probabilistic analysis, which significantly alleviates the effects of transformation. Gao et al. [9] propose a camera constraint-free view-based (CCFV) 3-D object retrieval algorithm which is not under the restriction of the camera array.

2.2 Semantic Understanding

The similarity between objects depends on subjective human observation and judgement. Thus, the basic object of semantic understanding is to bridge the gap between high-level semantic information and low-level model features.

The first semantic retrieval strategy was designed as an iterative and interactive algorithm. Since then, a great deal of related work has been performed and classified in three subdivisions. Leng and Qin [24] developed a relevance feedback concentric retrieval system which linearly combines a set of weighted descriptors. Learning from labeled models, this method dynamically updates the weights of each descriptor for improvements.

Annotation is a way of encoding semantic features in the form of textual messages or tags. Goldfeder and Allen [15] developed a method for the propagation of annotations attached to 3D models from Google 3DWarehouse. Later, Goldfeder et al. [16] continued their research on tag based propagation of annotations. Wen et al. [38] adopts a bipartite graph matching method to measure the semantic similarity between two objects based on multiple views.

Compared with 3D model retrieval, fewer algorithms for 3D model classification have been brought forward. One of the important reasons dues to "curse of dimensionality". In [37], a method is shown that how to learn a Mahanalobis distance metric for k-nearest neighbor classification by semi-definite programming. Li et al. [29] present a 3D model classification method based on nonparametric discriminant analysis combined with geometry projection based histogram models. Gao et al. [12] propose a semi-supervised method for 3D objects classification by constructing multiple hyper-graphs. Ji et al. [19] propose a hyperspectral image classification method to address both the pixel spectral and spatial constraints by learning from the relationship among pixels.

3 Deep Learning Based Classification Model

In this section, we implement an unsupervised learning procedure using deep belief networks (DBN) to extract features of 3D models. Some high-level feature can be got from this procedure [21]. And then, a semi-supervised learning process is carried out based on graph learning for 3D model recognition.

3.1 Extracting Feature Using Deep Belief Networks

The DBN are multi-layers networks constituted by several layers of Restricted Boltzmann machines (RBM). Hinton et al [18] proposed a greedy layer-wise algorithm for training the DBN, which uses RBM to model the input data. In practice, this greedy algorithm often learns deep belief networks that appear to fit input data well [35].

The DBN is a network of symmetrically coupled stochastic binary units. It contains a set of visible units $v \in [0, 1]^D$, and a sequence of layers of hidden units $h^1 \in \{0, 1\}, h^2 \in \{0, 1\}, \ldots, h^l \in \{0, 1\}$, There are connections only between hidden units and adjacent layer, as well as between visible units in the input layer and hidden units in the first hidden layer [2]. When given the visul v, we can obtain the hidden layer's output.

$$Q(h_i|v) = \frac{1}{1 + e^{-b_i - W_i v}} = sigm(b_i + W_i v) \tag{1}$$

Where $h_i \in \{0, 1\}$ and W is the linked weight matrix between visual layer and hidden layer, b_i is the bias of the unit i in the hidden layer.

When given the hidden layer h_i, we can also get the reconstruction of visual layer.

$$P(v_i|h) = \frac{1}{1 + e^{-c_i + W_{.j} h}} = sigm(c_i + W_{.j} h) \tag{2}$$

Where $W_{.j}$ is the j-th column of W. It needs to be stated that our input data is the depth image of all views from a model, and the real value is among the region $[0, 1]$.

Greedy layer-wise training is an unsupervised greedy strategy to train one layer at a time in the DBN. A k steps contrastive divergence (CD-K) method is applied when training DBN. The derivative of the log-likelihood of $P(v_0)$ with respect to the model parameters w can be written as (between input layer and first hidden layer):

$$\frac{\partial log P(v^0)}{\partial w_{ij}} = < v_i^0 h_j^0) > - < v_i^\infty h_j^\infty > \tag{3}$$

When training the parameters between layer l and $l + 1$, Markov chain and Monte-Carlo method is used to generate samples of layer l, beginning from input layer associated with trained parameters in prior layer. Then treat the samples of layer l as an visual layer v_0, and generate the next hidden layer with the probability $Q(h_0 = 1|v_0)$. The next step can be called reconstruction: re-sampling the visual layer v_1 with probability $P(v_{1j} = 1|h_0)$, and then compute the probability $Q(h_{1j} = 1|v_1)$ according to v_1. At last, update the parameters w and bias b^l and b^{l+1}.

$$W = W - \epsilon(h_0 v_0' - Q(h_1. = 1|v_1) v_1')$$
$$b^{l+1} = b^{l+1} - \epsilon(h_0 - Q(h_1. = 1|v_1)) \tag{4}$$
$$b^l = b^l - \epsilon(v_0 - v_1^l)$$

Where ϵ is the rate of learning, which can be set dynamically according to Hinton et al.[17].

The upper step is the special case of k-steps contrastive divergence method when $k = 1$, which has been experimentally proved to be efficient in [5]. We use $Q(h_1. = 1|v_1)$ to replace h_j^0, which has been illustrated in [17]. The pseudo-code to accomplish the task mentioned above has been provided in [3]. We just take the output of the last layer as the feature of a 3D object.

3.2   Semi-supervised Algorithm for 3D Model Recognition

Semi-supervise is also a natural choice in hard AI tasks witch involve several sub-tasks as well Deep architectures when labeled data is not enough[5]. In this subsection, we employee a semi-supervised method which can learn from local and global consistency.

Considering a general case of problem respecting with unlabeled data and labeled data, given a data set $X = \{x_1, x_2, \ldots, x_l, x_{l+1}, \ldots, x_n\}$ and a label set $L = \{1, \ldots, c\}$, the first l data have labels $\{Y_1, Y_2, \ldots, Y_l\}$, and the remains are unlabeled data,which could be set as zero. our goal is to get the information of the unlabeled data.

Define the membership vector of model i

$$Y_i = (m_{1i}, m_{2i}, \ldots, m_{ci}) \tag{5}$$

where

$$m_{ij} = \begin{cases} 1, \text{ if label model j belongs to class i} \\ 0, \text{ otherwise} \end{cases} \tag{6}$$

Where c is the number of categories.

We have obtained the feature of 3D model via unsupervised DBN learning. Suppose V_i is the feature of model i , and we can get an affinity matrix M, the element of M can be defined as:

$$M_{i,j} = \begin{cases} e^{-\frac{|V_i - V_j|^2}{2\sigma^2}} & \text{if i} \neq j \\ 0 & \text{otherwise} \end{cases} \tag{7}$$

Here M is symmetrical matrix with the diagonal elements being zero.M_{ij} can symbol the affinity between model i and model j. Then a regulation framework associated the intrinsic structure and label information is given as follow.

$$Q(F) = \sum_{i=0}^{N} \sum_{j=0}^{N} M_{ij} |\frac{F_i}{\sqrt{D_{ii}}} - \frac{F_j}{\sqrt{D_{jj}}}|^2 + \lambda \sum_{i=0}^{N-1} |F_i - Y_i|^2 \tag{8}$$

Where D is a diagonal matrix, whose (i, i) element equals the sum of elements in i th row of W. ie, $D_{ii} = \sum_j M_{ij}$, and its function is to normalize M. F_j is the model j's membership vector. The right hand of this equation contains two items, which represent two types of cost. The first is the smoothness constrains, which indicates the semantic difference between two models referring the similarity between models. The second is fitting constraint, which symbols the scale of change of semantic matrix from initial Y. The trade-off of the two constraints is captured by parameter λ.

Our target is to find F to satisfy the following condition:

$$F = arg \min_{F} Q(F). \tag{9}$$

The goal is to find the F that minimize the cost function respecting both the two constraints. This optimality problem can be solved by Lagrange multiplier method. By computing partial derivative of $Q(F)$ with respect to F, Equation (8) can be converted as follows:

Make the derivation over F, we have

$$\frac{\partial Q}{\partial F}|_{F=F^*} = F^* - LF^* + \frac{1}{\lambda}(F^* - Y), \tag{10}$$

where $L = D^{-\frac{1}{2}}MD^{\frac{1}{2}}$. And let Equation (10) to be zero and this equation can be transformed to be as the following form [39]:

$$F = (I + \frac{1}{\lambda}L)^{-1}Y. \tag{11}$$

The procedure mentioned above can be understood intuitively in term of spreading label information in a network [4]. As shown in Equation (11) models can get information from the neighbors and its initial label information. In the procedure of propagation, the self-reinforcement is avoided since diagonal elements in affinity matrix are set to zeros. The information spreads symmetrically since L is a symmetric matrix.

3.3   Classification

According to the label propagation procedure formulated above, full information of every model could be got. Each element in the membership vector $F_i = \{F_{ij}, \ldots, F_{ic}\}$ represents the membership of the responding category. So we can use the following method to distinguish which category a model belongs to.

$$c_m = arg \max_{i}\{F_{m1}, \ldots, F_{mi}, \ldots, F_{m}n\} \tag{12}$$

Where c_m is the classification result of model m, and n is the number of categories.

4   Experiment Setups

4.1   Model Database and Evaluation Measurement

It is very important to compare different algorithms using publicly available benchmark databases, containing models classified in different categories. To demonstrate the effectiveness of the proposed method, the experiments are conducted on the famous benchmark Princeton shape benchmark (PSB) [34].

The PSB model database is provided by the Shape Retrieval and Analysis Group from Princeton University, and it contains 1814 objects in general categories like human, building and vehicle, as illustrated by Fig. 1. We only choose the test model in the experiment, and the test set contains 907 3D models which classified into 92 classes.

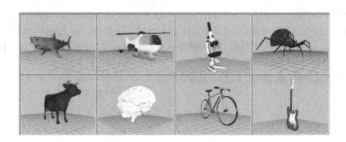

Fig. 1. Sample models in PSB

To estimate the effectiveness of different algorithms, we apply mean classification precision (*MCP*) as the criterion. Let N to represent the number of models in a class and let R be the number of right classification of a class. So we define $CP = R/N$, and *MCP* is the average of all classes' *CP*. We also use *accuracy* as criterion, which is the ratio between the number of right classification and the number of test models.

4.2 Compared Methods

In the experiments, there are two methods are applied to compare with the proposed method.

1. CMVD with graph learning (CMVD-GL)

    The compact multi-view descriptor (CMVD) method [6], combining 2D Polar-FFT, 2D Zernike Moments and 2D Krawthcouk moments, is an example of early descriptors fusion for 3D model retrieval. Then, the CMVD method is combined with the graph learning for 3D model classification, as a compared approach.
2. Equal weight using graph fusion (EW-GF)

    This method combines 2D Polar-FFT, 2D Zernike Moments, 2D Krawthcouk moments and the Desire descriptor [36] with equal weight fusion, and the equal weight fusion can be viewed as a fusion method based on graph learning with all elements in linked weight parameter α assigned to be 0.25, without considering the distinction of graphs.

5   Experimental Results

This section will discuss the experimental results of the proposed model and the compared methods in PSB.

Fig. 2 shows the comparison results of accuracy in PSB. It is obvious that the classification accuracy of all the algorithms is growing with the number of labeled samples increasing, and the proposed method is the best among these approaches. With the increasing of labeled number, the superiority of our method become more and more remarkable. From the figure, when the label number is 92, the discrepancy between our method and EW-GF method is 20.1%, and the difference between our method and CMVD-GL method is 22.1%. Once the label number is increased to 508, the discrepancy becomes 14.5% and 25.9%respectively.

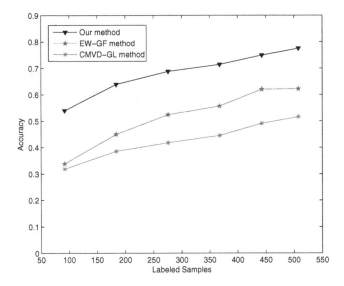

Fig. 2. Comparison results of accuracy in PSB

Fig. 3 indicates the MCP performance of these methods in PSB. It is apparent that our method has acquired the best performance among these approaches. The MCP value of the proposed model is 0.783, while that of the EW-GF and CMVD-GL method are 0.714 and 0.661 respectively. In other words, the performance of our method is 9.24% and 18.00% better than that of the EW-GF and CMVD-GL in the measurement of MCP.

Table. 1 represents a comprehensive comparison among several classes in PSB. The number of labeled samples is 276, while 3 samples labeled per category. From this table, our method surpasses other methods in most classes.

Therefore, the proposed model is better than the EW-GF and CMVD-GL methods, and is more suitable to 3D model classification.

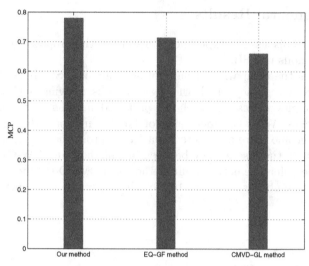

Fig. 3. MCP performance in PSB

Table 1. The comparison among different classes in PSB (the number of labeled samples is 276, 3 samples labeled per category)

Class name	Our method	EW-GF method	CMVD-GL method
biplane	0.727	0.636	0.594
commercial	0.875	0.500	0.508
fighter_jet	0.915	0.362	0.547
glider	1.000	0.750	0.700
stealth_bomber	1.000	1.000	0.783
hot_air_balloon	1.000	0.667	0.663
human	0.766	0.723	0.705
human_arms_out	0.765	0.676	0.676
sword	0.615	0.615	0.556
face	1.000	0.846	0.795
head	0.923	0.692	0.841
chess_set	1.000	1.000	0.996
computer_monitor	1.000	0.797	0.797
door	1.000	0.867	0.927
fireplace	1.000	0.333	0.370
bench	0.625	0.500	0.463
desk_chair	0.917	0.667	0.674
geographic_map	0.889	0.667	0.533
handgun	0.857	0.429	0.494
hourglass	0.667	0.667	0.724
streetlight	0.800	0.600	0.551
electrical_guitar	1.000	0.200	0.423
large_sail_boat	0.667	0.333	0.297
submarine	0.833	0.833	0.562
hammer	1.000	1.000	0.906
shovel	1.000	0.667	0.727

6   Conclusion

This paper introduces a novel 3D model classification model using deep learning technique. The deep belief networks have been proved to have powerful ability to represent the distribution of input data. Thus, the deep belief networks are applied to extract the features from different projected depth-buffer images of 3D model. After that, a contrastive divergence method is conducted, and then a trained-well deep belief networks has been constructed with great power of representing the input data. The features from the output of last layer is acquired. The unsupervised procedure utilizes lots of unlabeled data to obtain the high-level feature of 3D objects. Because of the limitation of labeled data, we use a semi-supervised learning method based on graph for label information propagation. The graph learning method is used to train the labeled and unlabeled data using the feature obtained from the trained deep belief networks, and finally get the label label information of unlabeled data, which is the basis for classification. The experiments are conducted in the publicly available Princeton Shape Benchmark (PSB), and the experimental results demonstrate the effectiveness of our method.

Acknowledgments. The 3D model database PSB is from the Shape Retrieval and Analysis Group at the University of Princeton. This work is supported by the National Natural Science Foundation of China (No. 61103093), the New Teachers' Fund for Doctor Stations from Ministry of Education (No.20111102120017) and the Fundamental Research Funds for the Central Universities.

References

1. Ansary, T.F., Daoudi, M., Vandeborre, J.P.: A bayesian 3-d search engine using adaptive views clustering. IEEE Transaction on Multimedia 9(1), 78–88 (2007)
2. Bengio, Y.: Learning deep architectures for ai. Foundations and Trends® in Machine Learning 2(1), 1–127 (2009)
3. Bengio, Y., Lamblin, P., Popovici, D., Larochelle, H.: Greedy layer-wise training of deep networks. In: Proceedings of the Advances in Neural Information Processing Systems, pp. 153–160 (2007)
4. Blum, A., Chawla, S.: Learning from labeled and unlabeled data using graph mincuts. In: Proceedings of the Eighteenth International Conference on Machine Learning, pp. 19–26 (2001)
5. Carreira-Perpinan, M.A., Hinton, G.E.: On contrastive divergence learning. In: Proceedings of the Tenth International Workshop on Artificial Intelligence and Statistics, pp. 33–40 (2005)
6. Daras, P., Axenopoulos, A.: A 3D shape retrieval framework supporting multimodal queries. International Journal of Computer Vision 89(2-3), 229–247 (2010)
7. Daras, P., Zarpalas, D., Tzovaras, D., Strintzis, M.G.: Efficient 3D model search and retrieval using generalized 3D radon transforms. IEEE Transactions on Multimedia 8(1), 101–114 (2006)
8. Gao, Y., Dai, Q.H., Zhang, N.Y.: 3D model comparison using spatial structure circular descriptor. Pattern Recognition 43(3), 1142–1151 (2010)

9. Gao, Y., Tang, J.H., Hong, R.C., Yan, S.C., Dai, Q.H., Zhang, N.Y., Chua, T.S.: Camera constraint-free view-based 3-d object retrieval. IEEE Transactions on Image Processing 21(4), 2269–2281 (2012)
10. Gao, Y., Tang, J.H., Li, H.J., Dai, Q.H., Zhang, N.Y.: View-based 3D model retrieval with probabilistic graph model. Neurocomputing 73(10), 1900–1905 (2010)
11. Gao, Y., Wang, M., Ji, R.R., Wu, X.D., Dai, Q.H.: 3D object retrieval with hausdorff distance learning. Accepted for Publication in IEEE Transactions on Industrial Electronics (2013)
12. Gao, Y., Wang, M., Tao, D.C., Ji, R.R., Dai, Q.H.: 3-d object retrieval and recognition with hypergraph analysis. IEEE Transactions on Image Processing 21(9), 4290–4303 (2012)
13. Gao, Y., Wang, M., Zha, Z.J., Tian, Q., Dai, Q.H., Zhang, N.Y.: Less is more: efficient 3-d object retrieval with query view selection. IEEE Transactions on Multimedia 13(5), 1007–1018 (2011)
14. Gao, Y., Yang, Y., Dai, Q., Zhang, N.: 3D object retrieval with bag-of-region-words. In: Proceedings of the ACM International Conference on Multimedia, Firenze, Italy, pp. 955–958 (2010)
15. Goldfeder, C., Allen, P.: Autotagging to improve text search for 3D models. In: ACM/IEEE-CS Joint Conference on Digital Libraries, Pittsburgh, PA, USA, pp. 355–358 (2008)
16. Goldfeder, C., Feng, H., Allen, P.: Shrec08 entry: Training set expansion via autotags. In: Proceedings of the IEEE International Conference on Shape Modeling and Applications, Stony Brook, NY, USA, pp. 233–234 (2008)
17. Hinton, G.E.: A practical guide to training restricted boltzmann machines. In: Montavon, G., Orr, G.B., Müller, K.-R. (eds.) Neural Networks: Tricks of the Trade, 2nd edn. LNCS, vol. 7700, pp. 599–619. Springer, Heidelberg (2012)
18. Hinton, G.E., Osindero, S., Teh, Y.W.: A fast learning algorithm for deep belief nets. Neural Computation 18(7), 1527–1554 (2006)
19. Ji, R.R., Gao, Y., Hong, R.C., Liu, Q., Tao, D.C., Li, X.L.: Spectral-Spatial Constraint Hyperspectral Image Classification. Accepted for Publication in IEEE Transactions on Geoscience and Remote Sensing (2013)
20. Ji, R.R., Yao, H., Liu, W., Sun, X., Tian, Q.: Task-dependent visual-codebook compression. IEEE Transactions on Image Processing 21(4), 2282–2293 (2012)
21. Le Roux, N., Bengio, Y.: Representational power of restricted boltzmann machines and deep belief networks. Neural Computation 20(6), 1631–1649 (2008)
22. Leng, B., Li, L., Qin, Z.: MADE: A composite visual-based 3D shape descriptor. In: Gagalowicz, A., Philips, W. (eds.) MIRAGE 2007. LNCS, vol. 4418, pp. 93–104. Springer, Heidelberg (2007)
23. Leng, B., Qin, Z.: Automatic combination of feature descriptors for effective 3D shape retrieval. In: Gagalowicz, A., Philips, W. (eds.) MIRAGE 2007. LNCS, vol. 4418, pp. 36–46. Springer, Heidelberg (2007)
24. Leng, B., Qin, Z.: A powerful relevance feedback mechanism for content-based 3D model retrieval. Multimedia Tools and Applications 40(1), 135–150 (2008)
25. Leng, B., Qin, Z., Cao, X.M., Wei, T., Zhang, Z.X.: Mate: a visual based 3D shape descriptor. Chinese Journal of Electronics 18(2), 291–296 (2009)
26. Leng, B., Qin, Z., Li, L.Q.: Support vector machine active learning for 3D model retrieval. Journal of Zhejiang University SCIENCE A 8(12), 1953–1961 (2007)
27. Leng, B., Xiong, Z.: Modelseek: an effective 3D model retrieval system. Multimedia Tools and Applications 51(3), 935–962 (2011)
28. Leng, B., Xiong, Z., Fu, X.W.: A 3D shape retrieval framework for 3D smart cities. Frontiers of Computer Science 4(3), 394–404 (2010)

29. Li, J.B., Sun, W.H., Wang, Y.H., Tang, L.L.: 3D model classification based on nonparametric discriminant analysis with kernels. Neural Computing and Applications 22(3-4), 771–781 (2013)
30. Papadakis, P., Pratikakis, I., Perantonis, S., Theoharis, T.: Efficient 3D shape matching and retrieval using a concrete radialized spherical projection representation. Pattern Recognition 40(9), 2437–2452 (2007)
31. Papadakis, P., Pratikakis, I., Theoharis, T., Perantonis, S.: Panorama: A 3D shape descriptor based on panoramic views for unsupervised 3D object retrieval. International Journal of Computer Vision 89(2), 177–192 (2010)
32. Park, Y.S., Yun, Y.I., Choi, J.S.: A new shape descriptor using sliced image histogram for 3D model retrieval. IEEE Transactions on Consumer Electronics 55(1), 240–247 (2009)
33. Patane, G., Spagnuolo, M., Falcidieno, B.: A minimal contouring approach to the computation of the reeb graph. IEEE Transactions on Visualization and Computer Graphics 15(4), 583–595 (2009)
34. Shilane, P., Min, P., Kazhdan, M., Funkhouser, T.: The princeton shape benchmark. In: Proceedings of Shape Modeling and Applications, Palazzo Ducale, Genova, Italy, pp. 167–178 (2004)
35. Sutskever, I., Hinton, G.E.: Deep, narrow sigmoid belief networks are universal approximators. Neural Computation 20(11), 2629–2636 (2008)
36. Vranic, D.V.: Desire: a composite 3D-shape descriptor. In: Proceedings of IEEE International Conference on Multimedia and Expo, Amsterdam, Netherlands, pp. 962–965 (2005)
37. Weinberger, K.Q., Saul, L.K.: Distance metric learning for large margin nearest neighbor classification. The Journal of Machine Learning Research 10(6), 207–244 (2009)
38. Wen, Y., Gao, Y., Hong, R.C., Luan, H.B., Liu, Q., Shen, J.L., Ji, R.R.: View-based 3D object retrieval by bipartite graph matching. In: Proceedings of the ACM Multimedia, Nara, Japan, pp. 897–900 (2012)
39. Zhou, D., Bousquet, O., Lal, T.N., Weston, J., Schölkopf, B.: Learning with local and global consistency. In: Proceedings of the Advances in Neural Information Processing Systems, pp. 321–328 (2004)

Pursuing Detector Efficiency for Simple Scene Pedestrian Detection

De-Dong Yuan, Jie Dong, Song-Zhi Su*, Shao-Zi Li, and Rong-Rong Ji

Dept. of Cognitive Science, Xiamen University, Xiamen, Fujian, China, 361005 Fujian
Key Laboratory of the Brain-like Intelligent Systems, Xiamen, Fujian, China, 361005
ssz@xmu.edu.cn

Abstract. Detector accuracy is by any means the key focus in most existing pedestrian detection algorithms especially for clutter scenes. However, it is not always necessary, while sometimes over-fitted, to directly leverage such detectors in scenarios with simple scene compositions. To this end, limited work has done on a systematic detector simplification towards balancing its speed and accuracy. In this paper, we study this problem by investigating two mutually correlated issues, i.e. fast edge-based feature extraction and detector score computation. For handling the first issue, a simple Structured Local Edge Pattern (SLEP) is proposed to extract and encode local edge cues, extremely effectively, into a histogram. For the second, an integral image based acceleration is proposed toward fast classifier score computation by transforming the classifier score into a linear sum of weights. Experimental results on CASIA gait recognition dataset show that our proposed method is highly efficient than most existing detectors, which even faster than the practical OpenCV pedestrian detector.

Keywords: Pedestrian detection, structured local edge pattern, linear Support Vector Machine, Integral image.

1 Introduction

Pedestrian detection has been widely studied in computer vision community and beyond, with emerging applications ranging from driver assistant system[1], human motion analysis [2-4], intelligent video surveillance [5] to object retrieval[6,7]. The major challenge in pedestrian detection lies in the variations caused by lighting, pose, occlusion and viewpoint. Recent years have witnessed tremendous progress in building robust pedestrian detectors, most of which mainly focus on improving the detector accuracy especially for cluttered scenes. This can be proven from the wide variety of complex features and classifiers proposed in the past decade. For instances the systematic review of state-of-the-art detectors by Dollar et al. in [8] and others [1,17]. Well known works include the feature designs based upon various local patch statistics or patterns for example Haar[9],Histogram of Oriented Gradient (HOG) [10], Covariance descriptor

---

* Corresponding author.

C. Gurrin et al. (Eds.): MMM 2014, Part II, LNCS 8326, pp. 140–150, 2014.

[11],Local Binary Pattern (LBP) [12], etc. In a standard pipeline, sliding window scanning is first deployed, within each sliding window the pedestrian detection is regarded as a binary classification problem based on the designed features and classifiers. The core of this method is to build a robust classifier.Among the supervised classifiers,neural networks [11],support vector machines [10,13], AdaBoost[14], and Random Forest [15] are extensively used.

However, few pedestrian descriptors are specified to detect pedestrian in simple scenes with limited background compositions and static or smoothed viewing angle changes, which is however important for scenarios such as real-time surveillance where the efficiency is with a higher priority.In some applications such as surveillance, the background is often relatively simple as shown in Fig.1. In such a case, pursuing the detector efficiency is the major bottleneck beyond its accuracy.

A lazy way is to directly reusing the pedestrian detector trained on cluttered scenes to simple scenes. Unfortunately, this is infeasible due to the data variance between test and training sets. On the other hand, re-training complex detectors for simple scenes are also problematic, due to the "over fitted" design on complex features and classifiers. For example as shown in Fig.1, where the OpenCV detector using HOG as descriptor and trained in complex scenes, gets some false alarms even the scene is simple. The major issues lie in two facets, i.e. the time consuming feature extraction and classifier score computing within each sliding window.

As for accelerating the sliding window search, efficient search techniques are exploited in the literature with the principle of search space pruning. Different from exhaustive sliding window search schemes [10], efficient search space pruning usually adopts optimization schemes to exam only subset of the potential solutions. For example, Efficient Sub-window Search (ESS) [16] scans the sub-images containing pedestrian in a branch-and-bound scheme and can converge to a globally optimal solution. Other similar works includes Implicit Shape Model [17], Hopping Window Strategy [18]. As for accelerating the feature computing and classifier scoring, Wu et al. proposed C4 detector , which using the CENTRIST visual descriptor and a cascade classifier to achieve real-time performance.

However, none of above approaches exploits the actual cues about simple scenes. Under the simple scenes, the background contains less texture. Therefore features that are computationally light can be employed to encode the contour information of pedestrian. Second, it is easier to collect the negative training set, and then building the corresponding classifiers.

Our proposed scheme deals with the fast pedestrian detection by proposing two mutually correlative components, i.e. a fast feature computing and a fast classifier scoring. As for the first one, we propose a Structured Local Edge Pattern (SLEP) by using the image edge information. For the second one, we propose to concatenate the histogram of SLEP over each image sub-region without normalization. Therefore, the classifier score calculation can be transformed into a sum of weights that can be seen as pixels of a response image, from which

Fig. 1. Pedestrian detectors in simple scenes. Green bounding boxes are the results returned by OpenCV detector, which is trained in complex scenes. Red bounding boxes are the results of our proposed method, using a simpler feature and trained in simple scenes.

perspective we accelerate the classification stage by using integral image,which is described in details in Section 3.

The rest of this paper is organized as follows, Section 2 analysis the bottleneck of fast pedestrian detector, Section 3 presents our proposed method, including an edge-based feature and a fast classifier score prediction method based on integral image. Section 4 describes the implementation details and experimental results. And finally, Section 5 concludes this paper and prospects our future work.

2 Bottleneck of Fast Pedestrian Detector

Sliding window search is extensively used in object recognition, in which feature is computed over each scanning window and then fed into a binary classifier to make a decision whether the current scanning window contains a pedestrian or not. In terms of window design, multi-scale scanning is typically used, which allows dense overlapping with a non-maxima suppression to fuse individual detection results.

Overall, a typical sliding window search based object detector is decomposed into three components i.e. feature extraction, classifier score prediction, and non-maxima suppression. Among them, the most time-consuming component comes from the feature extraction, serving as the performance bottleneck. We show this in Fig.2 under different scanning window strides and numbers of windows fed to NMS procedure. For example, consider building a pedestrian detector using HOG with Linear SVM, the process of HOG feature calculation in each detection window, including binned gradient orientation, tri-linear interpolation, Gaussian weighted for histogram, and feature normalization, will take up about 90% of the feature computing time. Note that no matter how we change the number of sliding windows and its scanning steps, feature extraction always accounted more than 90% of the whole detection time and is the most time-consuming procedure. To this end, a fast feature calculation, is available to boost the detection speed to real-time, which we investigated in Section 3. Based on the proposed feature,

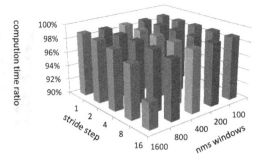

Fig. 2. The time proportion of feature extraction accounted for the whole pedestrian detection progress. We generate a 640 × 1280 image randomly and compute the processing time of feature extraction, binary classification prediction, and non-maximum suppression. Our experiment based on different stride steps with 1,2,4,8,16 and different numbers of windows (100, 200, 400, 800, 1600) fed into NMS procedure.

we find that the classifier score can be transformed into a weighted sum of each block in a scanning window, so the integral image technology is used to speed up the classifier score computation, which is presented in Section 3.2.

3   Proposed Method

3.1   Structured Local Edge Pattern

Local Edge Pattern (LEP) is a local image feature descriptor firstly proposed by Cheng [19] and widely used for image classification and retrieval. LEP is computed on binary edge images output from Canny detections in a manner similar to LBP, which encodes the edge pixels as well as center pixel as a 0/1 string. LEP histogram of a region R can be computed using Eq.1,

$$LEP_h[i] = \frac{N_i}{N}, i = 0, 1, ..., 511 \qquad (1)$$

where N_i is the number of pixels with LEP value i and N is the total number of pixels in region R, where i ranges within [0,511] and LEP_h within [0,1] In the sliding window based pedestrian detection, suppose the size of scanning-window is 64 × 128 with block size 16 × 16, and block stride size 8 × 8, then the final dimension of LEPh descriptor will be $(\frac{64-16}{8} + 1) \times (\frac{128-16}{8} + 1) \times 512 = 53760$, which is high-dimensional and inefficient to be directly used.

Given most of the pedestrian local edge features can be represented as following structure " $-$ / \ |", we proposed a feature called Structured Local Edge Patten(SLEP) , which reduces the dimension by capturing only the higher-order correlations among LEP structures:

$$SLEP = 16a_0 + 8(a_4\|a_8) + 4(a_3\|a_7) + 2(a_2\|a_6) + 1(a_1\|a_5) \qquad (2)$$

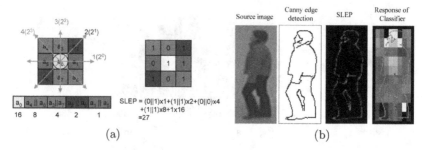

Fig. 3. SLEP feature for pedestrian detection. (a) demonstrates the computation of SLEP value, (b) the canny edge image, SLEP image and responses of the linear SVM classifier of the source image.

Fig. 4. Response image of SLEP classifier weights. The top is original images, the bottom is the response value of each pixel, provided by the classifier score. The red rectangles denote positive images, the blue rectangles denote negative images.

where a_i, i=0, ... ,8 denotes the corresponding binary edge value in a 3×3 window. As shown in Fig.3(a), SLEP encoding center pixel and edge pixel in five-directions, with scores ranging within [0,31]. As shown in Equ.2, as long as there is a pixel is encoded as 1, the direction of this pixel is coded as 1. To some extent this method overcomes the problem of noise disturbing in edge detection.

We further compute individual SLEP responses into a histogram. As another advantage, the SLEP histogram encodes direction between the edge pixels, which can well depict the contour of the pedestrian, as shown in Fig.3(b), where the consistent color denotes the same edge direction, meanwhile can be seen from the classifier response images, classifier response value is higher in the edge, as shown in Fig.4.

3.2   Fast Classifier Score Prediction Based on Integral Images

Histogram based feature is typically normalized with L1-norm, L2-norm, or L1-sqrt [10]. But in our experiments, we construct SLEP histogram without normalization, which will bring out two advantages. First, the feature computation

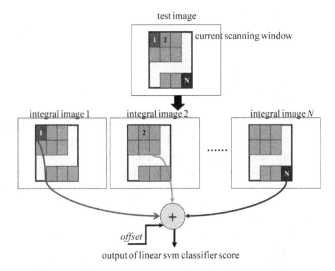

Fig. 5. Fast classifier score prediction based on integral images

time is reduced by omitting the normalization step. Second, the classifier score can be rapidly computed by integral images, as shown in Fig.5.

Let x represent the input image, $\{d_j\}_{j=1}^n$ denote the set of input image pixels, where n is the pixel number, $x|_y$ is the region y in the image x, $h = h(x|_y)$ is the SLEP histogram in the region y, K denotes histogram bins, set as 32 in this paper. Let h_k be the frequency of edge pattern with SLEP value k in $x|_y$. In the case of linear classifier such as Linear SVM, the classification can be formulated as:

$$f(h) = \beta + \sum_{b=1}^{N} \sum_{i=1}^{L} \alpha_i < h_b, h_b^i >$$

$$= \beta + \sum_{b=1}^{N} \sum_{i=1}^{L} \alpha_i \sum_{j=1}^{K} h_{b,j} h_{b,j}^i \qquad (3)$$

$$= \beta + \sum_{b=1}^{N} \sum_{j=1}^{K} h_{b,j} (\sum_{i=1}^{L} \alpha_i h_{b,j}^i)$$

in which $w_{b,j} = \sum_{i=1}^{L} \alpha_i h_{b,j}^i$, such that

$$f(h) = \beta + \sum_{b=1}^{N} \sum_{j=1}^{K} h_{b,j} w_{b,j}$$

$$= \beta + \sum_{b=1}^{N} \sum_{d \in x|_y} w_{b,c(d)} \qquad (4)$$

where N denotes the number of blocks in the scanning window, L is the total number of support vectors, $w_{b,j}$ represents a value contributed by the j-th pixel in the b-th block, $c(d)$ is the SLEP value corresponding to the pixel d, β is the bias of the trained SVM model. Thus the classification score is transformed into a weighted sum of each pixel.

In order to accelerate the classifier score calculation within a given detection window, N integral images corresponding to the N blocks is built as shown in Fig.5. Therefore, classifier score computation in the detection window only needs to accumulate the weights in each block and then sum them up. The accumulation weights can be fast computed by referring to its corresponding integral image, which is formulated as w_b in Eq.4.

Integral image is a data structure and algorithm for rapidly computing the sum of values in a rectangular region. In our case, the weight vector of SVMs can be seen as a concatenation of the block weight vectors. In each block, the SLEP pattern of pixel d is first computed, and then this pattern value is used as the index of the block weight vectors to obtain the pixel's response value. After the response image of a block is built, the corresponding integral image can easily get. Thus, classifier score of currently scanning window is just the sum of each block's accumulating response, which can rapidly return by the integral image.

4   Experiment Results

4.1   Dataset and Performance Measurements

We choose gait recognition as the testbed for our fast pedestrian detector, which typically constitutes of scenes with simple backgrounds. CASIA gait dataset [20] is used in our experiments, whose statistics is shown in Table.1[1]. The negative training subimages are randomly sample from the 307 training positive images. We randomly crop 20 windows per image from which only the ones with overlapping scores (to the ground truth) less than 0.4 are retained. This results in 5,499 negative cropped subimages, about 17 windows per image. The testing set includes 2,287 positive images with size 240×320. The recall-precision curve is used to evaluate the performance of our proposed method. A detection is considered true only if its overlapping score (OS) with a ground truth bounding box exceeds a fixed threshold. We follow the so-called OS-based PASCAL VOC evaluation criteria. OS is computed as Eq.5:

$$OS = \frac{dt \bigcap gt}{dt \bigcup gt} > thr, \tag{5}$$

where dt, gt denotes bounding boxes of detected samples and ground truth respectively. Average Precision (AP) is used to assess the ranked list of our proposed method and the baseline detectors.

---

[1] "- "means the negative images are not provided.

Table 1. Statistical information of training and testing set

	Train		Test		Size
	Positive	Negative	Positive	Negative	
#Images	307	-	2287	-	240x320
#Sub-images	307	5499	2287	-	48x120

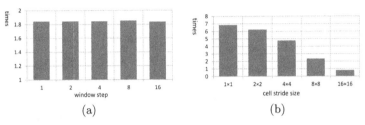

(a)   (b)

Fig. 6. (a) Comparing of the feature computation time in a full image with the sliding window step 1, 2, 4, 8, 16. Y-axis denotes the value of SLEP computation time divided by HOG computation time. (b) Computation time of score prediction without using integral image divided by that of using integral image, with different cell stride size 1x1, 2x2, 4x4, 8x8 and 16x16.

4.2   Detection Speed

As depicted before, SLEP and the fast classifier make our pedestrian method very efficient. To evaluate the efficiency gain of using SLEP feature, we evaluate the time complexity of these two steps. Let T_{SLEP} and T_{HOG} denote the computation time of SLEP and HOG in a scanning window respectively. Fig. 6(a) shows the value of T_{HOG} divided by T_{SLEP} with different steps of sliding windows. No matter how the step changes, the computation time of HOG in a full image is about 1.8 times than that of SLEP.

To evaluate the efficiency gain on the classifier stage, we compare the total computation time of all the cells with size 12×24 in a full image, with or without integral image, where the size of cell stride varies from 1×1, 2×2, 4×4, 8×8, to 16×16. As shown in Fig.6(b), when the stride size is 4×4, the computation time of score prediction without using integral image is about 4.4 times longer than that of using integral image. In our sybsequent implementation, we set the stride size of scanning window as 4×4. As for the detail implementation, we set the test image size as 320×240, the start scan scale of MSS as 0.6, the scan step as 1.05, the maximal scan levels as 15, the sliding window stride size as 4×4, the cell size of SLEP as 12×24, the scan window size as 48×120, and the total dimensions of the SLEP histogram in a scanning window as 2016 (less than that of HOG, i.e.,3780). Based on the SLEP feature and fast score prediction with integral image, the experiments on CASIA90 dataset shows that the computation time of our detector is only 1/7 to that of HOG based detector, which verifies the efficiency of our proposed method.

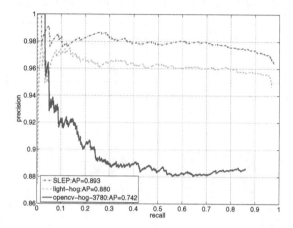

Fig. 7. ROC Curve and Average Precision of our proposed method (SLEP), light_HOG, and OpenCV_HOG

Fig. 8. Detection results of SLEP (Red), OpenCV_HOG (Green), and light_HOG (Blue).The red bounding boxes is the detections returned by our proposed SLEP feature, the green ones and blue ones are the results returned by HOG and light_HOG, respectively.

4.3 Detection Accuracy

As for the baseline of the proposed fast pedestrian detector, an accelerated version of HOG (called light HOG) is implemented in which three steps are omitted: Gaussian weighting, tri-linear interpolation, using the gradient magnitude as voting weights when building the histogram of a block. Intuitively, light HOG is the frequency of binned edge orientation essentially, encodes the image edge information. Recall-precision curve (RPC) is used to compare the performance of SLEP, light HOG, and OpenCV detector (using HOG as descriptor). An SVM with linear kernel trained by the libsvm [21] using the default parameter is selected as a classifier. As shown in Fig.7, the accuracy of SLEP is better than that of light HOG and HOG. Some detection results are shown in Fig.8. As can be seen in Fig.7 and Fig.8, light HOG feature also better than the original HOG proposed by Dalal et al. due to the non-iid issue of dataset. It also demonstrates that it is better to retrain the classifier using the recollected train positive and negative examples cropped from the target domain, instead of directly transferring the classifier from source domain. Since the poses, viewpoints, and lighting condition in the CASIA90 dataset is different from those of INRIA dataset, the OpenCV detector gets the worst testing performance. Our proposed achieves the best performance among these three features.

5   Conclusions and Future Work

Feature extraction is the bottleneck of pursuing a fast pedestrian detector. Aiming at tackling this problem, in the feature extraction stage, a simple edge based feature descriptor called Structured Local Edge Pattern (SLEP) is proposed. SLEP performs better than HOG in the simple scenes. In the classifier calculation stage, an accelerated classifier output prediction using integral image is proposed. Experiments on CASIA gait recognition dataset show its efficiency. The study shown in this paper serves as the start of our subsequent research. In our future work, we would further investigate how to integrate learning based feature design instead of the hand craft SLEP feature design for fast feature extraction. On the other hand, the possibility to extending the proposed fast detector into complex scenes with frequent occlusions and background variations would be also studied.

Acknowledgements. This work is supported by the Nature Science Foundation of China (No. 61373076, No.61202143), the Fundamental Research Funds for the Central Universities (No.201 3121026, No.2011121052), the 985 Project of Xiamen University, the Natural Science Foundation of Fujian Province (No.2013J05100, No.2010J01345 and No.2011J01367), the Key Projects Fund of Science and Technology in Xiamen (No. 3502Z20123017), the Research Fund for the Doctoral Program of Higher Education of China (No.20110 1211120024), the Special Fund for Developing Shenzhen's Strategic Emerging Industries (No. JCYJ20120614164600201), the Hunan Provincial Natural Science Foundation (12JJ2040), and the Hunan Province Research Foundation of Education Committee(09A046).

References

1. Geronimo, D., Lopez, A.M., Sappa, A.D., et al.: Survey of pedestrian detection for advanced driver assistance systems. Pattern Analysis and Machine Intelligence (PAMI) 32, 1239–1258 (2010)
2. Aggarwal, J.K., Ryoo, M.S.: Human activity analysis: A review. ACM Computing Surveys (CSUR) 43, 16 (2011)
3. Poppe, R.: A survey on vision-based human action recognition. Image and Vision Computing 28, 976–990 (2010)

4. Weinland, D., Ronfard, R., Boyer, E.: A survey of vision-based methods for action representation, segmentation and recognition. Computer Vision and Image Understanding 115, 224–241 (2011)
5. Hu, W., Tan, T., Wang, L., et al.: A survey on visual surveillance of object motion and behaviors. Systems, Man, and Cybernetics, Part C: Applications and Reviews 34, 334–352 (2004)
6. Gao, Y., Wang, M., Tao, D., et al.: 3D Object Retrieval and Recognition with Hypergraph Analysis. IEEE Transactions on Image Processing 21, 4290–4303 (2012)
7. Gao, Y., Tang, J., Hong, R., et al.: Camera Constraint-Free View-Based 3D Object Retrieval. IEEE Transactions on Image Processing 21, 2269–2281 (2012)
8. Dollar, P., Wojek, C., Schiele, B., et al.: Pedestrian detection: An evaluation of the s-tate of the art. Pattern Analysis and Machine Intelligence 34, 743–761 (2012)
9. Oren, M., Papageorgiou, C., Sinha, P., et al.: Pedestrian detection using wavelet templates. Computer Vision and Pattern Recognition, 193–199 (1997)
10. Dalal, N., Triggs, B.: Histograms of oriented gradients for human detection. Computer Vision and Pattern Recognition 1, 886–893 (2005)
11. Szarvas, M., Sakai, U., Ogata, J.: Real-time pedestrian detection using lidar and convolutional neural networks. In: Intelligent Vehicles Symposium, pp. 213–218 (2006)
12. Wang, X., Han, T.X., Yan, S.: An hog-lbp human detector with partial occlusion handling. In: 2009 IEEE 12th International Conference on Computer Vision, pp. 32–39 (2009)
13. Maji, S., Berg, A.C., Malik, J.: Classification using intersection kernel support vector machines is efficient. Computer Vision and Pattern Recognition, 1–8 (2008)
14. Zhu, Q., Yeh, M.C., Cheng, K.T., et al.: Fast human detection using a cascade of histograms of oriented gradients. Computer Vision and Pattern Recognition 2, 1491–1498 (2006)
15. Gall, J., Yao, A., Razavi, N., et al.: Hough forests for object detection, tracking, and action recognition. Pattern Analysis and Machine Intelligence 33, 2188–2202 (2011)
16. Lampert, C.H., Blaschko, M.B., Hofmann, T.: Efficient subwindow search: A branch and bound framework for object localization. Pattern Analysis and Machine Intelligence 31, 2129–2142 (2009)
17. Leibe, B., Leonardis, A., Schiele, B.: Robust object detection with interleaved categorization and segmentation. International Journal of Computer Vision 77, 259–289 (2008)
18. Moritz, C.T., Farley, C.T.: Human hopping on damped surfaces: strategies for adjusting leg mechanics. Proceedings of the Royal Society of London. Series B: Biological Sciences 270, 1741–1746 (2003)
19. Cheng, Y.C., Chen, S.Y.: Image classification using color, texture and regions. Image and Vision Computing 21, 759–776 (2003)
20. Wang, L., Hu, W., Tan, T.: A new attempt to gait-based human identification. Pattern Recognition 1, 115–118 (2002)
21. Chang, C.C., Lin, C.J.: Libsvm: a library for support vector machines. ACM Transactions on Intelligent Systems and Technology 2, 27 (2011)

Multi-view Action Synchronization in Complex Background

Longfei Zhang[1], Shuo Tang[1], Shikha Singhal[1,2], and Gangyi Ding[1]

[1] School of Software, Beijing Institute of Technology, Beijing, China
[2] School of Computer Science, Amity University Rajasthan, India
{longfeizhang,tang_shuo,dgy}bit.edu.cn,
shikhasinghal22@gmail.com

Abstract. This paper addresses temporal synchronization of human actions under multiple view situation. Many researchers focused on frame by frame alignment for sync these multi-view videos, and exploited features such as interesting point trajectory or 3d human motion feature for event detecting individual. However, since background are complex and dynamic in real world, traditional image-based features are not fit for video representation. We explore the approach by using robust spatio-temporal features and self-similarity matrices to represent actions across views. Multiple sequences can be aligned their temporal patch(Sliding window) using the Dynamic Time Warping algorithm hierarchically and measured by meta-action classifiers. Two datasets including the Pump and the Olympic dataset are used as test cases. The methods are showed the effectiveness in experiment and suited general video event dataset.

Keywords: Multi-view, Human action Synchronization, Video alignment, MoSIFT.

1 Introduction

The synchronization of human action in multiple views videos is a critical challenge due to different viewpoints, different frame rates, camera motions or even varying appearances of the moving objects[1,2,5].

In addition to the inter-view, inter-scene variability is another source of concern when trying to synchronize sequences from two similar though not identical dynamic scenes. As the same time, it is necessary to synchronize the actions in multi-view videos because that the temporal alignment of these videos always not clear for further analysis. This is of a lot of applications of video event recognition, detection, event knowledge base building and other multimedia and security usages. Therefore, lots of computer vision methods for action alignment and action representation in multi-view are proposed by researchers.

Several problems are needed to be solved in this challenging task[1,2]. The first challenge is how to measure the difference of appearance when an action happened in critical situations. The second problem is how to get rid of the noise affection. Many researchers also employ multi-view learning approaches to solving the appearance measuring problem. However, it is hard to find a good approach to build these models before the first problem is solved.

C. Gurrin et al. (Eds.): MMM 2014, Part II, LNCS 8326, pp. 151–160, 2014.

Fig. 1. Examples of human actions with MoSIFT feature points in multi-view video sequences in Pump dataset[23] and Olympic dataset[21]. Arrows in green and red circles in Images are MoSIFT points.

In this work, we address on action synchronization, including the sequences captured by multiple cameras under one three-dimensional background and video sequences belong to one action class but performed by different performers in different background. We mine the structure of the action patches first, and train the model of meta-actions. These action patches are paring by sliding windows. Meta-action models can specify the structure of action and build the self-similarity matrix [12] to show the difference between these multi-view actions. DTW [4] is employed to synchronize these sequences. We exploit the spatio-temporal features and bag-of-features to build the action model. We testified the effectiveness of our approach in the Olympic dataset [21] and the Pump dataset [9] (which are shown in Fig.1).

The rest of the paper is organized as follows. Section 2 overviews some of the related works. Section 3 describes the main algorithm. We present experimental validation in Section 4 and conclude the paper in Section 5.

2   Related Works

Action synchronization is to align a pair of videos or 3d tracking points which content same action frame by frame or sequence by sequence. 3d tracking points based action alignment is well researched in motion capture systems [4,5,6]. To account for the variation of human actions performed by specific subjects, the most popular method is the Dynamic Time Warping(DWT) with regression models [4]. However, it is a challenge to get the 3d tracking points in real video sequences.

Many researchers focused on frame by frame alignment for sync these multi-view videos captured in the same 3d background or same action in different 3d scenario. Image-based temporal features without correspondences as in [15,16], where authors investigate a temporal descriptor of image sequence based on co-occurences of appearance changes. [4] used canonical time warping to align facial actions, [8] used the correspondence of interest point features to be the clue for alignment. At the same time, many

3d matching methods were proposed, such as [20,24,17] exploited the Hyper Graph to measure the correlation of the 3d scenario, [26,22] proposed new codebook method for 3d object measurement. However, it is still a challenge to match the content of videos with different view port in sequence since the complex background in real action video. Since action in video contains as a sequence, therefore, it is a good thinking to align the action video sequence by sequence.

Feature based methods [14,11] used the correspondence of spatio-temporal features to be the clue for alignment. [7] presented a method training a discriminative aspect model to handle the view invariance but it required good parameter initialization. [12] presented an effective action representation named Self-Similarity Matrix, which is constructed by computing the pairwise similarity between pairs of figure-centric frames. HoG feature and trajectory analyzing were exploited in his work. [3] and [13] extended the work. However, in their experiments, the approach performed poor when views were quite different from other views, and real point trajectories are not stable for real actions in complex background and multi-view situations.

Our motivation is related to the works of [19] and [21] to exploit temporal structure of the human actions. Since [19] and [21] were working for action classification, their works could not be used for aligning video directly.

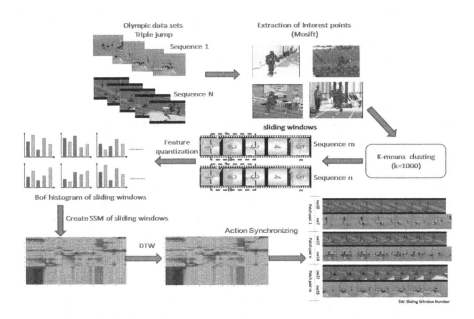

Fig. 2. The framework of the human action synchronizing in complex background

3   Action Synchronization Framework

In this section, we describe the common framework for synchronizing human actions. In first, we present the adaptive temporal descriptors of video sequences based on temporal self-similarities. After that, we describe the Dynamic Time Warping algorithm used in order to synchronize action video sequences.

3.1   Feature for Meta-action

We exploit MoSIFT descriptor and bag-of-words framework [10]to represent human action in video sequence. The MoSIFT feature is originated from the SIFT [9]. It is used to detect and describe the spatio-temporal interest points. MoSIFT point detection is basically based on the optical flow and DoG. Its descriptor is mainly used these two features. Rather than joining a complete HoG classifier with a complete HoF classifier, it builds a single feature descriptor which adds both HoF and HoG into one vector and it is called early fusion. The motivation of the MoSIFT feature is to joint the information of appearance and motion to be a sptio-temporal robust feature.

For each key frame, the number of extracted interesting points is different. In order to obtain a good video representation, bag-of-words (BoW) approach, which is based on a selective sampling k-means clustering, is used to quantize the combination of motion and appearance features to a fixed size vector for each frame.

In order to compare with previous works, we follow the configuration of the Olympic [21] and Pump datasets in [9]. A χ^2 kernel SVM classifier is also applied to train the meta-action classifiers because it has been shown to be better for measure feature distances and compares to previous work [9,10] and [23].

3.2   Self-similarity Measurement

Computing temporal descriptors requires two steps: (i) building for each sequence a self-similarity matrix (SSM) which captures similarities and dissimilarities along the video sequences and (ii) computing a temporal descriptor which captures the main structures of the SSM.

Considering a sequence of video, denoted $S_K = (S_0, S_1..S_k)$, the self-similarity matrix, $D(S) = d_{i,j}$, $(i, j=1...k)$, is a square symmetric matrix where each entry $d_{i,j}$ represents a distance between some features extracted from video sequence S_i and S_j. In this work, we use the cosine distance on each meta-action model.

3.3   Action Synchronization

Dynamic Time Warping is a good known technique which is used to find an optimal alignment between time dependent sequences. These sequences are curved in a non-linear way to map each other. Dynamic time warping is automatically applied to handle it with the time deformations and different speeds associated with time de-pendent data. It is an algorithm for measuring the similarity between two sequences which differ in time speed or time. In this paper, we use DTW to align two sliding windows sequences.

Our approach works as Algorithm 1:

Algorithm 1. Multi-view action synchronization algorithm

The input videos are M human action video sequences $V_M = (V_0, V_1..V_k)$. The output is a responds of operation $R(Y, N)$. Define $M = (M_0, M_1..M_m)$ as category of action. current video sequence is S;

Initialize $q = 1$, H; **repeat**

    Step1. construct Sliding window W_0^q with size λ_w and by step λ_s, where q_0^q belong to A, $A = (a_0^q, a_1^q..a_k^q)$.

    Step2. extract MoSIFT points p_0^q inside a_0^q.

    Step3. construct Bag-of-Words b_0^r inside temporal sliding window t_0^r, $r = 1..k - 1$.

    Step4. classify the b_0^r as a M_i.

    Step5. get $R(Y, N)_k$ if M_i is $m_t h$ meta-action.

    Step6. build self-similarity matrix by $R(Y, N)_k$.

    Step7. using DTW to make action synchronization in one action(local sequence of a action).

until $q = k$;

Step6. connect $R(Y, N)_k$ to $R(Y, N)_k$ to get the finial sequence assignments: $R(Y, N)$

Return $R(Y, N)$

In our approach, we converts the video from a volume of pixels to compact but descriptive interest points. We employ MoSIFT detector [10] to detect and describe spatio-temporal interest points, which method outperforms Laptev's STIP method [14] on the KTH dataset [10]and the Pump dataset [9,23].

For each key frame, the number of extracted key points can be different. A bag-of-words (BoW) approach, which is based on a selective sampling k-means clustering, is used to quantize the combined of motion and appearance features to a fixed size vector for each frame. In order to compare with previous works, We do not change the configuration in KTH dataset and Pump dataset. A χ^2 kernel SVM classifier is also applied because it has been shown to be better for calculating histogram distances and directly compares to previous work [9].

4 Experiments

In our experiments, we use two behavior datasets to verify our algorithm: the Olympic dataset[21], which is for measuring that one kind of action multiple view video synchronizing task, and the Pump [9] dataset, which is for measuring that one action with multiple view cameras video synchronizing task.

4.1 The Same Perspective Sequence

Olympic dataset contains videos of athletes practicing different sports. We can see a pair of sequences from this datasets each time which are shown in Fig.3.

Firstly, we extracted interest points described by MoSIFT algorithm from the original sequences and using the K-means to create the codebook containing 1000 visual words. Secondly, we create the size of sliding window with 8 frames (It is the minmum size of sliding window of meta-action a priori) and set the step size with 3 frames, so we

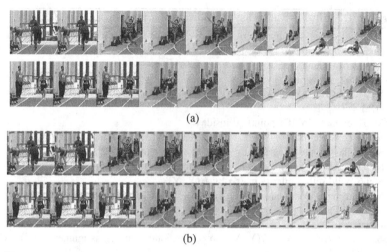

(a)

(b)

Fig. 3. Two sequences from the same perspective and the sliding windows. Fig.3(a) shows two sequences have same perspectives and Fig.3(b) shows the same sequences with the sliding windows.

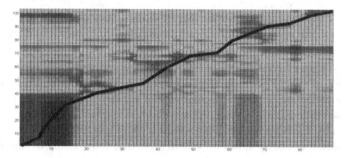

Fig. 4. The SSM of the two same perspective sequences

get 89 sliding windows from the first video sequence and 103 sliding windows from the second one as Fig.3(b) (the red dashed rectangles represent the sliding windows). In the third place, we create BOWs (Bag-of-Word) for each of the sliding window. The dimension of BOF is 1000 which is same as the number of cluster centers.

After feature is ready, we cluster these BOWs to N classes. Each class builds a meta-action classifier. The probabilities of classification are using to create the SSM (Self-Similarities matrix) from these two sequences which is like the Fig.4. As we can see from this figure, the SSM of the two video sequences X (vertical axis) containing 89 sliding windows and Y (horizontal axis) containing 103 sliding windows. Regions of low cost are indicated by dark colors and regions of high cost are indicated by light colors.

In order to find the optimal warping path of the SSM, we use the DTW algorithm in the last step. The DTW result of the Fig3 is like Fig4, we get the accumulated cost matrix with optimal warping path p* represented by the black line. From the figure of result, we note that only cells of SSM that exhibit low costs are covered by the optimal warping path. Tab1 shows the synchronization result of two sliding window sequences.

4.2   The Different Perspective Sequence

We also test our method using two different perspective sequences as shown in Fig.5(a). Because of the different viewpoints, varying relative positions of cameras and human beings in videos and individual variations of people in posture, motion and camera motions, and also because the DTW algorithm is not robust under noisy conditions [4], all of those make some mismatch between the two different perspective sequences as depicted in Fig.5(b), but using our framework of synchronization, most of the key action sliding windows such as hog, bound and jump are synchronized by our framework, as we can see from the Fig.5.

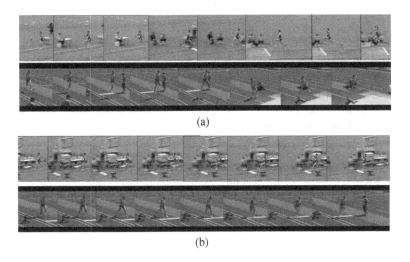

(a)

(b)

Fig. 5. Two sequences from the same perspective and the sliding windows. Fig.5(a) shows two video sequences from different viewpoints. and Fig.5(b) shows Two mismatch result of sliding windows .

The experiment was designed as a recognition task: a predefined event period was given to the system and the system was asked to determine its class of action. To achieve this, we aggregated all visual words over the duration of a single event. Thus, each event is represented by a visual word histogram.

In figure 6, where sw_i is the ID of sliding windows of different sequences. The first sliding window is the location of the first event. There are six event totally. We can find that video records start in different time in different frame rates.

Comparison to alternative methods: We compare the accurate of synchronization and miss frame rate for measuring the performance of SSM feature based method [12] and voting space feature and RANSAC based method[11].

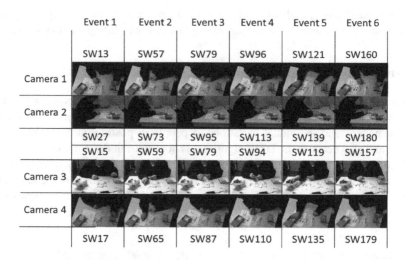

Fig. 6. One of the synchronization results of video sequences on the Pump dataset

Alignment accuracy was evaluated by measuring the average temporal misalignment. This is the average difference between the computed time of each frame and the frame's "ground-truth time. Since our method and [11] are based on sequence by sequence, and [12] is frame to frame, we use the Miss Frame Rate to measure the miss matching frame rate in Table1.

Table 1. Compare of different methods in the Pump and Olympic datasets

	Accurate	Miss Frames Rate
Pump dataset with our method	0.734	0.243
Pump dataset with [12]	0.735	0.212
Pump dataset with [11]	0.667	0.322
Olympic dataset with our method	0.651	0.326
Olympic dataset with [12]	0.232	0.615
Olympic dataset with [11]	0.437	0.31

From the table, we find that our method is perform good in both datasets. At the same time, [11] and [12] failed in the Pump dataset or Olympic dataset separately. Especially in Olympic dataset, our method announces 20% better than other two methods.

5 Conclusion

This paper proposed a action synchronization approach for multi-view action analysis in complex background. Instead of synchronizing the frames of video sequences, we focus on the sliding windows, which is more fit the human action synchronization and online action detection. The small scale actions in these sliding window is named as mata-action. We believe all actions are composed by meta-action sequences which can be mining unsupervised. Therefore, We don't use still features such as HOG but flexible spatio-temporal appearance feature, MoSIFT, to build the Bag-of-Words and Self-Similarity Matrix to represent video actions.

Public benchmark datasets, the Olympic and the Pump datasets, are employed to verify the performance of our approach. In one action with multiple view cameras video synchronizing task, our approatch is as good as [12]. At the same time, in one kind of action multiple view video synchronizing task, our approach yield 20% better than [12]and [11].

On the other hand, our approach can be improved at least in two points. One is that the sliding window size can be flexible. If we use more efficient cluster methods or automatic annotation method to mining video structure more accuracy, we will synchronize the action more accuracy. Another one is that we still need more effective features to represent the action in multiple view video sequence to break the action isolation and recognition.

Acknowledgments. This material is based upon work supported by the Key Technologies Research and Development Program of China Foundation under Grants No. 2012BAH38F01-05, 2012BAH38F05, and by the Research Fund for the Doctoral Program of Higher Education of China under Grants No. 20121101120033. Any opinions, findings, and conclusions or recommendations expressed in this material are those of the author(s) and do not necessarily reflect the views of the Key Technologies Research and Development Program of China Foundation or the Research Fund for the Doctoral Program of Higher Education of China.

References

1. Weinland, D., Ronfard, R., Boyer, E.: A survey of vision-based methods for action representation, segmentation and recognition. Computer Vision and Image Understanding 115(2), 224–241 (2011)
2. Poppe, R.: A survey on vision-based human action recognition. Image and Vision Computing 28(6), 976–990 (2010)
3. Dexter, E., Prez, P., Laptev, I.: Multi-view Synchronization of Human Actions and Dynamic Scenes. In: Proc. of BMVC 2009, pp. 1–11 (2009)
4. Zhou, F., Frade, F.: Generalized Time Warping for Multi-modal Alignment of Human Motion. In: IEEE Conference on Computer Vision and Pattern Recognition, CVPR (June 2012)
5. Zhou, F., Frade, F., Hodgins, J.: Hierarchical Aligned Cluster Analysis for Temporal Clustering of Human Motion. IEEE Transactions on Pattern Analysis and Machine Intelligence (PAMI) 35(3), 582–596 (2013)
6. Hsu, E., Pulli, K., Popovic, J.: Style translation for human motion. ACM Trans. Graph. 24(3), 1082–1089 (2005)
7. Farhadi, A., Tabrizi, M., Endres, I., Forsyth, D.: A latent model of discriminative aspect. In: International Conference on Computer Vision - ICCV, pp. 948–955 (2009)
8. Wedge, D., Huynh, D., Kovesi, P.: Using space-time interest points for video sequence synchronization. In: Proc. IAPR Conf. on Machine Vision Applications, pp. 190–194 (2007)

9. Gao, Z., Detyniecki, M., Chen, M.-Y., Hauptmann, A.G., Wactlar, H.D., Cai, A.: The Application of Spatio-temporal Feature and Multi-Sensor in Home Medical Devices. International Journal of Digital Content Technology and its Applications (IJDCTA) 4(6), 69–78 (2010)

10. Chen, M.Y., Hauptmann, A.: MoSIFT: Recognizing human actions in surveillance videos. CMU-CS-09-161, Carnegie Mellon University (2009)

11. Padua, F.L.C., Carceroni, R.L., Santos, G.A.M.R., Kutulakos, K.N.: Linear sequence-to-sequence alignment. IEEE Transactions on Pattern Analysis and Machine Intelligence (PAMI) 32(2), 304–320 (2010)

12. Junejo, I.N., Dexter, E., Laptev, I., Pérez, P.: Cross-view action recognition from temporal self-similarities. In: Forsyth, D., Torr, P., Zisserman, A. (eds.) ECCV 2008, Part II. LNCS, vol. 5303, pp. 293–306. Springer, Heidelberg (2008)

13. Junejo, I.N., Dexter, E., Laptev, I., Prez, P.: View-Independent Action Recognition from Temporal Self-Similarities. IEEE Trans. Pattern Anal. Mach. Intell. 33(1), 172–185 (2011)

14. Laptev, I., Belongie, S.J., Perez, P., Wills, J.: Periodic motion detection and segmentation via approximate sequence alignment. In: Proc. Int. Conf. on Computer Vision, vol. 1, pp. 816–823 (2005)

15. Ukrainitz, Y., Irani, M.: Aligning sequences and actions by maximizing space-time correlations. In: Leonardis, A., Bischof, H., Pinz, A. (eds.) ECCV 2006. LNCS, vol. 3953, pp. 538–550. Springer, Heidelberg (2006)

16. Ushizaki, M., Okatani, T., Deguchi, K.: Video synchronization based on co-occurrence of appearance changes in video sequences. In: Proc. International Conference on Pattern Recognition (ICPR), pp. III:71–III:74 (2006)

17. Gao, Y., Wang, M., Ji, R., Wu, X., Dai, Q.: 3D Object Retrieval with Hausdorff Distance Learning. IEEE Transactions on Industrial Electronics (2013)

18. Wolf, L., Zomet, A.: Wide baseline matching between unsynchronized video sequences. International Journal of Computer Vision 68(1), 43–52 (2006)

19. Xu, D., Chang, S.F.: Video Event Recognition Using Kernel Methods with Multilevel Temporal Alignment. IEEE Trans. Pattern Anal. Mach. Intell. 30(11), 1985–1997 (2008)

20. Gao, Y., Wang, M., Tao, D., Ji, R., Dai, Q.: 3D Object Retrieval and Recognition with Hypergraph Analysis. IEEE Transactions on Image Processing 21(9), 4290–4303 (2012)

21. Niebles, J.C., Chen, C.-W., Fei-Fei, L.: Modeling Temporal Structure of Decomposable Motion Segments for Activity Classification. In: Daniilidis, K., Maragos, P., Paragios, N. (eds.) ECCV 2010, Part II. LNCS, vol. 6312, pp. 392–405. Springer, Heidelberg (2010)

22. Ji, R., Duan, L., Chen, J., Xie, L., Yao, H., Gao, W.: Learning to distribute vocabulary indexing for scalable visual search. IEEE Transactions on Multimedia (2013)

23. Zhang, L.F., Guan, Z.Y., Hauptmann, A.: Co-Attention model for tiny activity analysis. Neurocomputing 105(1), 51–60 (2013)

24. Gao, Y., Tang, J.H., Hong, R.C., Yan, S.C., Dai, Q.H., Zhang, N., Chua, T.S.: Camera Constraint-Free View-Based 3D Object Retrieval. IEEE Transactions on Image Processing 21(4), 2269–2281 (2012)

25. Gao, Y., Wang, M., Zha, Z., Tian, Q., Dai, Q., Zhang, N.: Less is More: Efficient 3D Object Retrieval with Query View Selection. IEEE Transactions on Multimedia 11(5), 1007–1018 (2011)

26. Ji, R., Yao, H., Liu, W., Sun, X., Tian, Q.: Task-dependent visual-codebook compression. IEEE Transactions on Image Processing 21(4), 2282–2293

Parameter-Free Inter-view Depth Propagation for Mobile Free-View Video[*]

Binbin Xiong[1], Weimin Wu[2], Haojie Li[3], Hongtao Yu[4], and Hanzi Mao[1]

[1] Dept. of Electronics and Information Engineering,
Huazhong University of Science and Technology
[2] Wuhan National Laboratory of Optoelectronics, Dept. of Electronics and Information
Engineering, Huazhong University of Science and Technology
[3] Sch. of Software, Dalian University of Technology
[4] Library, Huazhong University of Science and Technology
viclol36@gmail.com, {wuwm,hanzimao}@hust.edu.cn,
hjli@dlut.edu.cn, leader_bin@hotmail.com

Abstract. As a result of the rapidly improving performance of mobile devices, there is an inevitable trend for the development of applications with stereoscopic functions. For these applications, especially those can be seen from free angles, a set of depth images are usually needed. To fit for bit-rate limited working conditions such as for mobile terminals, depth propagation is always exploited thus to decrease the transfer consumption. Traditionally, accurate propagation needs camera parameters to warp one view to another, which are usually unavailable. This paper proposes a parameter-free depth inter-view propagation method for free-view video with multi-view texture videos plus single-view depth video. Firstly, the depth is estimated for each view by the corresponding texture video. Then, inter-view relationships of depth views are investigated by the neighboring estimated depths. Finally, the obtained inter-view relationship is used to propagate the depth from one given depth with high quality to other views. Experimental results demonstrate that our scheme has better quality on both depth and virtual viewpoint generation.

Keywords: depth propagation, free-view video, mobile video, depth estimation.

1 Introduction

As a new trend in consumer electronics, 3D video on mobile devices is emerging recently with the mushroom growing of mobile capacities. Free viewpoint vision can provide immersive viewing experience by multi-view videos [1-5], which usually call for a huge volume of bandwidth. Therefore, how can the system designers bring free-view video to terminals on bandwidth limited wireless channel is a challenging

[*] This work was supported in part by National Science and Technology Major Project (No. 2013ZX03003015-005), 863 Hi Tech R&D Program of China (No. 2012AA121604), and International S& T Cooperation Program of China (No.2012DFG12010).

C. Gurrin et al. (Eds.): MMM 2014, Part II, LNCS 8326, pp. 161–169, 2014.

problem. Usually, the information needed to be transferred through the internet includes texture images and depth images. Multi-view depth is urgently needed since other views can be rendered with the help of these depth maps [6, 16, 19]. Therefore if we can decrease the number of depth images needed for transmission, the bandwidth consumption can certainly be lowered. There are many ways to obtain multi-view depth maps, such as capturing via depth camera [7-10], depth estimation [14] and so on. Beyond these methods, the depth propagation is an efficient way to generate multi-view depth from a limited number of viewpoints depth maps.

The depth propagation methods can be divided into two categories: temporal propagation and inter-view propagation. For the temporal propagation, it is assumed that the temporal correlation of depth is similar as that of the corresponding texture video. Therefore, the temporal correlation of texture video was investigated at first by temporal prediction technologies, such as motion estimation, optical flow and so on. For the inter-view depth propagation, it is assumed that the views of video including texture and depth video follow the constraint of epipolar geometry. Therefore, the depth of a certain view can be mapping to another view by the epipolar geometry [12, 17, 18]. Usually, camera parameters are essential for the description of epipolar geometry. This is the shortage of these methods since the camera parameters are absent for many 3D video applications [20-23].

On this background, we propose a new method to realize inter-view depth propagation without camera parameters, which will improve the 3D free view display in wireless transmission environment and especially fit for mobile platforms. To resolve the challenges mentioned above, we first estimate the depth for each view only by the corresponding texture video. Then, the inter-view relationship of depth is investigated by the neighboring estimated depths. After that, the obtained inter-view relationship is used to propagate the depth from one given depth with high quality to other views.

The contribution of this paper has two aspects: (a) we propose a parameter-free depth inter-view mapping method. The multi-view depth videos with high quality can be obtained as long as a single-view high quality depth is given; even the camera parameters are absent. It fills the gap between applications of single-view depth and multi-view depth. (b) A framework for mobile free-view video is built by our proposed method. The resource consumption of this framework is comparable with video only have a limited number of views, but it can provide smooth free-view user experience.

This paper is organized as follow: In Section 2, we discuss our scheme in details. After that, we present an application of mobile free-view video based on our proposed inter-view depth propagation in Section 3. Experiments and conclusions are presented in Sections 4 and 5 respectively.

2   Parameter-Free Depth Interview Propagation

Inter-view relationship is crucial for inter-view depth propagation and multi-view depth generation. Therefore, in our proposed scheme, we focus our efforts to exploit the inter-view relationship. The workflow of our proposed inter-view depth propagation scheme for free-view video is illustrated in Fig.1. As can be found in this workflow, there are three steps including depth estimation, inter-view relationship generation, and inter-view depth propagation. These three steps are specified in the following subsections.

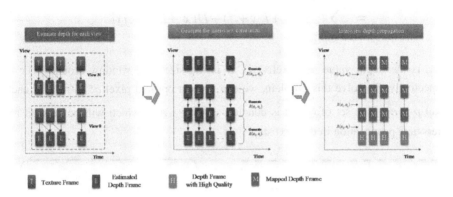

Fig. 1. The workflow of the proposed parameter-free depth inter-view propagation scheme

2.1   Depth Estimation for Each View

In this step, the depth for each view is generated in order to obtain the inter-view relationship of depth views.   It requires the accuracy of depth relationship rather than that of the depth itself. To this end, we do not predict accurate depth information for one viewpoint in this step, but find a logical distance relationship and then use it for inter-view relationship computation.

The depth estimation method in 2D-to-3D method in the previous work [8] is employed. In this method, depth cues are detected firstly, mainly including salience map and occlusion between objects. After that, the logical distance relationship is setup and a rough depth map can be obtained. Finally, the rough depth map is normalized to describe the relative distance among objects in the corresponding scene. Fig. 2 illustrates the workflow of depth estimation proposed by [8]. These estimated depth maps are not accurate enough for free-view vision, while it also contains the inter-view correlation. Therefore, these estimated depth is used to deduce the inter-view correlation of depth maps, which is described in the subsection 2.2

2.2   Inter-view Relationship Generation

After obtaining the estimated depth map for each viewpoint, the inter-view correlation can be generated with the help of these depth maps. Suppose $p(x,y)$ is a pixel in the current viewpoint e_k, $q(x,y)$ is a pixel in window $N(p')$ with the center of p' in adjacent viewpoint e_l, where p' is the pixel in adjacent view with the same coordinate of pixel $p(x,y)$ in e_k. The objective function is described as:

$$\min_{p\in e_k, q\in e_l, q\in N(p')} f(p,q) \tag{1}$$

where

$$f(p(x,y), q(x',y'))$$
$$= \sum_{-\frac{w}{2}\leq i,j\leq\frac{w}{2}} |D(x+i, y+j) - D(x'+i, y'+j)| \tag{2}$$

$D_{(x,y)}$ is the depth value on pixel (x,y), w is the size of a window around p or q. respectively. To solve this problem, we can find an optimal pixel q^* in e_l for each pixel p in e_k. The set of (p,q^*) is denoted as $R(e_k, e_l)$, which will be utilized for inter-view propagation in the next step.

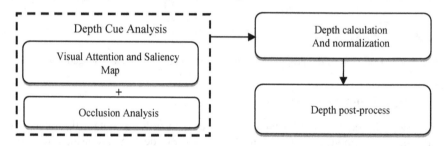

Fig. 2. The depth cue detection based depth estimation method

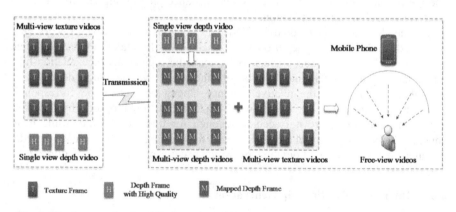

Fig. 3. The framework of mobile free-view video based on inter-view depth propagation

2.3   Inter-view Depth Propagation

It is supposed that a single-view depth with high accuracy is obtained in the streaming. The problem is how to propagate the single-view depth to multi-view depth. Here, we use the inter-view correlation $R(e_k, e_l)$ to generate the depth of another view-angle. It is assumed that we use the *kth* view H to propagate the *lth* view M, so that it has

$$R(H, M) = R(e_k, e_l) \qquad (3)$$

then,

$$D(q) = D(p), q \in M, p \in H \qquad (4)$$

where (p, q) should satisfy the function of $R(H, M)$. For each $q \in M$, we can get its depth value from its best matching pixel $p \in H$. In this way, the depth with high accuracy is propagated from H to M.

3 Mobile Free-View Video Application Based on Depth Inter-view Propagation

The proposed application is based on the encoding scheme with N-view texture videos but single-view depth video in wireless transmission environment. In this case, in order to save the network bandwidth and resources of mobile device, the number of provided views should be limited, while the distance between the neighboring views are rather long. Therefore, a large amount of virtual views are needed for better user experience. In order to provide this attractive function to mobile users, multi-view depth is desired to supply the abundant space information of 3D scenes. As illustrated in Fig.3, we propose that only single or a small number of depths are needed to be generated and transmitted. Before the free-view display, the viewpoint angle and numbers of virtual views are obtained according to the user interaction. Therefore, the corresponding multi-view depth is obtained from a single view of depth, which can be generated with high accuracy, with our proposed inter-view depth propagation.

The advantages of the framework illustrated in Fig.3 include three aspects:

(a) It is that only a small number of texture videos plus a single view of depth video is transmitted. Therefore, the bit-rate consumption is much lower than the traditional system with both multi-view texture and depth videos.

(b) It is very efficient to generate multi-view depth and render virtual views according to the user interaction, rather than compressing and transmitting all the possible required depth and texture videos.

(c) The single view depth, which is used to generate other views of depth, can be obtained and transmitted with high accuracy. Therefore, the quality of multi-view depth is higher than the method that it generates all the depth by the video with low quality.

4   Experiments and Discussions

In order to reveal the performance of our proposed scheme, we compare our method to other two methods. The details of the experiment settings are as following.

1. Scheme A: The proposed scheme with inter-view depth propagation. In this scheme, the original texture videos of viewpoints #A and #B accompanied by the original depth video of viewpoint #A are used as input. The depth video of viewpoint #B will be propagated by our scheme. Finally, viewpoint #C which is in the middle position between viewpoints #A and #B is synthesized by the above texture and depth videos.
2. Scheme B: The original texture and depth videos of viewpoints #A and #B are used as input to synthesize viewpoint #C.
3. Scheme C: The original texture videos of viewpoints #A and #B are used as input. The 2D-to-3D conversion in [8] is employed to generate depth videos of viewpoints #A and #B. Finally, viewpoint #C which is in the middle position between viewpoints #A and #B is synthesized by the above texture and depth videos.

We select standard test sequences Lovebird1 (viewpoints 4 and 6 are selected for viewpoint 5 synthesis), Street (viewpoints 27 and 29 are selected for viewpoint 28 synthesis) and Ghost Fly (viewpoints 1 and 3 are selected for viewpoint 2 synthesis) for our experiments from JVC standardization group [9, 11-15]. View synthesis reference software v.3.5 [10] is selected for virtual view synthesis. The experimental results are shown by Figs. 4 to 7. In Fig. 4, the quality of depth maps and synthesized virtual viewpoint are given. As can be found on the results of these schemes to both test sequences, the quality for the propagated depth is about 30dB, leading the quality of corresponding virtual viewpoint of Scheme B, in which the original depth maps are involved. The performance of Scheme A comes from high quality of the proposed inter-view propagation.

We further compare the quality of synthesized virtual viewpoint by three schemes. In Fig. 5 to 7, (a) is the original captured viewpoint, (b) is the synthesized viewpoint by the original depth maps of neighboring viewpoints (Scheme B), (c) is the synthesized viewpoint by the propagated depth maps obtained by the proposed scheme (Scheme A), (d) is the synthesized viewpoint by the propagated depth maps obtained by 2D-to-3D conversion (Scheme C). As can be found in Fig. 5(d), Scheme C has ghost effects on both the near and far objects, leading worse perceptual quality. The ghost effect is due to inaccurate depth maps in synthesizing procedure. On the other hand, as depicted Fig. 5(c), the ghost effect is eliminated in Scheme A due to the high quality of the propagated depth maps.

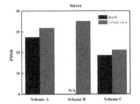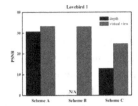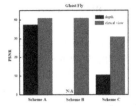

Fig. 4. Quality performance comparisons of three schemes with test sequence Steer, Lovebird1 and Ghost Fly respectively

Fig. 5. Enlarged details for quality comparisons for three schemes with Lovebird1

Fig. 6. Enlarged details for quality comparisons for three schemes with Street

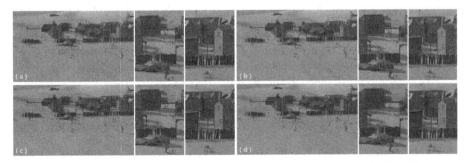

Fig. 7. Enlarged details for quality comparisons for three schemes with Ghost Fly

5   Conclusions

In this paper, we propose a parameter-free inter-view depth propagation scheme for mobile applications. In this scheme, multi-view depth maps can be generated with high quality efficiently. The proposed scheme is helpful in saving communication bandwidth, which is the most concern for wireless transmission and mobile devices. The experimental results have shown that the quality of propagated depth maps and the consequent virtual viewpoints is more than 30dB, and they are close to the quality performance that obtained by the original depth maps. Therefore, our scheme has great potentials in future mobile 3D video applications.

References

1. Smolic, A.: An overview of 3D video and free viewpoint video. In: Jiang, X., Petkov, N. (eds.) CAIP 2009. LNCS, vol. 5702, pp. 1–8. Springer, Heidelberg (2009)
2. Fehn, C.: Depth-image-based rendering (DIBR), compression, and transmission for a new approach on 3D-TV. Electronic Imaging 2004, 93–104 (2004)
3. Wilson, A.D., Benko, H.: Combining multiple depth cameras and projectors for interactions on above and between surfaces. In: Proceedings of the 23rd Annual ACM Symposium on User Interface Software and Technology, pp. 273–282 (2010)
4. Gudmundsson, S.A., Aanaes, H., Larsen, R.: Fusion of stereo vision and time-of-flight imaging for improved 3d estimation. International Journal of Intelligent Systems Technologies and Applications 5(3), 425–433 (2008)
5. Smisek, J., Jancosek, M., Pajdla, T.: 3D with Kinect. In: Consumer Depth Cameras for Computer Vision, pp. 3–25. Springer, London (2013)
6. Torralba, A., Oliva, A.: Depth estimation from image structure. IEEE Transactions on Pattern Analysis and Machine Intelligence 24(9), 1226–1238 (2002)
7. Zhu, J., Wang, L., Gao, J., et al.: Spatial-temporal fusion for high accuracy depth maps using dynamic MRFs. IEEE Transactions on Pattern Analysis and Machine Intelligence 32(5), 899–909 (2010)
8. Zhang, J., Yang, Y., Dai, Q.: A novel 2D-to-3D scheme by visual attention and occlusion analysis. In: 3DTV Conference: The True Vision-Capture, Transmission and Display of 3D Video, pp. 1–4 (2011)
9. Ohm, J.R.: Call for proposals on 3d video coding technology, ISO/IEC JTC1/SC29/WG11, Doc. W12036
10. Lee, C., Ho, Y.S.: View Synthesis Tools for 3D Video, ISO/IEC JTC1/SC29/WG11, Doc M15851
11. Gao, Y., Wang, M., Tao, D., Ji, R., Dai, Q.: 3D Object Retrieval and Recognition with Hypergraph Analysis. IEEE Transactions on Image Processing 21(9), 4290–4303 (2012)
12. Gao, Y., Tang, J., Hong, R., Yan, S., Dai, Q., Zhang, N., Chua, T.-S.: Camera Constraint-Free View-Based 3D Object Retrieval. IEEE Transactions on Image Processing 21(4), 2269–2281 (2012)
13. Gao, Y., Wang, M., Zha, Z., Tian, Q., Dai, Q., Zhang, N.: Less is More: Efficient 3D Object Retrieval with Query View Selection. IEEE Transactions on Multimedia 11(5), 1007–1018 (2011)
14. Gao, Y., Dai, Q., Zhang, N.: 3D Model Comparison using Spatial Structure Circular Descriptor. Pattern Recognition 43(3), 1142–1151 (2010)

15. Shen, J., Shepherd, J., Cui, B., Tan, K.L.: A novel framework for efficient automated singer identification in large music databases. ACM Transactions on Information Systems (TOIS) 27(3), 18

16. Ji, R., Gao, Y., Hong, R., Liu, Q., Tao, D., Li, X.: Spectral-Spatial Constraint Hyperspectral Image Classification. IEEE Transactions on Geoscience and Remote Sensing (2013)

17. Ji, R., Duan, L.Y., Chen, J., Yao, H., Yuan, J., Rui, Y., Gao, W.: Location discriminative vocabulary coding for mobile landmark search. International Journal of Computer Vision 96(3), 290–314

18. Liu, Q., Yang, Y., Ji, R., Gao, Y., Yu, L.: Cross-View Down/Up-Sampling Method for Multi-View Depth Video Coding. IEEE Signal Processing Letters 19(5), 295–298

19. Yang, Y., Liu, Q., Ji, R., Gao, Y.: Dynamic 3D Scene Depth Reconstruction via Optical Flow Field Rectification. PLoS ONE 7(11), e47041

20. Yang, Y., Liu, Q., Ji, R., Gao, Y.: Remote Dynamic Three-Dimensional Scene Reconstruction. PLoS ONE 8(5), e55586

21. Yang, Y., Jiang, G., Yu, M., Zhu, D.: Parallel process of hyper-space-based multiview video compression. In: IEEE International Conference on Image Processing, pp. 521–524 (2006)

22. Yang, Y., Dai, Q., Jiang, G., Ho, Y.S.: Comparative Interactivity Analysis in Multiview Video Coding Schemes. ETRI Journal 32(4), 566–576

23. Liu, Q., Yang, Y., Gao, Y., Hong, R.: Texture-adaptive hole-filling algorithm in raster-order for three-dimensional video applications. Neurocomputing 111, 154–160

24. Liu, Q., Yang, Y., Gao, Y., Ji, R., Yu, L.: A Bayesian Framework for Dense Depth Estimation Based on Spatial-temporal Correlation. Neurocomputing 104, 1–9

Coverage Field Analysis
to the Quality of Light Field Rendering

Changjian Zhu, Li Yu, and Peng Zhou

Department of Electronics and Information Engineering,
Huazhong University of Science and Technology,
430074 Wuhan, P.R. China
{changjianzhu,hustlyu,zhoupeng}@hust.edu.cn

Abstract. In light field rendering (LFR), the geometric configuration of cameras concerns the rendering quality of virtual views. A mathematical model of coverage field (CF) is proposed in this paper to quantify the relationship between the rendering quality and the geometric configuration of cameras. We analyze the impact of changes in CF with the rendering quality by a set of positions of the virtual views and the geometric configuration of cameras. An optimization algorithm is also presented to optimize the geometric configuration of cameras with the help of CF. The experimental results show that the proposed CF can effectively quantify the quality of LFR, and can be used to optimize the geometric configuration of cameras.

Keywords: Light field rendering, coverage field, geometric configuration, rendering quality.

1 Introduction

Light field rendering (LFR) system aims to synthesize arbitrary virtual views of a three-dimensional (3D) scene from a set of captured images which are obtained by cameras [1]. When synthesizing a virtual view, the rendering quality can be improved by optimizing the geometric configuration of cameras, and it is needed to establish a relationship between the rendering quality and the geometric configuration of cameras [2]. For the optimization of the geometric configuration of cameras, many optimization algorithms have been proposed [2]-[6]. At the same time, analyzing the effects of different the geometric configuration of cameras on the rendering quality is a fundamental issue for LFR.

The rendering quality analysis of LFR is associated with the 3D scene representation, reconsitution, and so on. For the research of 3D scene representation, there are a lot of research results [7]-[17]. The purpose of our work is promoting these research results. For the rendering quality analysis, Nguyen et al. [18] pointed out that image-based rendering (IBR) quality can be quantified by using IBR configurations such as the depth and intensity estimate errors, the scene geometry and the texture, the number of actual cameras, and their positions and resolutions. They proposed an

C. Gurrin et al. (Eds.): MMM 2014, Part II, LNCS 8326, pp. 170–180, 2014.
© Springer International Publishing Switzerland 2014

algorithm that per-pixel depth was used to quantitatively analyze the rendering quality of IBR. Liu et al. [19] used a proper number of images for rendering and a 3D surface to describe the relation between multi-view data capturing and quality of the rendered view. Yang et al. [17] proposed an elegant model by taking the contourlet transform and neighboring references together, and can be used for the case when the original reference and ground-truth geometric information are both absent. Shidanshidi et al. [20] proposed a quantitative approach for comparison and evaluation of LFR algorithms. These works have a great contribution on the analysis of the rendering quality. Based on these works, we will be analyzing the effects of the geometric configuration of cameras on the rendering quality.

For the analysis of the geometric configuration of cameras, Shidanshidi et al. [21] proposed an effective sampling density (ESD) of a scene, and used it to establish an optimization model to calculate the optimum configuration of cameras including the positions and the orientations. Zhang et al. [2] proposed an active rearranged capturing approach which used the distance between light ray of rendering and camera to quantify the rendering quality. Our work is inspired by the above research results, and we propose an model called coverage field (CF) to quantify the relationship between the rendering quality and the geometric configuration of cameras. We derived a formula to express the rendering quality with an area of CF, and the rendering quality of virtual views is assessed by CF. At the same time, the CF is further used to establish an optimization algorithm for the geometric configuration of cameras.

2   Analysis for Coverage Field and Rendering Quality

2.1   The Definition of Coverage Field

We assume that virtual views and cameras are placed along a straight line. The cameras are used to capture images, and the virtual views are rendered by these captured images. Fig. 1 shows the top view of a plane associated with one row of pixels. As shown in Fig. 1, the dotted line is for the range of virtual view, and the solid line is for the camera. For clarity, only two light rays of field of view (FOV) on the edge are shown in 2D space. The coverage range of the FOV of virtual view or camera is the area between the left edge and the right edge. The CF describes the topology of the intersecting coverage range of the FOV between camera and virtual view. In Fig. 1, the coverage range of the bold black dotted line is the CF. Generally, the geometric configuration of cameras consisted of position and orientation. If the position or orientation of cameras is changed, the area of CF will also be changed. We will study the relationship between geometric configuration of camera and CF with position or orientation of cameras being fixed. In this paper, we assume that the orientation of cameras and virtual views is fixed, and they are pointing to the same orientation. This assumption does not hinder to demonstrate effectiveness of the CF.

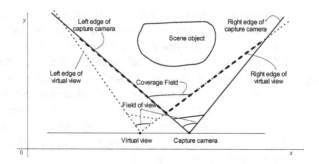

Fig. 1. The coverage field of light field rendering

2.2   Derivation of the Coverage Field Area

First, we formulate the relationship between the area of CF and the position of cameras. As shown in Fig. 2, the straight line $y = Z_p$ is the camera plane, the straight line $y = Z$ is a constant depth line which make the area of CF is limited, c_i is the camera center, and v_k is the virtual view center. Assuming the FOV of camera is θ, and it is consisted by M light rays in 2D space. $r_{c_i}^n$ is the nth light ray of the camera c_i, and the equation of $r_{c_i}^n$ could be written as follows

$$y_n(c_i) = \frac{(M-1)\cdot(x-c_i)}{\tan(\theta/2)\cdot(2\cdot(n-1)-M+1)} + Z_p,$$ (1)

where $n = 1, 2, ..., M$, $i = 1, 2, ..., N$, and N is the number of cameras. Similarly, assuming the FOV of virtual view is θ, and it is consisted by M light rays in 2D space. Therefore, $r_{v_k}^m$ is the mth light ray of the virtual view v_k, and the equation of $r_{v_k}^m$ is

$$y_m(v_k) = \frac{(M-1)\cdot(x-v_k)}{\tan(\theta/2)\cdot(2\cdot(m-1)-M+1)} + Z_p,$$ (2)

where $m = 1, 2, ..., M$, $k = 1, 2, ..., K$, and K is the number of virtual views.

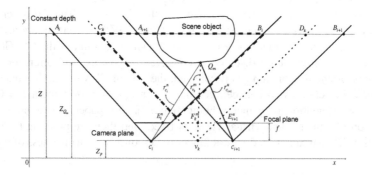

Fig. 2. The formulation of coverage field and light ray interpolation

After that, we calculate the area of CF with $Z_p \leq y \leq Z$. As shown in Fig. 2, the topology model of the bold black dotted line is a triangle, and it can be calculated easily. For the calculation of the area of CF, it is associated with the relation between the virtual view v_k and the camera c_i. There are two kinds of the relations, which are $c_i \geq v_k$ and $c_i < v_k$. Therefore, the area of CF is

$$S(c_i, v_k) = \begin{cases} S_{\mathrm{I}}, & v_k \leq c_i < v_k + \mathrm{B} \\ S_{\mathrm{II}}, & v_k - \mathrm{B} < c_i < v_k \\ 0, & \text{otherwise} \end{cases} \quad (3)$$

where $\mathrm{B} = \dfrac{2 \cdot (Z - Z_p)}{\tan((\pi - \theta)/2)}$, and S_{I}, S_{II} is

$$S_{\mathrm{I}} = \frac{1}{4} \cdot \tan\left(\frac{\pi - \theta}{2}\right) \cdot c_i^2 - \left(\frac{1}{2} \cdot \tan\left(\frac{\pi - \theta}{2}\right) \cdot v_k + Z - Z_p\right) \cdot c_i + (Z - Z_p) \cdot v_k + \frac{1}{4} \cdot \tan\left(\frac{\pi - \theta}{2}\right) \cdot v_k^2 + \frac{(Z - Z_p)^2}{\tan\left(\frac{\pi - \theta}{2}\right)}, \quad (4)$$

$$S_{\mathrm{II}} = \frac{1}{4} \cdot \tan\left(\frac{\pi - \theta}{2}\right) \cdot c_i^2 + \left(Z - Z_p - \frac{1}{2} \cdot \tan\left(\frac{\pi - \theta}{2}\right) \cdot v_k\right) \cdot c_i - (Z - Z_p) \cdot v_k + \frac{1}{4} \cdot \tan\left(\frac{\pi - \theta}{2}\right) \cdot v_k^2 + \frac{(Z - Z_p)^2}{\tan\left(\frac{\pi - \theta}{2}\right)}. \quad (5)$$

2.3   Derivation of the Rendering Quality

Given a Lumigraph, one can generate a new image from an arbitrary camera by ray interpolation. For each ray, the value is properly interpolated using the chosen basis functions [22]. Therefore, the rendering method is based on light ray interpolation. As shown in Fig. 2, the light ray $r_{v_k}^m$ is rendered by the light rays $r_{c_i}^n$ and $r_{c_{i+1}}^u$, where $u = 1, 2, ..., M$. The light ray $r_{v_k}^m$ intersects the focal plane on point $F_k^m\left(q_{v_k}^m, f + Z_p\right)$, where f is the focal length of camera. The light ray $r_{c_i}^n$ intersects the focal plane on point $E_i^n\left(q_{c_i}^n, f + Z_p\right)$, and $r_{c_{i+1}}^u$ intersects the focal plane on point $E_{i+1}^u\left(q_{c_{i+1}}^u, f + Z_p\right)$. Therefore, the rendering result can be calculated as

$$\hat{p}_m(v_k) = \frac{q_{v_k}^m - q_{c_{i+1}}^u}{q_{c_i}^n - q_{c_{i+1}}^u} \cdot p_n(c_i) + \frac{q_{v_k}^m - q_{c_i}^n}{q_{c_{i+1}}^u - q_{c_i}^n} \cdot p_u(c_{i+1}), \quad (6)$$

where $\hat{p}_m(v_k)$ is the rendering result of the point F_k^m, $p_n(c_i)$ is the pixel of the point E_i^n, and $p_u(c_{i+1})$ is the pixel of the point E_{i+1}^u, where

$$q_{v_k}^m = \frac{f \cdot \tan(\theta/2) \cdot (2 \cdot (m-1) - M + 1)}{M - 1} + v_k, \quad (7)$$

$$q_{c_i}^n = \frac{f \cdot \tan(\theta/2) \cdot (2 \cdot (m-1) - M + 1)}{M - 1} + \frac{f \cdot v_k + (Z_{Q_m} - Z_p - f) \cdot c_i}{Z_{Q_m} - Z_p}, \quad (8)$$

where Z_{Q_m} is the depth of the scene. Therefore, the rendering error is calculated as follows

$$e_m(v_k) = \frac{\ddot{p}_m(v_k)}{(2)!} \cdot \left(\frac{Z_{Q_m} - Z_p - f}{Z_{Q_m} - Z_p}\right)^2 \cdot (v_k - c_i) \cdot (v_k - c_{i+1}). \quad (9)$$

where $p_m(v_k)$ is the expectation of the point F_k^m, $e_m(v_k)$ is the rendering error of v_k, By the formula (9), we can have the relationship between rendering error and position of cameras.

2.4   The Quantification of Rendering Quality

By the formula (3), we get a formula of relationship between area of CF and position of camera, and substituting it to (9) to obtain the formula of quantify rendering quality by the area of CF. Therefore, the formula of $e_m(v_k)$ with the area of CF as follows

$$e_m(v_k) = \begin{cases} e_{\mathrm{I}}, & c_i < v_k \leq c_{i+1} \\ e_{\mathrm{II}}, & c_i \geq v_k, c_{i+1} \geq v_k \\ e_{\mathrm{III}}, & c_i < v_k, c_{i+1} < v_k \end{cases}, \tag{10}$$

where $c_i, c_{i+1} \in \left[v_k - \dfrac{2 \cdot (Z - Z_p)}{\tan((\pi - \theta)/2)}, v_k + \dfrac{2 \cdot (Z - Z_p)}{\tan((\pi - \theta)/2)} \right]$, and e_{I}, e_{II}, e_{III} is calculated as follows

$$e_{\mathrm{I}} = \frac{4 \cdot p_m^{'}(v_k) \cdot \left(\dfrac{Z_p - Z_{Q_m} + f}{Z_{Q_m} - Z_p} \right)^2}{(2)!} \cdot \frac{\left(Z_p - Z + \sqrt{S_{\mathrm{II}} \cdot \tan((\pi - \theta)/2)} \right) \cdot \left(Z_p - Z + \sqrt{S_{\mathrm{I}} \cdot \tan((\pi - \theta)/2)} \right)}{\tan((\pi - \theta)/2)^2} \tag{11}$$

,

$$e_{\mathrm{II}} = \frac{4 \cdot p_m^{'}(v_k)}{(2)!} \cdot \left(\frac{Z_{Q_m} - Z_p - f}{Z_{Q_m} - Z_p} \right)^2 \cdot \frac{\left(Z_i - Z + \sqrt{\tan((\pi - \theta)/2) \cdot S_{\mathrm{I}}} \right)^2}{\tan((\pi - \theta)/2)^2}, \tag{12}$$

$$e_{\mathrm{III}} = \frac{4 \cdot p_m^{'}(v_k)}{(2)!} \cdot \left(\frac{Z_{Q_m} - Z_p - f}{Z_{Q_m} - Z_p} \right)^2 \cdot \frac{\left(Z_p - Z + \sqrt{\tan((\pi - \theta)/2) \cdot S_{\mathrm{II}}} \right)^2}{\tan((\pi - \theta)/2)^2}. \tag{13}$$

The formula (10) is the basic equation of the LFR that describes the quantitative relationship the area of CF and rendering quality. From the above analysis, the CF is considered as a measurement of rendering quality for LFR system, where imaging process depends on the position of between camera and virtual view, and it depends on the constant depth assumption and depth of object.

3   Optimization of Camera Position by Coverage Field

We use the CF to optimize the position of cameras. Assume that there are N cameras for a static scene capturing, and there are K virtual views to be rendered through light ray interpolation from the N captured images. The position of these K virtual views can be fixed in advance. Our goal is to improve the rendering quality of the K virtual views by optimizing these N cameras position. Here both N and K are positive finite integers, and we assume that $N < K$.

These N cameras can be optimized by a modified fuzzy c-means (FCM) clustering algorithm [23]. As we defined previously, the CF is associated with the virtual view v_k and camera c_i, so that these K virtual views can be divided into N groups by the area of the corresponding CF. Therefore, the best position of cameras can be obtained by,

$$\hat{c}_i = \arg\max_{c_i} \sum_{k=1}^{K}\sum_{i=1}^{N}\left(\left(u_{i,k}\right)^b \cdot S\left(c_i,v_k\right)\right), \tag{14}$$

where $i = 1, 2, ..., N$, b is a weighted index, $u_{i,k}$ is a membership function. We can use the iteration method of modified FCM to determine the best position of cameras.

4   Experimental Results and Discussions

4.1   Simulation of the Coverage Field Area and the Rendering Quality

In order to analyze the area of CF with respect to positions of camera, the formula (3) is used for the simulation. In all cases, we set $Z = 10$ cm, and $\theta = \pi/3$ radian. A virtual view is fixed on the point 0.0 cm or 1.0 cm. The positions of camera are the points between -1.0 cm and 2.0 cm.

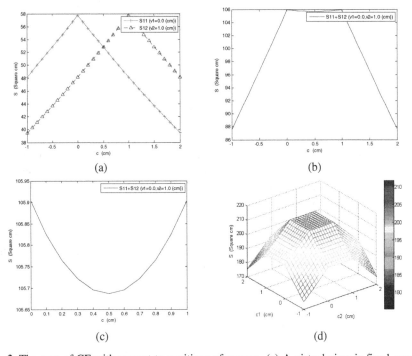

Fig. 3. The area of CF with respect to positions of camera. (a) A virtual view is fixed, camera movements and the area of CF is calculated. (b) Sum of two CF for a camera and two virtual views. (c) A larger view of figure (b) for camera position changes from 0.0 cm to 1.0 cm. (d) Sum of four CF for two cameras and two virtual views.

As shown in Fig. 3(a), a curve of the CF is composed of two parabolas, the maximum value of the area is that the camera moves to the point 0.0 cm or 1.0cm. Another phenomenon is that the distance between camera and virtual view is the smaller, the area of CF is the larger.

As shown in Figs. 3(b) and (c), the curve of the sum of two CF is composed of three parabolas. The maximum of the sum of the area is that the camera moves to the point 0.0 cm or 1.0 cm. If these two virtual views are fixed, and two cameras are moved, then the maximum values of the sum of the area are that the two cameras respectively move to the points 0.0 cm and 1.0 cm. As shown in Fig. 3(d), there are four peaks.

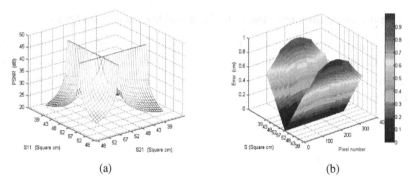

(a)   (b)

Fig. 4. The simulation results of our CF. (a) The simulation of PSNR. (b) Rendering error with respect to the depth and the area of CF, and the object is a circle.

The variables of formula (9) contain not only position of camera and virtual view, but also contain the area of CF and depth information. Fig. 4(a) shows the peak signal to noise ratio (PSNR) with respect to the area of CF, and the PSNR calculated by the formula (9). A phenomenon is that the area of CF is the larger, the rendering quality is the better, and the distance between camera and virtual view is the smaller. As long as there is an area of CF of camera has a maximum, then the PSNR is the maximum, and the PSNR is no longer as another the area of CF varies.

If the scene is a circle, and we assume that a virtual view is fixed on the point 0.0 cm, and the positions of camera are the points between -2.0 cm and 2.0 cm, then simulation results as shown in Fig. 4(b). A phenomenon is that the rendering error will become smaller when the depth is the larger.

4.2   Synthetic Experimental Results and Discussions

We verify the effectiveness of the proposed CF with one synthetic scene, namely, Church, and all rendered from POV-Ray, which creates 3D photo-realistic images using ray tracing. We use 2 cameras on the camera plane to capture the scene. The positions of the two cameras are points between -1.0 cm and 2.0 cm. Two virtual views are fixed on the points 0.0 cm and 1.0 cm. There are 31 capture positions of cameras, and 961 rendering results of each virtual view.

Fig. 5. shows one group of rendering results for Church scene. When the first camera on the point -0.5 cm, the second camera on the point 1.0 cm, and the images are shows by Fig. 5(a) and Fig. 5(b). The virtual views which on the points 0.0 cm and 1.0 cm are rendered, the rendering results as shows in Fig.5(c)(d), respectively. Intuitively, Fig.5(d) is better than Fig.5 (c), that is because the Fig.5 (c) has distortion which was caused by the distance between the camera and the virtual view is too big.

| (a) | (b) | (c) | (d) |

Fig. 5. The synthetic results of the Church scene. (a) The images captured on the point -0.5. (b) The images captured on the point 1.0 cm. (c) Synthetic result of the virtual view on the point 0.0 cm. (d) Synthetic result of the virtual view on the point 1.0 cm.

Fig. 6(a) shows the PSNR of the rendered image of the Church scene with respect to the area of CF. A phenomenon is that the area of CF is larger can result in better rendering quality, and the distance between camera and virtual view is smaller. As long as there is a camera the area of CF has a maximum value, then PSNR is the maximum, and the PSNR is no longer as another the area of CF varies. The conclusion is the same as the conclusion which is obtained by Fig. 4(a) and Fig. 6(b).

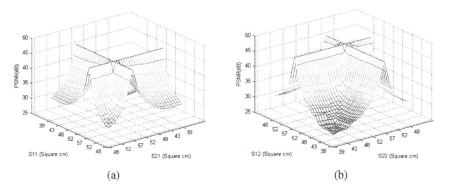

| (a) | (b) |

Fig. 6. Results of the synthetic experiment. (a) PSNR of rendered images, and the virtual view on the point 0.0 cm. (b) PSNR of rendered images, and the virtual view on the point 1.0 cm.

From the synthetic experiment results, we found that the relationship between rendering quality and the area of CF the same as the simulation results in section 4.1. The relationship is that the area of CF is the larger, and the rendering quality is the better, and the distance between camera and virtual view is the smaller.

4.3   Optimization Experimental Results

We verify the effectiveness of proposed optimization algorithm by the CF with one synthetic scene. The scene is named Church and is captured by 2 cameras. Seven virtual views are rendered which positions are at the camera plane, and two group virtual views were rendered by 2 group cameras capture images. Table.1 shows the position of virtual views. The Table.2 shows the initial position and the best position of cameras. The modified FCM algorithm is used to optimize position of the cameras. The threshold is set to 0.002. The algorithm is converged when iteration to nine.

Table 1. The position of virtual views. The unit is cm.

The position of virtual views	1	2	3	4	5	6	7
First group	-1.0	-0.5	0.0	0.5	1.0	1.5	2.0
Second group	-1.0	-0.3	0.3	0.9	1.4	1.5	2.0

Table 2. The position of cameras. The unit is cm.

The position of cameras	The first camera	The second camera
The initial position of the cameras	0.0	1.0
The best position of the cameras for the first group	-0.4658	1.1882
The best position of the cameras for the second group	-0.2790	1.4179

5   Conclusions

This paper proposes a CF to describes the topology of the intersecting coverage range of the FOV between the camera and the virtual view. We find that the CF can provide a mathematical model to quantify the relationship between the rendering quality and geometric configuration of cameras. The CF can be used to optimize the position of cameras in the LFR system.

Acknowledgments. This work was supported in part by the National Natural Science Foundation of China (NSFC) (No. 61170194, 61202301, 61231010), the Fundamental Research Funds for the Central Universities (HUST2012QN076), The research Funds for the Doctoral Program (20120142110015).

References

1. Levoy, M., Hanrahan, P.: Light Field Rendering. In: Proceedings of the 23rd Annual Conference on Computer Graphics and Interactive Techniques, pp. 31–42. ACM Press, New York (1996)

2. Zhang, C., Chen, T.: Active rearranged Capturing of Image-Based Rendering Scenes-Theory and Practice. IEEE Transactions on Multimedia 9(3), 520–531 (2007)
3. Werner, T., Hlavac, V., Leonardis, A., Pajdla, T.: Selection of Reference Views for Image-based Representation. In: 13th International Conference on Pattern Recognition, pp. 73–77. IEEE Press, Vienna (1996)
4. Fleishman, S., Cohen-Or, D., Lischinski, D.: Automatic Camera Placement for Image-Based Modeling. Computer Graphics Forum 19(2), 101–110 (2000)
5. Vaish, V., Wilburn, B., Joshi, N., Levoy, M.: Using Plane+Parallax for Calibrating Dense Camera Arrays. In: Proceedings of the 2004 IEEE Computer Society Conference on Computer Vision and Pattern Recognition, pp. 1–2. IEEE Press, Washington (2004)
6. Lin, Z., Wong, T.T., Shum, H.Y.: Relighting with the Reflected Irradiance Field: Representation, Sampling and Reconstruction. International Journal of Computer Vision 49(2-3), 229–246 (2002)
7. Gao, Y., Wang, M., Tao, D., Ji, R., Dai, Q.: 3D Object Retrieval and Recognition with Hypergraph Analysis. IEEE Transactions on Image Processing 9(21), 4290–4303 (2012)
8. Gao, Y., Tang, J., Hong, R., Yan, S., Dai, Q., Zhang, N., Chua, T.S.: Camera Constraint-Free View-Based 3D Object Retrieval. IEEE Transactions on Image Processing 4(21), 2269–2281 (2012)
9. Gao, Y., Wang, M., Zha, Z., Tian, Q., Dai, Q., Zhang, N.: Less is More: Efficient 3D Object Retrieval with Query View Selection. IEEE Transactions on Multimedia 5(11), 1007–1018 (2011)
10. Gao, Y., Dai, Q., Zhang, N.Y.: 3D Model Comparison Using Spatial Structure Circular Descriptor. Pattern Recognition 43(3), 1142–1151 (2010)
11. Shen, J., Shepherd, J., Cui, B., Tan, K.L.: A Novel Framework for Efficient Automated Singer Identification in Large Music Databases. ACM Transactions on Information Systems (TOIS) 27(3), 18 (2009)
12. Ji, R., Yao, H., Liu, W., Sun, X., Tian, Q.: Task-dependent Visual-codebook Compression. IEEE Transactions on Image Processing 21(4), 2282–2293 (2012)
13. Ji, R., Duan, L.Y., Chen, J., Yao, H., Yuan, J., Rui, Y., Gao, W.: Location Discriminative Vocabulary Coding for Mobile Landmark Search. International Journal of Computer Vision 96(3), 290–314 (2012)
14. Liu, Q., Yang, Y., Ji, R., Gao, Y., Yu, L.: Cross-view Down/up-sampling Method for Multiview Depth Video Coding. IEEE Signal Processing Letters 19(5), 295–298 (2012)
15. Yang, Y., Liu, Q., Ji, R., Gao, Y.: Dynamic 3d Scene Depth Reconstruction Via Optical Flow Field Rectification. PLoS ONE 7(11), e47041 (2012)
16. Yang, Y., Liu, Q., Ji, R., Gao, Y.: Remote Dynamic Three-Dimensional Scene Reconstruction. PLoS ONE 8(5), e55586 (2013)
17. Yang, Y., Dai, Q.: Contourlet-based Image Quality Assessment for Synthesized Virtual Image. Electronics Letters 46(7), 492–494 (2010)
18. Nguyen, H.T., Do, M.N.: Error Analysis for Image-Based Rendering with Depth Information. IEEE Transactions on Image Processing 18(4), 703–716 (2009)
19. Liu, S.X., An, P., Zhang, Z.Y., Zhang, Q., Shen, L.Q., Jiang, G.Y.: On the Relationship Between Multi-View Data Capturing and Quality of Rendered Virtual View. Imaging Science Journal 57(5), 250–259 (2009)
20. Shidanshidi, H., Safaei, F., Li, W.: A Quantitative Approach for Comparison and Evaluation of Light Field Rendering Techniques. In: 2011 IEEE International Conference on Multimedia and Expo (ICME), pp. 1–4. IEEE Press, Pennsylvania (2011)

21. Shidanshidi, H., Safaei, F., Li, W.: Objective Evaluation of Light Field Rendering Methods Using Effective Sampling Density. In: 13th International Workshop on Multimedia Signal Processing (MMSP), pp. 1–6. IEEE Press, Hangzhou (2011)
22. Gortler, S., Grzeszczuk, R., Szeliski, R., Cohen, M.F.: The Lumigraph. In: Proceedings of the 23rd Annual Conference on Computer Graphics and Interactive Techniques, pp. 43–54. ACM Press, New York (1996)
23. Bezdek, J.C., Coray, C., Gunderson, R., Watson, J.: Detection and Characterization of Cluster Substructure I. Linear Structure: Fuzzy c-lines. SIAM Journal on Applied Mathematics 40(2), 339–357 (1981)

Personalized Recommendation by Exploring Social Users' Behaviors

Guoshuai Zhao, Xueming Qian, and He Feng

SMILES LAB, Xi'an Jiaotong University, China
zgs2012@stu.xjtu.edu.cn, qianxm@mail.xjtu.edu.cn

Abstract. With the popularity and rapid development of social network, more and more people enjoy sharing their experiences, such as reviews, ratings and moods. And there are great opportunities to solve the cold start and sparse data problem with the new factors of social network like interpersonal influence and interest based on circles of friends. Some algorithm models and social factors have been proposed in this domain, but have not been fully considered. In this paper, two social factors: interpersonal rating behaviors similarity and interpersonal interest similarity, are fused into a consolidated personalized recommendation model based on probabilistic matrix factorization. And the two factors can enhance the inner link between features in the latent space. We implement a series of experiments on Yelp dataset. And experimental results show the outperformance of proposed approach.

Keywords: recommender system, entropy, rating behaviors, social networks.

1 Introduction

Recommender system (RS) is an emerging research orientation in recent years, and it has been demonstrated to solve information overload to a certain extent. In E-Commerce, such as Amazon, it also has been utilized to provide attractive and useful products' information for users from mass scales of information. A survey shows that at least 20 percent of the sales in Amazon come from the work of the RS. The traditional collaborative filtering algorithms [7-9] could be deemed to the first generation of recommender systems [6,19,20] to predict user interest. However, with the rapidly increasing number of registered users and more and more new products hit store shelves, the problem of cold start for users (new users into the RS with little historical behavior) and sparsity of datasets (the proportion of rated user-item pairs in all the user-item pairs of RS) have been increasingly intractable. And with the popularity and rapid development of social network, more and more users enjoy sharing their experiences, such as reviews, ratings and moods. So we can mine the information we are interested in from social networks to make the prediction ratings more accurate. In this paper, we propose personalized recommendation approach by exploring social users' behavior.

The main contributions of this paper are as following: 1) Propose a personalized recommendation model based on probabilistic matrix factorization combining two factors: interpersonal rating behaviors similarity, and interpersonal interest similarity.

C. Gurrin et al. (Eds.): MMM 2014, Part II, LNCS 8326, pp. 181–191, 2014.

And both of them make connections between user's latent feature vectors and his/her friends'. 2) In social circle, we utilize entropy which is based on the same category with rated history in users' circles of friends, to describe interpersonal rating behaviors similarity. 3) Experimental results and discussions show the effectiveness of the proposed approach.

The rest of this paper is organized as following: Related work on personalized recommendation system and probabilistic matrix factorization model for rating and adoption prediction problem is reviewed in section 2. The proposed personalized recommendation model combining interpersonal interest similarity and interpersonal rating behaviors similarity is introduced in detail in section 3. The experiments and results are given in section 4. And at last in section 5 conclusions are drawn.

2   Related Work

The traditional collaborative filtering algorithms [7-9] could be deemed to the first generation of recommender systems [6,19,20,21] to predict user interest. And the model we proposed is based on probabilistic matrix factorization with consideration of factors of social network.

To introduce various complicated approaches and models [1, 2, 3, 5], we firstly review the basic probabilistic matrix factorization (BaseMF) approach [4] briefly, which doesn't take any factors into consideration. It utilizes user latent feature vector and item latent feature to predict the ratings user to item, and then the task of this model is minimizing the objective function which involve the prediction errors and the Frobenius norm of matrix. This objective function can be minimized efficiently using gradient descent method in [3], which is also implemented in this paper.

Nowadays with the popularity of internet, more and more people enjoy the social networks as Facebook, Twitter, Yelp[1], Douban[2], Epinions[3], etc. The interpersonal relationships become transparent and opened, especially the circles of friends, which bring opportunities and challenges for recommender system (RS) to solve cold start and sparsity problem of datasets. Many models based social network [3, 11-14, 17, 18,19,20,21] have been proposed to improve the performance of the RS. Java et al. [11] had analyzed a large social network in a new form of social media known as micro-blog. Such networks were found to have a high degree correlation and reciprocity, indicating close mutual acquaintances among users. And they had identified different types of user intentions and studied the community structures. And we can believe that the ability to categorize friends into groups (e.g. family, co-workers) would greatly benefit the adoption of micro-blog platforms based on author's analysis of user intentions. That is to say user's friends' interest and categories could reflect user intentions and interest. In [21], a personalized product recommendation system is proposed by mining user-contributed photos in existing

---

[1] http://www.yelp.com
[2] http://www.douban.com
[3] http://www.epinions.com

social media sharing website such as Flickr[4]. Both visual information and the user generated content are fused to improve recommendation performances. They have shown that the more information we obtained from users' sides, the better performances are achieved. Chen et al. [12] explored three separate dimensions in designing such a recommender: content sources, topic interest models for users, and social voting. They implemented 12 algorithms in the design space they formulated, and demonstrated that both topic relevance and the social voting process were helpful in providing recommendations. Piao et al. [14] proposed an entropy-based recommendation algorithm to solve cold start problem and discover users' hidden interests. A hierarchical user interest mining method is proposed to explore user's potential shopping needs based on user-contributed photos in her/his social media sites [21]. We recommend personalized products according to the mined user interests. Mehta et al. [13] had calculated entropy-based similarity between users to achieve solution for scalability problem. Iwata et al. [15] proposed a model for user behaviors in online stores that provide recommendation services, and estimated the probability of purchasing an item given recommendations for each user based on the maximum entropy principle. In [17], authors proposed a context-aware recommender system, which proceeded contextual information by utilized random decision trees to group the ratings with similar contexts. At the same time Pearson correlation coefficient was proposed to measure user similarity, and then their model could learn user latent factor vectors and item latent factor vectors by matrix factorization.

Recently, Yang et al. [1] proposed using the concept of 'inferred trust circle' based on the circles of friends to recommend user favorite items. Their approach not only refined the interpersonal trust in the complex networks, but also reduced the load of big data. Meanwhile, besides the interpersonal influence, Jiang et al. [2] demonstrated that individual preference is also a significant factor in social network. Just like the idea of interpersonal influence Yang et al. [1] proposed, according to the preference similarity, users latent features should be similar to their friends' based on the probabilistic matrix factorization model [4]. Qian et al. propose to fuse three social factors: personal interest, interpersonal interest similarity, and interpersonal influence, into a unified personalized recommendation model based on probabilistic matrix factorization [19, 20]. They represent personality by user-item relevance of user interest to the topic of item by mining the topic of item based on the natural item category tags of rating datasets. Moreover, each item is denoted by a category/topic distribution vector. The user-user relationship of social network contains two factors: interpersonal influence and interpersonal interest similarity.

3   The Approach

In this paper, two social factors are fused into the proposed personalized recommendation approach: interpersonal interest similarity, and interpersonal rating behaviors similarity. And we will introduce two factors in detail respectively. And

---

[4] http://www.flickr.com

then the objective function of the proposed algorithm based on the probabilistic matrix factorization model is inferred at last.

3.1 Interpersonal Interest Similarity

User interest is a significant factor to affect users' decision-making process, which has been proved by psychology and sociology studies [10]. Moreover, Jiang et al. [2] demonstrated the effect of ContextMF model with consideration of both individual preference and interpersonal influence. However, there is a main difference between user interest factor in our model and individual preference in ContextMF [2]: we utilize friends' interest in same category to link user latent feature vector, that is to say, user latent feature should be similar to his/her friends' latent feature according to the similarity of their interests.

According to natural item category tags of rating datasets, we can get category distribution of the item, which can be seen as the naive topic distribution D_i of item i.

Just like the item *Steakhouses Argentine* in New York in Yelp dataset belongs to the sub-category **Steakhouses**, meanwhile it certainly belongs to the first-level category **Restaurants**, and in this paper, we just put user into a distinct group according to the first-level, that means, we analyse user interest similarity and the rating behaviors similarity just in single category because the item naive topic distribution is different from other categories, and there are sufficient sub-categories in each category to describe item naive topic distribution, such as the 114 sub-categories in **Restaurants**. According to user's historical rating data, we summarize the number of all the rated items to measure user interest, that is to say, the more rated items are, the more user interest is:

$$D_u^c = \frac{1}{\left|H_u^c\right|} \sum_{i \in H_u^c} D_i \tag{1}$$

where H_u^c is the set of items rated by user u in c.

And we denote the interest similarity between user u and his/her friend v by $W_{u,v}$, and each of the rows is normalized to unity $\sum_v W_{u,v}^* = 1$.

$$W_{u,v} = Sim(D_u, D_v) \tag{2}$$

where the similarity function is measured by cosine similarity as:

$$Sim(D_u, D_v) = \frac{D_u \bullet D_v}{\left|D_u\right| \times \left|D_v\right|} \tag{3}$$

Then the basic idea of this factor is that user latent feature should be similar to his/her friends'.

3.2   Interpersonal Rating behaviors Similarity

Besides the category tags information, user's ratings are more helpful to be utilized to describe user's rating behavior habits and his/her rating standards. As we all know, the higher probability of occurrence of certain information, the easier we predict the user behaviors including ratings. So we can mine user's interest information for predictions by comparing the ratings similarity in same sub-category by entropy algorithm.

There are some existed approaches which describe the similarities and behaviors analysis between users by entropy [13-16], but there are two main differences of our approach: 1) Unlike [13-16], they utilize entropy to calculate the similarity among all users, even there are no connections among some users, while we utilized entropy algorithm in social circle of friends to calculate the similarity of rating behaviors. One of advantages of our approach is with lower computational cost because we confine the calculation by social circle. Another advantage of our approach is that better performances are achieved by filtering out the insignificant information. 2) We extend the scope of entropy to fit the comparability and pervasiveness of ratings between user and his/her friends. Because the ratings of a user and his/her friends to the same item are very few, we replace ratings of the same item with average ratings in same sub-category. Thus we calculate the ratings similarity as follows:

$$E(U_u, U_v) = -\sum_{c'=1}^{n} p(d_{c'}) \log_2 p(d_{c'}) \tag{4}$$

where U_u and U_v denotes user u and his/her friend v, $p(d_{c'})$ denotes the frequency of the errors $d_{c'}$, which is calculated by the average ratings between user u and his/her friend v in same sub-category c'. To solve sparsity problem of ratings to the same item in social network, we represent $d_{c'}$ as following:

$$d_{c'} = \left|K_{u,v}^{c'}\right| \times \left|R_{u,c'} - R_{v,c'}\right| \tag{5}$$

where $\left|K_{u,v}^{c'}\right|$ is the indicator function, and if both of user u and v have rated item in sub-category c', $\left|K_{u,v}^{c'}\right|$ is equal to 1, otherwise, it is 0. $R_{u,c'}$ denotes u's average rating in c' and $R_{v,c'}$ denotes v's average rating in c'.

As we all know, the higher entropy is, the smaller user ratings similarity becomes. So we denote ratings similarity between user u and his/her friend v by $E_{u,v}$, which is the reciprocal of entropy, and each of the rows is normalized to unity $\sum_v E_{u,v}^* = 1$.

$$E_{u,v} = \frac{1}{E(U_u, U_v)} \tag{6}$$

Then the basic idea of this factor is that user u's rating behaviors should be similar to its friend v's to some extent.

3.3   Personalized Recommendation Model

The personalized recommendation model contains these following aspects: 1) The Frobenius norm of matrix U and P, which is used to avoid over-fitting as [3]. 2)

Interest circle influence $W_{u,v}^{c*}$, which means the similarity degree between u and v. 3) User interpersonal ratings similarity $E_{u,v}^{c*}$, which has effects on understanding your rating behaviors and mining the users, whose ratings are similar to yours in circle of your friends.

With similarity to CircleCon Model [1] and Context Model [2], the objective function of our model is as following:

$$
\begin{aligned}
\Psi^c &\left(R^c, U^c, P^c, W^{c*}, E^{c*} \right) \\
&= \frac{1}{2} \sum_{u,i} \left(R_{u,i}^c - \hat{R}_{u,i}^c \right)^2 + \frac{\lambda}{2} \left(\left\| U^c \right\|_F^2 + \left\| P^c \right\|_F^2 \right) \\
&+ \frac{\beta}{2} \sum_u \left(\left(U_u^c - \sum_v W_{u,v}^{c*} U_v^c \right) \left(U_u^c - \sum_v W_{u,v}^{c*} U_v^c \right)^T \right) \\
&+ \frac{\gamma}{2} \sum_u \left(\left(U_u^c - \sum_v E_{u,v}^{c*} U_v^c \right) \left(U_u^c - \sum_v E_{u,v}^{c*} U_v^c \right)^T \right)
\end{aligned}
\tag{7}
$$

where $R_{u,i}^c$ is the real rating value and $\hat{R}_{u,i}^c$ is the predicted rating value in c as following:

$$
\hat{R}_{u,i}^c = r^c + U_u^c P_i^{cT}
\tag{8}
$$

where r^c is empirically set as user's average rating value in category c, U and P is user and item latent feature matrices in this model. And the factor of interpersonal interest similarity is enforced by the second term in the objective function, which denotes that user u's latent feature U_u should be close to the average of his/her friend v's latent feature with weight of $W_{u,v}^{c*}$ in c. The factor of interpersonal ratings similarity is enforced by the last term, which means that user u's latent feature U_u should be close to the average of his/her friend v's latent feature with weight of $E_{u,v}^{c*}$ in c.

3.4   Model Training

In this paper, we aim at the separate user latent feature U^c and item latent feature P^c in category c by the corresponding matrix factorization model as Eq. (7). And the objective function can be minimized by the gradient decent approach as [3]. More formally, the gradients of the objective function with respect to the variables U_u and P_i in c are shown as Eq. (9) and Eq. (10) respectively:

$$
\begin{aligned}
\frac{\partial \Psi^c}{\partial U_u^c} &= \sum_{i \in H_u^c} I_{u,i}^{R^c} \left(\hat{R}_{u,i}^c - R_{u,i}^c \right) P_i^c + \lambda U_u^c \\
&+ \beta \left(U_u^c - \sum_{v \in F_u^c} W_{u,v}^{c*} U_v^c \right) - \beta \sum_{v:u \in F_v^c} W_{v,u}^{c*} \left(U_v^c - \sum_{w \in F_v^c} W_{v,w}^{c*} U_w^c \right) \\
&+ \gamma \left(U_u^c - \sum_{v \in F_u^c} E_{u,v}^{c*} U_v^c \right) - \gamma \sum_{v:u \in F_v^c} E_{v,u}^{c*} \left(U_v^c - \sum_{w \in F_v^c} E_{v,w}^{c*} U_w^c \right)
\end{aligned}
\tag{9}
$$

$$\frac{\partial \Psi^c}{\partial P_i^c} = \sum_{i \in H_u^c} I_{u,i}^{R^c} \left(\hat{R}_{u,i}^c - R_{u,i}^c \right) U_u^c + \lambda P_i^c \qquad (10)$$

where $I_{u,i}^{R^c}$ is the indicator function which is equal to 1 if user u has rated item i in c, and equal to 0 otherwise. $\hat{R}_{u,i}^c$ is the predicted rating value user u to item i in c according to Eq. (8).

The initial values of U^c and P^c are sampled from the normal distribution with zero mean. We will set U^c and P^c the same initial values when comparing with each factor to insure the fairness, even it empirically has little effect on the latent feature matrix learning. In each iteration, the user and item latent feature vectors U^c and P^c are updated based on the previous values to insure the fastest decreases of the objective function. Note that the step size is a considerable issue. But in this paper, for each appropriate step size, it's always fair to each algorithm if it's set as the invariant, so we just adjust it to insure the decreases of the objective function in training.

Then the algorithm is shown as Table 1, where l is the step size, and t is the iteration time.

Table 1. Personalized recommendation algorithm based on rating behaviors

Algorithm of proposed personalized recommendation model
1)   initialization: $\Psi^c(t) = \Psi^c \left(U^c(t), P^c(t) \right)$, $t=0$.
2)   given: parameters $k, l, \lambda, \beta, \gamma$, average rating value r^c.

3)   iteration:
      while ($t<1000$)
        calculate $\dfrac{\partial \Psi^c(t)}{\partial U^c}, \dfrac{\partial \Psi^c(t)}{\partial P^c}$
        $U^c(t) = U^c(t) - l\dfrac{\partial \Psi^c(t)}{\partial U^c}$, $P^c(t) = P^c(t) - l\dfrac{\partial \Psi^c(t)}{\partial P^c}$
        t++
      end while

4)   return: $U^c, P^c \longleftarrow U^c(1000), P^c(1000)$
5)   prediction: $\hat{R}_{u,i}^c = r^c + U_u^c P_i^{c\mathrm{T}}$

4   Experiments

We implement a series of experiments to estimate the performance of proposed approach, and compare the factors by observing the performance and the effectiveness of each factor on Yelp dataset [19,20]. In this section, we will show you the introduction of dataset, the performance measures and results and discussion.

4.1   Yelp Dataset

Yelp is a local directory service with social networks and user reviews. It is one of the most popular consumer review websites and has more than 71 million monthly unique visitors as of January 2012. It combines local reviews and social networks to create a local online community with the slogan: "real people real review". And most of all, Yelp dataset contains the exact ratings without any subjective factor. It's the crucial problem to measure the performance of this algorithm with the objective authenticity of test collection. Meanwhile, Yelp dataset is similar to Epinions, which has been used in [1, 2, 3, 5,19,20].

We have crawled nearly 60 thousand users' circles of friends and their rated items from November 2012 to January 2013. And five categories are utilized to implement experiments and the statistics of them are shown in Table 2. More detail of this dataset can be found from website of SMILES LAB[5].

We experiment with 80% of each user's rating data as the training set and 20% of each user's rating data as the test set in each category to ensure all users' latent features are learned , and certainly sample the data randomly.

Table 2. Yelp Data: Statistics of the test categories

Category	User Count	Item Count	Rating Count	Sparsity	r^c
Home Services	2500	3213	5180	6.449e-4	3.707
Night Life	4000	21337	99878	1.170e-3	3.594
Pets	1624	1672	3093	1.139e-3	3.975
Restaurants	2000	32725	91946	1.405e-03	3.677
Shopping	3000	16154	33352	6.882e-04	3.819

4.2   Performance Measures

When we get user latent feature U^c and item latent feature P^c, the performance of our algorithm will be embodied by the errors. From [1-4] we can see Root Mean Square Error (RMSE) and Mean Absolute Error (MAE) as the most popular accuracy measures, which are defined as following:

$$RMSE = \sqrt{\frac{\sum_{(u,i)\in\Re_{test}}\left(R_{u,i} - \hat{R}_{u,i}\right)^2}{\left|\Re_{test}\right|}} \qquad (11)$$

$$MAE = \frac{\sum_{(u,i)\in\Re_{test}}\left|R_{u,i} - \hat{R}_{u,i}\right|}{\left|\Re_{test}\right|} \qquad (12)$$

---

[5] http://smiles.xjtu.edu.cn

Table 3. Performance comparison based on CircleCon2b of training on each category of Yelp

Category	BaseMF		CircleCon2b		ContextMF		URB	
	RMSE	MAE	RMSE	MAE	RMSE	MAE	RMSE	MAE
Home Services	3.26	2.57	2.14	1.68	1.72	1.34	**1.58**	**1.26**
Night Life	2.20	1.65	1.50	1.16	1.32	1.02	**1.18**	**0.93**
Pets	3.53	2.78	2.19	1.72	1.72	1.29	**1.46**	**1.16**
Restaurants	1.88	1.39	1.34	1.04	1.28	1.00	**1.15**	**0.91**
Shopping	2.52	1.90	1.73	1.34	1.41	1.09	**1.32**	**1.03**
Average	2.68	2.06	1.78	1.39	1.49	1.15	**1.34**	**1.06**

Table 4. Performance of the two independent factors on Restaurants of Yelp

Factors	BaseMF (without any factor considered)	
RMSE	1.854396	
MAE	1.361975	
Factors	**Interpersonal Interest Similarity**	**Ratings similarity**
RMSE	1.35943	1.24698
MAE	1.05077	0.97693
Factors	**Interpersonal Interest Similarity+ Ratings similarity**	
RMSE	1.14942	
MAE	0.91018	

where $R_{u,i}$ is the real rating value of user u on item i, $\hat{R}_{u,i}$ is the corresponding predicted rating value according to Eq. (8), and $\Re_{test}$ is the set of all user-item pairs in the test set.

4.3 Results and Discussion

In this paper, three existing models are compared with our social recommendation algorithm based on users' rating behaviors (URB) on Yelp dataset: BaseMF [3, 4], CircleCon2b [1] and ContextMF [2].

The performance of different algorithms including our algorithm are showed in Table 3 with the parameter $\lambda=0.1$ as [1], $\beta=30$, and $\gamma=50$, which are tradeoffs to adjust the strengths of different terms in the objective function. And the k which denotes the dimensionality of latent feature U and P, is set $k=10$ as [1]. Previous works [2, 3] had investigated the changes of performance with different k, but as an invariable, it is fair for all compared algorithms. And then we demonstrate the effectiveness and reliability of the proposed model according to experimental results shown in Table 3.

Considering the effectiveness of each factor, we compare the performance of the two independent factors in Restaurants of Yelp respectively. And the experimental results are shown in Table 4 from which we can see that both of the factors have effects on improving the accuracy of recommender system.

5   Conclusions

In this paper, a personalized recommendation approach is proposed by combining social network factors: interpersonal interest similarity and interpersonal rating behaviors similarity. In particular, the interpersonal rating behaviors similarity denotes user's rating behavior habits and his/her rating standards. We can mine user's interest information from comparing the ratings similarity in the same sub-category based on entropy algorithm. We conducted a series of experiments in five categories on Yelp dataset to compare existing approaches and the experimental results showed the significant improvements. At the moment, we just exploit user historical rating records and interpersonal relationship of social networks, but this only goes so far. In the future, we will take user location information and interpersonal influence into consideration to improve our algorithm.

Acknowledgments. This work is supported partly by NSFC No.60903121, No.61173109, and Microsoft Research Asia.

References

[1] Yang, X.-W., Steck, H., Liu, Y.: Circle-based recommendation in online social networks. In: KDD 2012, pp. 1267–1275 (August 2012)

[2] Jiang, M., Cui, P., Liu, R., Yang, Q., Wang, F., Zhu, W.-W., Yang, S.-Q.: Social contextual recommendation. In: CIKM 2012, pp. 45–54 (October 2012)

[3] Jamali, M., Ester, M.: A matrix factorization technique with trust propagation for recommendation in social networks. In: Proceedings of the Fourth ACM Conference on Recommender Systems, RecSys (2010)

[4] Salakhutdinov, R., Mnih, A.: Probabilistic matrix factorization. In: NIPS 2008 (2008)

[5] Koren, Y., Bell, R., Volinsky, C.: Matrix factorization techniques for recommender systems. Computer, 30–37 (August 2009)

[6] Adomavicius, G., Tuzhilin, A.: Toward the next generation of recommender systems: a survey of the state-of-the-art and possible extensions. IEEE Transactions on Knowledge and Data Engineering, 734–749 (June 2005)

[7] Bell, R., Koren, Y., Volinsky, C.: Modeling relationships at multiple scales to improve accuracy of large recommender systems. In: KDD 2007, pp. 95–104 (2007)

[8] Sarwar, B., Karypis, G., Konstan, J., Reidl, J.: Item-based collaborative filtering recommendation algorithms. In: Proceedings of the 10th International Conference on World Wide Web (WWW), pp. 285–295 (2001)

[9] Jahrer, M., Toscher, A., Legenstein, R.: Combining predictions for accurate recommender systems. In: KDD 2010, pp. 693–702 (2010)

[10] Bond, R., Smith, P.B.: Culture and conformity: a meta-analysis of studies using asch's (1952b, 1956) line judgment task. Psychological Bulletin, 111–137 (January 1996)

[11] Java, A., Song, X., Finin, T., Tseng, B.: Why we twitter: understanding microblogging usage and communities. In: WebKDD/SNA-KDD 2007: Proceedings of the 9th WebKDD and 1st SNA-KDD 2007 Workshop on Web Mining and Social Network Analysis, pp. 56–65. ACM, NY (2007)

[12] Chen, J., Nairn, R., Nelson, L., Bernstein, M., Chi, E.H.: Short and tweet: experiments on recommending content from information streams. In: Proceedings of the 28th International Conference on Human Factors in Computing Systems, pp. 1185–1194. ACM (2010)

[13] Mehta, H., Bhatia, S.K., Bedi, P., Dixit, V.S.: Collaborative Personalized Web Recommender System using Entropy based Similarity Measure. IJCSI 8(6) (November 2011)

[14] Piao, C.-H., Zhao, J., Zheng, L.-J.: Research on entropy-based collaborative filtering algorithm and personalized recommendation in e-commerce. SOCA 3, 147–157 (2009)

[15] Iwata, T., Saito, K., Yamada, T.: Modeling User Behavior in Recommender Systems based on Maximum Entropy. In: WWW 2007 Proceedings of the 16th International Conference on World Wide Web, pp. 1281–1282. ACM, NY (2007)

[16] Dou, Y., Liu, P., Chen, Y., Lei, Z.: Entropy based Broadband User Service Behavior Analysis. In: International Conference on Network Infrastructure and Digital Content, pp. 440–445 (2009)

[17] Liu, X., Aberer, K.: SoCo: a social network aided context-aware recommender system. In: WWW 2013, pp. 781–802 (2013)

[18] Shen, J., Wang, M., Yan, S., Cui, P.: Multimedia recommendation: technology and techniques. In: SIGIR 2013, p. 1131 (2013)

[19] Qian, X., Feng, H., Zhao, G., Mei, T.: Personalized Recommendation Combining User Interest and Social Circle. IEEE Trans. Knowledge and Data Engineering x(y), xx–yy (2013)

[20] Feng, H., Qian, X.: Recommendation via user's personality and social contextual. In: ACM CIKM 2013 (2013)

[21] Feng, H., Qian, X.: Mining User-Contributed Photos for Personalized Product Recommendation. Neurocomputing (2014)

Where Is the News Breaking? Towards a Location-Based Event Detection Framework for Journalists

Bahareh Rahmanzadeh Heravi[1], Donn Morrison[2], Prashant Khare[1], and Stephane Marchand-Maillet[3]

[1] Digital Enterprise Research Institute (DERI), National University of Ireland, Galway, Ireland
Bahareh.Heravi@deri.org, Prashant.Khare@deri.org
[2] Norwegian University of Science and Technology, Trondheim, Norway
donn.morrison@idi.ntnu.no
[3] University of Geneva, Switzerland
Stephane.marchand-maillet@unige.ch

Abstract. The rise of user-generated content (UCG) as a source of information in the journalistic lifecycle is driving the need for automated methods to detect, filter, contextualise and verify citizen reports of breaking news events. In this position paper we outline the technological challenges in incorporating UCG into news reporting and describe our proposed framework for exploiting UGC from social media for location-based event detection and filtering to reduce the workload of journalists covering breaking and ongoing news events. News organisations increasingly rely on manually curated UGC. Manual monitoring, filtering, verification and curation of UGC, however, is a time and effort consuming task, and our proposed framework takes a first step in addressing many of the issues surrounding these processes.

Keywords: Event Detection, Location extraction, Citizen Journalism, User Generated Content, Social News, Semantic News, Social Web, Semantic Web, Linked Data, Social Semantic Journalism.

1 Introduction

Social media platforms have recently become a prominent mode of sharing real-time information and in doing so have evolved into more than simply a user-to-user interaction medium, but an important asset to widespread source of newsworthy information being circulated every second. This has turned the former consumers of news and information - the audience - into potential broadcasters of breaking news.

The ubiquity of mobile technology combined with social media has made it more likely than ever that an individual or a community, not a professional journalist, will be the initial source of information for a breaking news event. This community-sourced data, or "citizen/social journalism", is a valuable source of information for news organisations.

Journalists are already monitoring social media for scoops, details, and images, but the process is laborious and provides inconsistent results. In the deadline-driven world of journalism, the need to process huge volumes of community-sourced data for extracting potential news stories is a universal problem. This data, known as

C. Gurrin et al. (Eds.): MMM 2014, Part II, LNCS 8326, pp. 192–204, 2014.
© Springer International Publishing Switzerland 2014

user-generated content (UGC), is mostly unstructured, unfiltered and unverified, and often lacks contextual information. Traditional approaches to newsgathering are quickly overwhelmed by the volume and velocity of information being produced.

User-generated content shared on social media plays a significant role in the process of capturing news events, classifying and verifying stories and also keeping the audience in the loop with timely and accurate news. Every minute over 350 new blog posts are created [7], 100 hours of new video is uploaded to YouTube [36], over 540,000 tweets are sent [29] and Facebook users share 684,478 pieces of content [7]. Hidden amongst this data is valuable information that the journalist can use to create breaking news stories. However, the scale of the data precludes manual processing and there exist no effective tools that can source, aggregate, filter and verify this content for news reportage.

Detection of newsworthy events is an area of research which can be readily applied to the early stages of the journalistic lifecycle. In the past, event detection from unstructured text has been used for applications from first story detection (FSD) [16], where novel news stories are detected from news organisations, to the more general sense of discovering anomalous patterns within large streams of data [20, 11]. In the journalistic context, event detection aims to decrease the time between the occurrence of a news event and the point at which a journalist is made aware of the event. Location-based event detection would then act as a geographic filter, further decreasing this time span by considering only UGC events occurring in areas where breaking news may be expected.

Figure 1 shows how, in the wider context, the framework described in this paper fits into the emerging topic of Social Semantic Journalism [18, 10], which addresses a universal problem experienced by media organisations: the combination of vast amount of UGC across social media platforms and the limited amount of time the journalist has to extract potential news stories from these mostly unstructured, unfiltered and unverified data. In this situation, there is evidently a need for solutions

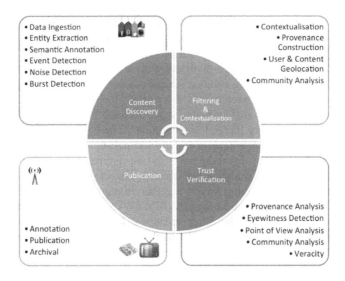

Fig. 1. Social Semantic Journalism Framework, adopted from [10]

that can help source, filter and verify social media content for media organisations who are now competing with the continuous flow of free content available on the web. Social Semantic Journalism also aims to address the chief obstacle facing news organisations: the vetting process, since the current manual process of verifying user-generated content is considered to be overwhelming and inadequate [21].

In this position paper we focus on the following objectives which we envisage a location-based event detection framework will have in relation to journalism involving UGC:

- Improve access to location-specific events from UGC;
- Decrease the time delay between the event and the reporting of the event by news organisations;
- Aid journalists in assessing the veracity of events reported in UGC; and
- Aid in efforts to trace back to first person reports.

The remainder of this paper is organised as follows. Section 2 describes the related work and existing approaches that could be adapted as components of the framework, both for unstructured UGC as well as associated metadata. Section 3 presents our proposed framework and examines the role of event detection in the journalistic lifecycle. Finally, Section 4 offers concluding remarks on the feasibility and impact of a location-based event detection framework in journalism.

2   Related Work

In the event detection literature, an event is defined as a real-world occurrence with an associated time period and a specific location. Considering the existence of a time-ordered stream of published messages relating to the real-world occurrence, the goal of event detection is to detect the occurrence based on the stream of messages [3]. Classical event detection algorithms can be broadly classified into two categories: message-pivot methods and feature-pivot methods. Message-pivot approaches detect events by clustering messages based on the semantic distance between them. An example is single-pass clustering algorithms [31,1].

A specialised form of message-pivot event detection is first story detection (FSD). FSD in a stream-based setting stores a stream of news stories, each represented as a vector of terms, and compares a new story, i.e., one not yet stored, to all stored stories. Those sufficiently different (by some distance metric or similarity measure such as cosine distance) are flagged "first stories". The FSD approach can be used to detect events as well as eyewitnesses to help journalists to identify the credibility of the tweets in order to report breaking news and this approach is used in [16]. Feature-pivot approaches involve studying the distributions of words in the messages and discovering events by grouping words together. Examples include Event Detection with Clustering of Wavelet-based Signals (EDCoW) [32] and algorithms for defining communities of keywords. The latter creates a keyword graph of documents or messages and uses community detection methods analogous to those used for social network analysis to discover and describe events [25]. Chen et al. [5] proposed a semi supervised system that crawled the data, specific to organisations and related users, using *fixed keywords* particular to the organisation, its key brands, and prominent people such as CEO. They developed a classifier which detected the temporally emerging topics from within the *fixed keywords* crawled data and formed the clusters of emerging topics. From emerging topic clusters, a supervised system detects those topics which are fast emerging based on certain features and can be considered as *hot emerging topics*.

Apart from analysing the temporal arrangement of words, there is a lot more information that can be retrieved from the text in terms of context of the messages. Natural language processing (NLP) has been used for event detection from an information extraction (IE) perspective [8]. While the use of NLP encourages the extraction of entities enclosed in text, a supervised learning approach can extract more information about entities and context. TwiCal is presented in [20] as the first open domain event extraction and categorisation tool for Twitter. The system is based on an annotated corpus of events in Twitter which are used as training data for sequence level models. Shallow linguistic analysis is used to create features for the classifier, thereby recognising event triggers as a sequence labeling task, using conditional random fields (CRF) for learning and inference.

Finally, event detection has been explored for other domains such as sports [34, 17, 13, 35, 27, 22] for purposes such as extracting highlights [22], detecting notable events such as point scores [35] and automatically structuring sports video [13]. However, these domains have a considerably smaller scale compared to that available from social media sources.

From a journalistic viewpoint event detection from social media stream relates to the discovery and filtering of UGC. Twitter has recently made advances in this with its updated search, which incorporates aspects of event detection by using Amazon's Mechanical Turk service to detect and verify trending topics and breaking news events by using human evaluators to categorise search queries and provide additional context. While the approach yields good results, it relies heavily on manual input from Mechanical Turk - an online crowdsourcing platform where the human workers perform the HIT (human intelligence tasks) that computers are unable to. Hence, a framework that reduces manual effort involved in the detection of the events would be a huge asset to the field of journalism.

A location-based event detection framework aims to address the "where" in the fundamental Big-5 information-gathering questions (who, what, where, when, why), depicted in Figure 2. Metadata such as GPS coordinates, user-specified location information (e.g., from Twitter, Facebook profiles), and even more advanced methods such as landmark detection from image and video data can all be exploited to associate the location of UGC in the event detection process. In this paper we focus on Twitter, where four types of location information are considered most relevant for detecting the location of an event as follows:

- *Geo-tagged tweets*: These are tweets which are tagged with GPS coordinates. These are the most straightforward to process for location identification. However, only a small fraction of tweets (~1%) include GPS coordinates [12].
- *User specified profile location*: These are the information that a user presents in his/her profile information, normally identifying their residential location. Studies show roughly 3% of Twitter users include location in their profiles

[14]. Time zone information can also provide an indication of where a user is located.

- *Entity extraction and NLP techniques*: These use entity extraction and natural language processing techniques for identification of 'place' type entities in microblog text.
- *Social network analysis*: This method leverages a user's social relationships and the spatial distribution of locations in his/her network for identification of potential locations [12].

Further in this section, different approaches to detect the location from the types of available location information (as mentioned above) are briefly explained and reviewed:

Geotagged Data and User Profile

User-generated content may contain geographic coordinates in its meta-data. Sakaki et al. [24] used Twitter to specifically detect an earthquake. They relied on the GPS location attached to tweets as well as user profile locations, both of which are in latitudinal and longitudinal format. Location approximation techniques are deployed using Kalman filtering (a Bayes filter variant that uses a set of signals to determine an estimation which is much more precise than single observation). The derived location is then queried using the Google Maps API to check the location on the map and establish the location of the user.

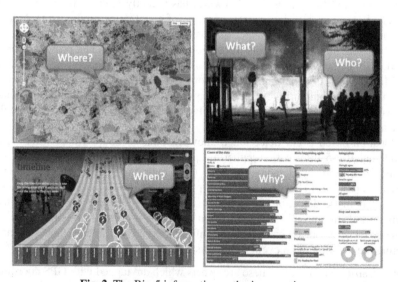

Fig. 2. The Big-5 information-gathering questions

Unankard et al. [30] followed a similar approach to extract event locations from tweets. The authors assigned a hotspot location to event clusters if sufficient correlation is found between the detected location and the event. Before calculating the correlation score they extracted the event location and user location. To find the user location, they depend on the geographic tags of the tweets and if that is not available they use the user profile location. The geographic tags are derived from coordinates logged by the devices of the users (e.g., smart phones), and profile location information is derived directly from profile text. The geographic tag data is then queried using Google API to find the location name, and in the case of user-profile location they queried gazetteer database- a list of locations downloaded from GeoNames[1] and stored in a local database. For event location, they followed an NLP-based approach to extract location entities from the text. Once both types of locations are determined, a correlation score is calculated for the event using both the locations, which is then assigned to the cluster.

Entity Extraction and NLP Techniques

NLP techniques assist in observing the events, sentiments, and to extract information such as variety of entities and tagging them. In order to extract the location for an event from the user-generated content, the text data is processed through NLP tools to determine the entities and their context with respect to parts of speech (POS). Ritter et al. [19] tested the performance of Stanford Named Entity Recognition tagger [9] named entity taggers[2], and parts of speech taggers[3] on Twitter streaming data. Unankard et al. [30], discussed earlier, also applied NLP techniques to determine the event location along with user location. To extract the event location they processed the textual data of the tweets with Named Entity Recognition (NER) for which they used the Standard Named Entity Recogniser to identify the location entities from the text messages. The most frequent location in each cluster of the detected event gets assigned as the location of the event. A correlation score is calculated between event location and user location (discussed above), by computing the level of granularity each location derives (whether it derives a country, state, city, or place name).

User Social Network Analysis

A user's social network plays an important role in determining the user's location. Often when the content-based approaches (geo-tagged data, user profile location) fail to determine the location of a user, it is the user's social network that can help in understanding from where the user is posting the content. Sadilek et al. [23], created a location prediction system, named Flap, which implements a probabilistic model of human mobility and generates a graph of people's fine grained location based on their friendship graph. First they recreate the friendship graph based on content similarity between users and redefine the edges. The location is predicted on a dynamic Bayesian network of the user and friends (from recreated friendship graph). The input sequence consists of locations visited by a user's friends (during supervised learning, the user's location was also given as input).

Jurgens [12] proposed a method based on a combination of spatial label propagation and a final location selection method to infer the geographical location of a user. In the same line, it is shown that an individual's location prediction is accessible to social network providers [2], where a relation is mapped between user's geographic locations and the friendship relationships on Facebook network. The same

---

[1] http://www.geonames.org/
[2] http://nlp.stanford.edu/software/segmenter.shtml
[3] http://nlp.stanford.edu/software/tagger.shtml

mapped relation is then reverted to infer a user's probable location based on his/her social relationships. Location-based social networks (LBSN) materialise directly the combination of social and geographical proximity and their study provide insight into how location and proximity impact social relationships [26].

The next section introduces the proposed framework that leverages the aforementioned techniques for inferring the location of an event.

3   A Framework for Location-Based Event Detection from UGC

Our framework for event detection focuses on detecting events from streams of disparate social media sources. Briefly, it aims to detect events as they happen from a stream of various social media modalities. We distinguish three types of events which require different methods:

- Breaking news - Events that are current and that were not precipitated or known a priori, e.g., a plane crash or the death of a prominent figure. Detecting breaking news from UGC requires real-time stream processing and analysis of popular social media sources.
- Running stories - This requires ongoing analyses of previously breaking news events or ongoing coverage of scheduled events.
- Scheduled events - These have known start and end dates (e.g., the Olympic Games) and can be followed by selecting certain topics or following users influential in those topics. However, unpredictable sub-events within these must be automatically detected in ways similar to breaking news.

A straightforward approach (disregarding the problem of data access), would consist of a stream processor that extracts named entities and hashtags from streaming UGC and monitors the frequency acceleration of these terms over time. The aim would be to flag the terms accelerating most frequently over time, as well as particular named entities, as potentially interesting events for journalists. Alternatively, first story detection (FSD) could be adapted to a more general framework for event detection from social media for breaking news, by treating tweets or other micro blog entries in a similar fashion. Being able to identify when something is happening in relation to a particular topic or location can be used as a first step for discovery, using entities detected through NLP.

There are other notable gaps in the literature of event detection. These include seamless interoperability between data sources and modeling different viewpoints of certain events, not only via different information streams. We propose a framework that will advance the state of the art by developing a linked data-based approach to event detection. This approach will build on previous work, e.g., [11], to develop a scalable and linked framework for flexible event detection from UGC streams. With respect to event detection in the information extraction context, recent efforts based on the work of Ritter [20] indicate that unsupervised approaches leveraging unlabeled data are promising. However, to our knowledge, NLP based event extraction for social media approaches do not leverage semantics. We envisage a framework that uses ontology-based IE expertise and improves on such semi-supervised approaches by exploiting existing linked data resources as well as available event ontologies.

Figure 3 depicts the proposed framework with two variations depending on where the location information is desired. Scenario A uses location information pre-specified by the journalist as a filter, thus it would discover events in that specified location. Scenario B clusters all detected events based on location information extracted from the media, leaving the journalist free to select among clusters. Scenario A is better suited to events where prior knowledge about the location is known, for example running stories or scheduled events as described above. In the case of general breaking news events, it is likely that any filter would be too restrictive unless such an event is expected at the specified location.

Scenario B, on the other hand, can be expected to detect a wider range of events, and is suitable for any of the three event types, with the caveat of a higher processing requirement and leaving the journalist with more manual intervention. In Scenario B, as shown in figure 4, the input stream is processed through various location detection techniques (as discussed above - geo tagged tweets, NLP techniques using Stanford NER or third party library AlchemyAPI, or social network analysis of the user), which results in cluster formation of various locations. However, there is one aspect that is not covered in the above techniques and that is leveraging the Semantic Web to refine the location cluster formation. Thus, this framework proposes using location knowledge relation mapping (from Linked Geo Data[4] or OpenGeoSpatial[5] data) to harvest the geographical proximities (two places from same country or from same city) so as to refine our location clusters, which means that if there are two different tweets: *"bombing at Boylston Street"* and *"Explosion at Boston marathon near finish line"*, we are likely to infer that *Boylston Street* is also conveying information about *Boston* or on a more macro level - *United States*, apparently both are conveying information about explosions in United States. However, this might be a computationally time costly process as it may call for querying the location knowledge data (graphs) each time a new text document is processed.

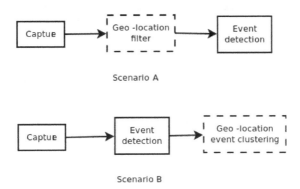

Fig. 3. Location-based event detection frameworks in two scenarios

---

[4] http://linkedgeodata.org/About
[5] http://www.opengeospatial.org/standards/geosparql

Once the location clusters are formed, the system (owner) has the leverage to mine and process the text data corresponding to each location cluster through event detection techniques. This can be achieved by breaking the input stream (of location cluster) into small time segments and analysing the *burst* keywords (meanwhile filtering out stop words) in the timeframes. The keywords whose frequency exceeds a *threshold* are considered and their corresponding text content (tweets) is taken into account. Further, *tweets* are matched against each other through vector cosine similarity (cosine similarity between text documents - converts text as vectors and evaluates the distance between vectors thus inferring the degree of relatedness between two text vectors) and if the matching output is more than a certain minimum value than they are cascaded into the same clusters, which signifies a particular *trend/event* against the *burst* keyword. This process is performed on incoming streaming data continuously and it results in clusters of events within the cluster of locations. That means, for any given location there are different clusters signifying different events.

Now that we have the location, event, and some named entities retrieved through an intermediate process (where the text was processed through Entity taggers), we can use those entities to explore and learn more about their co-occurrence significance by querying archived data from Linked Open Data (LOD) or other archival data. This would yield into a more insightful information about the event and generate more sense about its significance. Until this point, we have the events based on locations, related archival/historical data (archived stories relating to the entities of the current event), and users particular to events. What we do not have is a mechanism to verify the credibility and authenticity of the information.

In the next phase of the framework, we suggest ideas that can assist the journalists in determining the veracity. This is a multi-level process of filtering and reducing the tweets data from the clusters formed in the event detection process to a smaller set of tweets and its users, which can be further manually analysed. There are different approaches for determining the veracity of the content. Filtering the data based on multiple observations are likely to yield better results, for instance, merging the two different approaches - one based on a human centred design approach as proposed by Diakopoulos et al. [6] and other based on social network analysis.

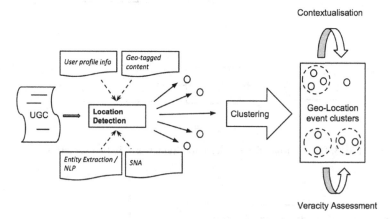

Fig. 4. Location-based event detection in Scenario B

A classifier is trained where tweets are analysed whether the user is a witness to the event he has posted or not. Text of the tweet is analysed against a dictionary[6], which contains a rich set of words across several language dimensions (for instance *affect, cognitive mechanism, health* etc.). It is hypothesised that such words are indicative of a user being witness and carrying an experiential authority pertaining to the event the user has posted. The classifier matches the tweets (which are not *re-tweets*) against the dictionary which may reflect the user's experience in time and space during an event. The tweets (with its user) which contain at least one such word from dictionary are marked as *eyewitness*. Now, all such users are further filtered through another check to narrow down to a more credible list of users. This can be achieved by analysing the social network/profile of the user, for example, we can guess users' location on the basis of geo-tagged information of the location in the profile, and also determine how reliable a profile is. Following the work by Castillo et al. [4], the *user-based* features in a data can be given a priority while assessing the authenticity of a user profile. The *user-based* features consider a user's characteristics based on the frequency of tweets, number of followers, and followee of the user. There are other features too, which can be considered while determining the trust value of the information, such as length of tweets, usage of special characters in text, number of re-tweets, and presence of a URL etc. A certain weight is given to the users who share the same location as that of the event, and even a higher weight is given to the users who have at least a threshold minimum number of followers, followees, and tweets (this threshold value can be based on a training data). Information propagated through credible users is seen to be as reliable, and the *user-based* features are indicative of users' reputation and hence credibility [4]. Once the user list and the corresponding tweets from an event are narrowed down to credible and authentic ones, the manual verification efforts can be greatly reduced.

The above described processes are likely to culminate into the following functionalities: *news search, popular/trending issues, related content (from archives), news by location, real time trends,* and *verified news.* This would result in a platform for a panoramic view of the real time data of the user generated content over various social media platforms, and provide a leverage to perform analysis and visualize the history of an event or related events.

Framework Evaluation

The effectiveness of the proposed location-based event detection framework must be evaluated through experiments on real data. Each module can be evaluated separately to measure suitability in the framework as a whole. For example, the event detection module can be tested using manually annotated event streams. The accuracy of the location extraction component can be similarly tested and evaluated using manually annotated data much like the TREC-2009 Blog Track where a retroactive event detection (RED) approach was taken [15], i.e., event streams are captured, annotated, and then used to train, validate, and test event detection accuracy. Metrics for accuracy range from classification accuracy (%) to information retrieval metrics such as *precision* and *recall.* For example, the evaluation of the Twitter-based event detection framework in [33] used precision, a measure of the proportion of relevant

---

[6] http://liwc.net/

events   detected   from   all   events   detected.   More   formally,

$$precision = \frac{|relevant\ events \cap retrieved\ events|}{|retrieved\ events|}.$$

Because location predictions can be approximate, metrics used to evaluate the module should be sufficiently flexible to reward high-precision, high-accuracy results and penalise low-precision results. Evaluation of the framework as a whole, on the other hand, is envisioned be more end-user focused, with hands-on use by real journalists. This stage of evaluation can only be carried out when a working prototype has been implemented comprising the necessary modules.

4   Impact and Conclusions

We have outlined a general framework for location-based event detection of UGC for journalists. The framework is part of a larger vision of Social Semantic Journalism which aims to address the many technological challenges facing journalists today by aiding in the sourcing, aggregation, filtering and verification of UGC for news reportage; the larger goal being to reduce the workload facing journalists today.

The proposed framework illustrates two potential scenarios that can be targeted. The first uses pre-specified location information as a filter for an event where that location may be known *a priori*, and lends itself to running stories or scheduled events. The second clusters all detected events by location, ultimately resulting in more manual work on the part of the journalist to inspect and identify event clusters of interest. The aim of the framework is to provide a platform and a tool to assist journalists as well as users to generate insights over several dimensions of UGC analysis.

The proposed framework is an attempt to knit together various level techniques which have till now been, to the best of our knowledge, researched or existing as modular services rather than being a part of a large information retrieval system. It further relies on the Semantic Web technologies to refine its inferences about locations and insights about the news/topic from several layers of mappings in Linked Open Data. However, the challenges which will need to be addressed and countered are pertaining to the immense magnitude of noise that comes with the information.

References

1. Allan, J., Papka, R., Lavrenko, V.: On-line new event detection and tracking. In: Proceedings of the 21st Annual International ACM SIGIR Conference on Research and Development in Information Retrieval, pp. 37–45. ACM (1998)
2. Backstrom, L., Sun, E., Marlow, C.: Find me if you can: improving geographical prediction with social and spatial proximity. In: Proceedings of the 19th International Conference on World Wide Web, pp. 61–70. ACM (2010)
3. Becker, H., Naaman, M., Gravano, L.: Beyond trending topics: Real-world event identification on Twitter. In: Proceedings of the Fifth International AAAI Conference on Weblogs and Social Media, ICWSM 2011 (2011)
4. Castillo, C., Mendoza, M., Poblete, B.: Information credibility on twitter. In: Proceedings of the 20th International Conference on World Wide Web, pp. 675–684. ACM (2011)

5. Chen, Y., Amiri, H., Li, Z., Chua, T.S.: Emerging topic detection for organizations from microblogs. In: Proceedings of the 36th International ACM SIGIR Conference on Research and Development in Information Retrieval, pp. 43–52. ACM (2013)
6. Diakopoulos, N., De Choudhury, M., Naaman, M.: Finding and assessing social media information sources in the context of journalism. In: Proceedings of the 2012 ACM Annual Conference on Human Factors in Computing Systems, pp. 2451–2460. ACM (2012)
7. DOMO, How Much Data is Created Every Minute? http://www.domo.com/blog/2012/06/how-much-data-is-created-every-minute/ (retrieved October 16, 2013)
8. Elloumi, S., Jaoua, A., Ferjani, F., Semmar, N., Besançon, R., Jaam, J., Hammami, H.: General Learning Approach for Event Extraction: Case of Management Change event. Journal of Information Sciences (2012)
9. Finkel, J.R., Grenager, T., Manning, C.: Incorporating non-local information into information extraction systems by gibbs sampling. In: Proceedings of the 43rd Annual Meeting on Association for Computational Linguistics, pp. 363–370. Association for Computational Linguistics (2005)
10. Heravi, B.R., McGinnis, J.: A Framework for Social Semantic Journalism. In: First International IFIP Working Conference on Value-Driven Social & Semantic Collective Intelligence (VaSCo), at ACM Web Science 2013, Paris, France (May 2013)
11. Hromic, H., Karnstedt, M., Wang, M., Hogan, A., Belák, V., Hayes, C.: Event Planning in a Stream of Big Data. In: LWA Workshop on Knowledge Discovery, Data Mining and Machine Learning (KDML). Workshop at LWA: Lernen, Wissen, Adaption (2012)
12. Jurgens, D.: That's What Friends are for: Inferring Location in Online Social Media Platforms Based on Social Relationships. In: Seventh International AAAI Conference on Weblogs and Social Media (2013)
13. Kijak, E., Gravier, G., Gros, P., Oisel, L., Bimbot, F.: HMM based structuring of tennis videos using visual and audio cues. In: Proceedings of the 2003 International Conference on Multimedia and Expo, ICME 2003, vol. 3, p. III-309. IEEE (2003)
14. Leetaru, K., Wang, S., Cao, G., Padmanabhan, A., Shook, E.: Mapping the global Twitter heartbeat: The geography of Twitter. First Monday 18(5) (2013) (n. pag. Web. August 9, 2013)
15. Macdonald, C., Ounis, I., Soboroff, I.: Overview of the TREC 2007 Blog Track. In: TREC, vol. 7, pp. 31–43 (2007)
16. Petrovic, S., Osborne, M., Lavrenko, V.: Streaming first story detection with application to twitter. In: Proceedings of NAACL, vol. 10 (2010)
17. Qian, X., Liu, G., Wang, H., Li, Z., Wang, Z.: Soccer video event detection by fusing middle level visual semantics of an event clip. In: Qiu, G., Lam, K.M., Kiya, H., Xue, X.-Y., Kuo, C.-C.J., Lew, M.S. (eds.) PCM 2010, Part II. LNCS, vol. 6298, pp. 439–451. Springer, Heidelberg (2010)
18. Rahmanzadeh Heravi, B., Boran, M., Breslin, J.: Towards Social Semantic Journalism. In: Sixth International AAAI Conference on Weblogs and Social Media (2012)
19. Ritter, A., Clark, S., Etzioni, O.: Named entity recognition in tweets: an experimental study. In: Proceedings of the Conference on Empirical Methods in Natural Language Processing, pp. 1524–1534. Association for Computational Linguistics (2011)
20. Ritter, A., Etzioni Mausam, O., Clark, S.: Open domain event extraction from Twitter. In: Proceedings of the 18th ACM SIGKDD International Conference on Knowledge Discovery and Data Mining (KDD 2012), pp. 1104–1112. ACM, New York (2012), doi:10.1145/2339530.2339704

21. Rosen, J.: Definition of Citizen Journalism (2008), http://www.youtube.com/watch?v=QcYSmRZuep4 (retrieved October 16, 2013)
22. Rui, Y., Gupta, A., Acero, A.: Automatically extracting highlights for TV baseball programs. In: Proceedings of the Eighth ACM International Conference on Multimedia, pp. 105–115. ACM (2000)
23. Sadilek, A., Kautz, H., Bigham, J.P.: Finding your friends and following them to where you are. In: Proceedings of the Fifth ACM International Conference on Web Search and Data Mining, pp. 723–732. ACM (2012)
24. Sakaki, T., Okazaki, M., Matsuo, Y.: Earthquake shakes Twitter users: real-time event detection by social sensors. In: Proceedings of the 19th International Conference on World Wide Web, pp. 851–860. ACM (2010)
25. Sayyadi, H., Hurst, M., Maykov, A.: Event detection and tracking in social streams. In: ICWSM 2009, pp. 311–314 (2009)
26. Scellato, S., Noulas, A., Lambiotte, R., Mascolo, C.: Socio-Spatial Properties of Online Location-Based Social Networks. In: ICWSM, vol. 11, pp. 329–336 (2011)
27. Shirazi, A., Rohs, M., Schleicher, R., Kratz, S., Müller, A., Schmidt, A.: Real-time nonverbal opinion sharing through mobile phones during sports events. In: Proceedings of the 2011 Annual Conference on Human Factors in Computing Systems, pp. 307–310. ACM (2011)
28. Sonderman, J.: One-third of adults under 30 get news on social networks now, http://www.poynter.org/latest-news/mediawire/189776/one-third-of-adults-under-30-get-news-on-social-networks-now/ (retrieved October 16, 2013)
29. Statisticbrain, Twitter Statistics, http://www.statisticbrain.com/twitter-statistics/
30. Unankard, S., Li, X., Sharaf, M.A.: Location-Based Emerging Event Detection in Social Networks. In: Ishikawa, Y., Li, J., Wang, W., Zhang, R., Zhang, W. (eds.) APWeb 2013. LNCS, vol. 7808, pp. 280–291. Springer, Heidelberg (2013)
31. Van Rijsbergen, C.J.: Information Retrieval (1979) ISBN 0-408-70929-4
32. Weng, J., Lee, B.: Event detection in Twitter. In: Proc. of ICWSM 2011, pp. 401–408 (2011)
33. Weng, J., Lee, B.-S.: Event Detection in Twitter. In: ICWSM (2011)
34. Xu, C., Zhang, Y.F., Zhu, G., Rui, Y., Lu, H., Huang, Q.: Using webcast text for semantic event detection in broadcast sports video. IEEE Transactions on Multimedia 10(7), 1342–1355 (2008)
35. Xu, M., Maddage, N.C., Xu, C., Kankanhalli, M., Tian, Q.: Creating audio keywords for event detection in soccer video. In: Proceedings of the 2003 International Conference on Multimedia and Expo, ICME 2003, Vol. 2, pp. II-281. IEEE (2003)
36. YouTube statistics (2013), http://www.youtube.com/yt/press/statistics.html (retrieved October 16, 2013)

Location-Aware Music Artist Recommendation

Markus Schedl[1] and Dominik Schnitzer[2]

[1] Johannes Kepler University Linz, Austria
http://www.cp.jku.at
[2] Austrian Research Institute for Artificial Intelligence, Vienna, Austria
http://www.ofai.at

Abstract. Current advances in music recommendation underline the importance of multimodal and user-centric approaches in order to transcend limits imposed by methods that solely use audio, web, or collaborative filtering data. We propose several hybrid music recommendation algorithms that combine information on the *music content*, the *music context*, and the *user context*, in particular integrating geospatial notions of similarity. To this end, we use a novel standardized data set of music listening activities inferred from microblogs (`MusicMicro`) and state-of-the-art techniques to extract audio features and contextual web features. The multimodal recommendation approaches are evaluated for the task of music artist recommendation. We show that traditional approaches (in particular, collaborative filtering) benefit from adding a user context component, geolocation in this case.

Keywords: Music Information Retrieval, Hybrid Music Recommendation, Personalization, Evaluation.

1 Introduction

Music Information Retrieval (MIR) is currently seeing a paradigm shift, away from system-centric perspectives towards user-centric approaches [4]. Incorporating user models and addressing user-specific demands in music retrieval and music recommendation is hence becoming more and more important.

Given the importance of user-centric and hybrid methods to MIR, we propose here several approaches that combine *music content*, *music context*, and *user context* aspects to build a music retrieval system [14]. Music content and music context are incorporated using state-of-the-art feature extractors and corresponding similarity estimators. The user context is addressed by taking into account *musical preference* and *geospatial data*, using a standardized collection of listening behavior mined from microblog data [13].

We make use of the best feature extraction and similarity computation algorithms currently available to model *music content* and *music context*. We then integrate these similarity models as well as a *user context* model into several novel user-aware music recommendation approaches that encompass all three modalities important to human music perception [14].

C. Gurrin et al. (Eds.): MMM 2014, Part II, LNCS 8326, pp. 205–213, 2014.
© Springer International Publishing Switzerland 2014

The remainder of the paper is organized as follows. Section 3 details the acquisition of the raw music (meta-)data, which serves as input to the feature extraction and data representation techniques presented in Section 4. Section 5 proposes several methods to incorporate geospatial information into music recommendation models. We further provide experimental evidence that adding a geospatial, user-aware component to a single-modality recommendation strategy is capable of improving recommendation results. Section 2 briefly reviews related literature. Finally, Section 6 draws conclusions and points to further research directions.

2   Related Work

Specific related work on geospatial music retrieval is very sparse, probably due to the fact that geospatially annotated music listening data is hardly available. Among the few works, Park et al. [7] use geospatial positions and suggest music that matches a selected environment, based on aspects such as ambient noise, surrounding, or traffic. Raimond et al. [10] combine information from different sources to derive geospatial information on artists, aiming at locating them on a map. Another possibility to link music to geographical information is presented by Byklum [2], who searches lyrics for geographical content like names of cities or countries. Zangerle et al. [17] use a co-occurrence-based approach to map tweets to artists and songs and eventually construct a music recommendation system. However, they do not take location into account.

On a more general level, this work relates to context-based and hybrid recommendation systems, a detailed review of which is unfortunately beyond the scope of the paper. A comprehensive elaboration, including a decent literature overview, can be found in [11].

3   Data Acquisition

The only standardized public data set of general microblogs, as far as we are aware of, is the one used in the TREC 2011 and 2012 Microblog tracks[1] [5]. Although this set contains approximately 16 million tweets, it is not suited for our task as it is not tailored to music-related activities, i.e. the amount of music-related posts is marginal.

We hence have to acquire multimodal data sets of *music items* and *listeners*, reflecting the three broad aspects of human music perception (*music content*, *music context*, and *user context*) [14]. Whereas the *music content* refers to all information that is derived from the audio signal itself (such as ryhthm, timbre, or melody), the *music context* covers contextual information that cannot be derived from the actual audio with current technology (e.g., meaning of song lyrics, background of a performer, or co-listening relationships between artists). The *user context.* encompasses all information that describe the listener. Examples range from musical education to spatiotemporal properties to physiological measures to current activities.

---

[1] http://trec.nist.gov/data/tweets

User Context. Only very recently a data set of music listening activities inferred from microblogs has been released [13]. It is entitled `MusicMicro` and is freely available[2], fostering reproducibility of social media-related MIR research. This data set contains about 600,000 listening events posted on `Twitter`[3]. Each event is represented by a tuple *¡twitter-id, user-id, month, weekday, longitude, latitude, country-id, city-id, artist-id, track-id¿*, which allows for spatiotemporal identification of listening behavior.

Music Content. Based on the lists of artist and song names in the `MusicMicro` collection, we gather snippets of the songs from `7digital`[4]. These serve as input to the music content feature extractors.

Music Context. To capture aspects of human music perception which are not encoded in the audio signal, we extract music-related web pages. Following the approach suggested in [15], we retrieve the top 50 web pages returned by the `Bing`[5] search engine for queries comprising the artist name[6] and the additional keyword "music", to disambiguate the query for artists such as "Bush", "Kiss", or "Hole".

In summary, we gathered raw data covering each of the three categories of perceptual music aspects [14]: *music content* (audio snippets), *music context* (related web pages), and *user context* (user-specific music listening events with spatiotemporal labels).

4 Data Representation

To represent the *music content*, we use state-of-the-art audio music feature extractors proposed in [8], which constitute a reference in music feature extraction for similarity-based retrieval. In particular, we extract auditory music features that combine various rhythmic information derived from the audio signal, e.g., "onset patterns" and "onset coefficients" (note onsets), with timbral features, e.g., "Mel Frequency Cepstral Coefficients" and the two handcrafted specialized descriptors for "attackness" and "harmonicness". The eventual output is pairwise similarity estimates between songs, which are later aggregated to the artist level.

We again employ a state-of-the-art technique to obtain features reflecting the *music context*. To describe the music items at the artist level, we follow the approach proposed in [15]. In particular, we model each artist by creating a "virtual artist documents", i.e. we concatenate all web pages retrieved for the artist. In accordance with findings of [12], we then use a dictionary of music-related terms (genres, styles, instruments, and moods) to index the resulting documents. From

---

[2] http://www.cp.jku.at/musicmicro
[3] http://www.twitter.com
[4] http://www.7digital.com
[5] http://www.bing.com
[6] Please note that issuing queries at the song level is not reasonable, as doing so typically yields only very few results.

the index, we compute term weights according to the best feature combination found in the large-scale experiments of [15]: TF_C3.IDF_I.SIM_COS, i.e. computing term weight vectors and artist similarity estimates according to Equations 1, 2, and 3, respectively for tf, idf, and *cosine similarity*; $f_{d,t}$ represents the number of occurrences of term t in document d, N is the total number of documents, $\mathcal{D}_t$ is the set of documents containing term t, F_t is the total number of occurrences of term t in the document collection, $\mathcal{T}_d$ is the set of distinct terms in document d, and W_d is the length of document d.

$$tf_{d,t} = 1 + \log_2 f_{d,t} \tag{1}$$

$$w_t = 1 - \frac{n_t}{\log_2 N}, \quad n_t = \sum_{d \in \mathcal{D}_t} \left(-\frac{f_{d,t}}{F_t} \log_2 \frac{f_{d,t}}{F_t} \right) \tag{2}$$

$$S_{d_1,d_2} = \frac{\sum_{t \in \mathcal{T}_{d_1,d_2}} (w_{d_1,t} \cdot w_{d_2,t})}{W_{d_1} \cdot W_{d_2}} \tag{3}$$

Rectifying the Similarity Space

Recent work has shown that "hubs" can be a problem in similarity spaces [9]. Hubs are data items that are frequently found among the nearest neighbors of many other data items, but cannot have all of these data items as nearest neighbors themselves. In recommendation systems, such hubs are usually undesired, because they are unjustifiably recommended much more frequently than any other data items, which strongly hinders serendipitous encounters, hence harms user satisfaction. To alleviate this problem, [16] suggests an approach called "mutually proximity" (MP), which rectifies high-dimensional similarity spaces in which the data set itself has low intrinsic dimensionality. This MP approach proved particularly beneficial for text features and music audio features, as shown in [16]. In the case of the audio features used here, this normalization is already included in the employed feature extraction algorithm. For the web-based music context features, we apply MP on the similarity matrix to suppress the formation of hubs in the ultimate recommendation approach.

Availability of the Data Sets

All components of the data set used in this paper are publicly available to allow researchers reproduce the results reported. The sole exception is the actual audio content of the songs under consideration. We cannot share them due to copyright restrictions. However, we provide identifiers by means of which corresponding 30-second-clips can be downloaded from 7digital. The MusicMicro set of geolocalized music listening events from microblogs [13] can be downloaded as well[7]. All other data (audio feature vectors and artist term weight vectors) can be shared upon request to the first author.

---

[7] http://www.cp.jku.at/musicmicro

5   Music Recommendation Models

Hybrid music retrieval and recommendation approaches, which base their similarity computation on more than one modality, are frequently suggested in literature, e.g. [3,1,4,14,6]. A systematic evaluation of approaches that integrate state-of-the-art music content (audio) and music context (web) similarity measures was only conducted very recently, though [omitted-due-to-review]. One finding of this study is that including a small amount of audio features in an otherwise solely web-based similarity measure (or vice versa) considerably improves retrieval performance. Given audio similarities $asim(i, j)$ and web similarities $wsim(i, j)$ between two artists i and j, Equation 4 shows the hybrid similarity model that performed best for artist retrieval according to [omitted-due-to-review][8]. It is hence used in the following as music content/music context-based model (CB).

$$sim(i, j) = 0.15 \cdot asim(i, j) + 0.85 \cdot wsim(i, j) \qquad (4)$$

Building user-aware recommendation systems obviously requires a user model. In our case, each user u is modeled by the set of artists $UM(u)$ he or she listened to. Based on this simple model, we implement the following recommendation strategies:

- CB: the hybrid (music content and music context) music retrieval model according to Equation 4
- CF: a standard user-based collaborative filtering model
- GEO: a model solely based on geospatial proximities
- GEO-CF: a model that combines GEO and CF by taking the union of the two recommended artist sets
- CF-GEO-LIN and CF-GEO-GAUSS: CF-based models that weight users according to their geospatial distance to the seed user, using either a linear or exponential geospatial distance measure
- RB: a random baseline model

In the CB model, the hybrid music similarity function (Equation 4) is used to determine the artists closest to $UM(u)$, which are then recommended. In the CF model, the K users closest to u are determined (using the Jaccard index between the user models), and the artists listened to by these nearest users are recommended. The GEO model defines user distance solely via the geospatial distance between users. To this end, we first compute a centroid of each user u's geospatial listening distribution $\mu_u[\lambda, \varphi]$[9]. For recommendation, the artists of the seed user's closest neighbors, measured via geodesic distance between the centroids, are suggested. The GEO-CF model simply recommends the union of the GEO and CF model's output.

---

[8] Audio similarities are aggregated on the artist level by computing the minimum of the distances between all pairs of tracks by i and j.

[9] It is common to denote longitude by λ and latitude by φ.

To integrate geospatial information into the CF model (CF-GEO-LIN and CF-GEO-GAUSS), we use the normalized geodesic distance $gdist(u,v)$ (Equation 5) between the seed user u and each other user v to weight the distance based on the user models. To this end, we propose two different weighting schemes: linear weighting and weighting according to a Gaussian kernel around $\mu_u[\lambda, \varphi]$. We eventually obtain a geospatially modified user similarity $sim(u,v)$ by adapting the Jaccard index between $UM(u)$ and $UM(v)$ via geospatial linear or Gauss weighting, according to Equation 6 (CF-GEO-LIN) or Equation 7 (CF-GEO-GAUSS), respectively. We recommend the artists listened to by u's nearest users v.

$$gdist(u,v) = \arccos\left(\sin(\mu_u[\varphi]) \cdot \sin(\mu_v[\varphi]) + \cos(\mu_u[\varphi]) \cdot \right.$$
$$\left. \cos(\mu_v[\varphi]) \cdot \cos(\mu_u[\lambda] - \mu_v[\lambda])\right) \cdot$$
$$\max(gdist)^{-1} \tag{5}$$

$$sim(u,v) = J(UM(u), UM(v)) \cdot gdist(u,v)^{-1} \tag{6}$$

$$sim(u,v) = J(UM(u), UM(v)) \cdot \exp(-gdist(u,v)) \tag{7}$$

For comparison, we further implemented a random baseline model (RB) that randomly picks K users from the filtered user set (filtering with respect to the parameter τ, see below) and recommends the artists they listened to. In addition, we ensure that all algorithms recommend approximately the same number of artists on average, to make results comparable. To this end, we use the number of artists recommended by the CF approach as baseline and adapt the parameters of the other approaches in such a way that they output a similar number of artists.

5.1 Experimental Setup

In order to ensure sufficient artist coverage of users, we evaluate our models using different thresholds τ for the minimum number of unique artists a user must have listened to in order to include him or her in the experiments. We vary τ between 30 and 200 using a step size of 10. Denoting as U_τ the number of users in the MusicMicro data set with equal or more than τ unique artists, $U_{30} = 881$, $U_{100} = 32$, and $U_{200} = 5$. We perform U_τ-fold leave-one-out cross-validation for each value of τ.

5.2 Results

Figure 1 shows accuracies for $K = 5$ nearest neighbors and $\tau = [30 \dots 200]$. We can see that all approaches significantly outperform the random baseline. Results for the CB approach and the CF approaches show an inverse characteristics over τ, which suggests a combination of both. As this is not trivial, it will be part of future work. The reason for CB outperforming CF for large numbers of τ is obviously the limited diversity among the K nearest neighbors (for $\tau > 150$), which seriously hampers CF-based approaches. Hence, "power users" benefit more from CB approaches than from CF approaches. The GEO approach performs rather poorly, being quite close to the baseline in most settings. Creating a recommender solely on location information hence seems not beneficial.

The hybrid approaches that use geospatial weighting to adapt CF-based similarities do not outperform the CF only approach. Possible explanations are (i) that using the centroid of a user's listening positions as summary of his or her overall location is too coarse a description, in particular for users who travel a lot, and (ii) that geodesic distance alone frequently does not reflect cultural distance, which seems more important for the recommendation task at hand. For instance, same geodesic distances between two users can have very different meaning in regions with different population density (e.g., Hong Kong versus Russia) or at cultural borders (North Korea versus South Korea, Spain versus Morocco, etc.). Future work will take a closer look at these aspects and investigate whether incorporating political and cultural information will improve results. In contrast, the hybrid approach GEO-CF that takes the set union of GEO- and CF-based recommendations performs superiorly. Considering both similar users and similar locations as equally important (GEO-CF) thus outperforms geospatial weighting of user similarities performed in CF-GEO-LIN and CF-GEO-GAUSS.

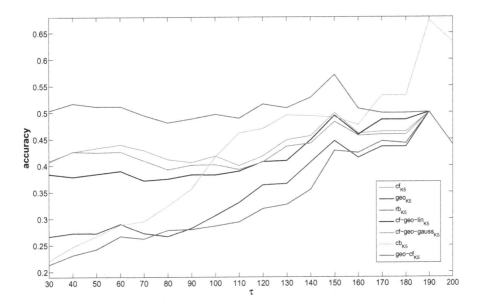

Fig. 1. Accuracy plots for different values of τ and $K = 5$

6 Conclusions and Outlook

We presented different hybrid music recommendation approaches that use state-of-the-art music content (audio) and music context (web) features, as well as contextual user information, more precisely a data set of geolocated music listening activities mined from microblogs. Experimental results indicate that hybrid (music content/music context/user context) strategies, in general, are capable of outperforming approaches using only one data source. However, the question of how to combine the different modalities is crucial. Among the hybrid approaches, we found that recommending the set union of a user-based collaborative filtering recommender and of a recommender based on geospatial proximity of users performed superior.

Future work includes investigating other data sources related to the user context, for instance, listening time or demographics. Furthermore, we presume that refining the notion of spatial proximity by taking into account political and cultural borders, for instance, defined by language or religion, will lead to better prediction accuracy, thus in turn to better user-aware music recommendation systems.

Acknowledgments. This research is supported by the Austrian Science Funds (FWF): P22856 and P24095, and by the EU FP7 project PHENICX: 601166.

References

1. Bogdanov, D., Serrà, J., Wack, N., Herrera, P., Serra, X.: Unifying Low-Level and High-Level Music Similarity Measures. IEEE Transactions on Multimedia 13(4), 687–701 (2011)
2. Byklum, D.: Geography and Music: Making the Connection. Journal of Geography 93(6), 274–278 (1994)
3. Coviello, E., Chan, A.B., Lanckriet, G.: Time Series Models for Semantic Music Annotation. IEEE Transactions on Audio, Speech, and Language Processing 19(5), 1343–1359 (2011)
4. Liem, C., Müller, M., Eck, D., Tzanetakis, G., Hanjalic, A.: The Need for Music Information Retrieval with User-centered and Multimodal Strategies. In: Proc. MIRUM, Scottsdale, AZ, USA (2011)
5. McCreadie, R., Soboroff, I., Lin, J., Macdonald, C., Ounis, I., McCullough, D.: On Building a Reusable Twitter Corpus. In: Proc. SIGIR, Portland, OR, USA (2012)
6. McFee, B., Lanckriet, G.: Heterogeneous Embedding for Subjective Artist Similarity. In: Proc. ISMIR, Kobe, Japan (2009)
7. Park, S., Kim, S., Lee, S., Yeo, W.S.: Online Map Interface for Creative and Interactive MusicMaking. In: Proc. NIME, Sydney, Australia (2010)
8. Pohle, T., Schnitzer, D., Schedl, M., Knees, P., Widmer, G.: On Rhythm and General Music Similarity. In: Proc. ISMIR, Kobe, Japan (2009)
9. Radovanović, M., Nanopoulos, A., Ivanović, M.: Hubs in Space: Popular Nearest Neighbors in High-dimensional Data. The Journal of Machine Learning Research, 2487–2531 (2010)

10. Raimond, Y., Sutton, C., Sandler, M.: Automatic Interlinking of Music Datasets on the Semantic Web. In: Proc. WWW: LDOW Workshop, Beijing, China (2008)
11. Ricci, F., Rokach, L., Shapira, B., Kantor, P.B. (eds.): Recommender Systems Handbook. Springer (2011)
12. Schedl, M.: #nowplaying Madonna: A Large-Scale Evaluation on Estimating Similarities Between Music Artists and Between Movies from Microblogs. Information Retrieval 15, 183–217 (2012)
13. Schedl, M.: Leveraging Microblogs for Spatiotemporal Music Information Retrieval. In: Serdyukov, P., Braslavski, P., Kuznetsov, S.O., Kamps, J., Rüger, S., Agichtein, E., Segalovich, I., Yilmaz, E. (eds.) ECIR 2013. LNCS, vol. 7814, pp. 796–799. Springer, Heidelberg (2013)
14. Schedl, M., Flexer, A.: Putting the User in the Center of Music Information Retrieval. In: Proc. ISMIR, Porto, Portugal (2012)
15. Schedl, M., Pohle, T., Knees, P., Widmer, G.: Exploring the Music Similarity Space on the Web. ACM Transactions on Information Systems 29(3) (July 2011)
16. Schnitzer, D., Flexer, A., Schedl, M., Widmer, G.: Local and Global Scaling Reduce Hubs in Space. Journal of Machine Learning Research 13, 2871–2902 (2012)
17. Zangerle, E., Gassler, W., Specht, G.: Exploiting Twitter's Collective Knowledge for Music Recommendations. In: Proc. WWW: #MSM Workshop, Lyon, France (2012)

Task-Driven Image Retrieval Using Geographic Information

Peixiang Dong[1], Kuizhi Mei[1], Ji Zhang[1], Hao Lei[1], and Jianping Fan[2]

[1] Institute of Artificial Intelligence and Robotics, Xi'an Jiaotong University, China
[2] Department of Computer Science, School of Information Science and Technology,
Northwest University, Xi'an 710069, China

Abstract. When large-scale online geo-tagged images come into view, it is important to leverage geographic information for web image retrieval. In this paper, a geo-metadata based image retrieval system is proposed using both textual tags and visual features. This image retrieval system is especially useful for tourism related tasks such as tourism recommendation and tourism guide. First, the requested image retrieval task is classified into one of the three different types according to the retrieval purpose, and then it can be handled with specific method. Second, a WordNet hierarchy based semantic similarity is developed to measure the similarity between different cities. This semantic similarity is somehow consistent with the visual similarity. Finally, a high-level image representation method is proposed to narrow the semantic gap between the low-level visual features and high-level image concepts. The proposed algorithm is evaluated on an image set which is consisted of totally 177,158 images of 120 most popular cities all over the world collected from Flickr, and the experiments have provided very positive results.

Keywords: content-based image retrieval, geographic information, semantic city similarity, high-level image representation.

1 Introduction

In recent years, as digital cameras and smart phones become more affordable and widespread, high-quality digital images are growing exponentially on the Internet. Thus there is an increasing need of new techniques to support more effective image search. Most existing content-based image retrieval (CBIR) algorithms mainly focused on the low-level visual features [1], but it is doubtful for their effectiveness because there is always a semantic gap between the low-level visual features and the high-level image concepts [2]. The keyword-based image retrieval is the most popular method for current image search tasks, and it is more suitable for most users. However, the main obstacle for supporting keyword-based image retrieval is that users may not be able to find the most suitable keywords to describe their image needs precisely or they may not even know what to look for (i.e., *I don't know what I am looking for, but I'll know when I find it*) [3].

C. Gurrin et al. (Eds.): MMM 2014, Part II, LNCS 8326, pp. 214–226, 2014.

owner: 7150652@N02
datetaken: 2011-12-14
tags: newyork city newyork
church harlem manhattan
methodist united methodist
methodist church
latitude: 40.811827
longitude: -73.946753
views: 82

owner: 67004864@N02
datetaken: 2010-12-08
tags: china beijing peking
great wall of china
latitude: 40.005527
longitude: 116.600646
views: 22

Fig. 1. Examples of multiple textual information for Flickr images

A good news is that there is a large amount of textual tagged images from popular web album services such as Flickr [4]. As shown in Fig. 1, people may not only share the photos, but also provide some relevant information, such as the owner, taken date, tags, taken location, and views. In particular, the photo's geographical information, which is also called *geotag*, plays an important role in the domain of image retrieval. The geographic information could be classified into two different types: the text keywords (especially the city names, e.g., "NewYork" and "Beijing") and the GPS coordinates(i.e., the numerical values of the latitude and the longitude).

With the emergence of large amounts of geographically tagged photo images, there has been much work employing geographic information to support image management [5–10]. However, in order to perform effective retrieval of geo-tagged images, the traditional approaches suffer from at least three main challenges. 1) The purely low-level visual features are not sufficient for the web image retrieval tasks due to the diversity of the images on the Internet. Some kind of high-level visual features would be much more powerful. 2) The weak and subjective textual tags are not able to describe an image accurately. 3) The geotag of a photo image is measured at the position where the picture is taken rather than where it is located.

In this paper, a task-driven image retrieval system is presented leveraging the geographic information. To evaluate the proposed system, we build a geo-tagged image set by collecting 177,158 photo images with geotags (both geographic keywords and GPS coordinates) from Flickr. The main contributions of this paper are listed as follows:

– leveraging the geographic information, a requested image search task is first classified into one of the three types due to the image search purpose and then handled by a specific method. And this is noted as "task-driven image retrieval" in this paper;
– based on the WordNet [11, 12] hierarchy, a "city similarity" that measures the semantic relationship between two cities is first proposed in this paper;
– based on the similarity with a number of representative categories, a high-level image representation method is developed to narrow the semantic gap between low-level visual features and high-level image concepts.

The rest of this paper is organized as follows. Section 2 reviews the most related work briefly. Section 3 presents the detail explanation of the three types of image retrieval tasks. The semantic city similarity measurement and the high-level image representation method are proposed in section 4. Section 5 describes the experimental results of our task-driven image retrieval system and we conclude this paper in section 6.

2   Related Work

In this section, we provide a brief review of the most related work with this paper from two aspects: *high-level image representation* and *geo-tagged image retrieval*.

2.1   High-Level Image Representation

Applying low-level visual features with the BoW(bag-of-words) image representation method has been proven to be a powerful tool for the object recognition and image classification tasks. However, the pixels or even local image patches carry little semantic meanings [13], so the high-level image representations would be more suitable for the image retrieval task. Lampert et al. [14] have proposed high-level attributes that allow the transfer of knowledge between object categories for object detection. This description consists of arbitrary semantic attributes, such as shape, color or even geographic information. However, these human-specified attributes could not provide universal descriptions for all categories. Li et al. [13] have proposed a so called "Object Bank" for high-level image representation, where an image is represented as a response map of a large number of pre-trained generic object detectors. Torresani et al. [15] have proposed a new descriptor for images using classemes. The descriptor is the output of a large number of weakly trained object category classifiers on the image. However, this method could not provide accurate descriptions for an image because of the low accuracy of the weakly trained object category classifiers.

2.2   Geo-tagged Image Retrieval

There has been many pioneer work employing geotags for image annotation, image collection, and tourist recommendation. Hays et al. [5] have proposed a system to estimate geographic information from a single image. A database was built over 6 million GPS-tagged images from the Flickr online photo collection, and various types of visual features are compared in that work. Kennedy et al. [16] have proposed a similar work with [5]. However, they used a combination of textual tags and image features to generate representative sets of images for location-driven features and landmarks. Both the global features (i.e., Gabor textures and color histograms) and local features (i.e., SIFT features) are combined to generate visual features. Crandall et al. [17] have used textual tags and image features for content analysis and geospatial information for structural analysis. The mean shift algorithm is employed to find the popular places at which people take photos. A significant characteristic is that they consider the geo-tagged image collection tasks on a very large database (about 35 million images collected from Flickr). Liang et al. [6] have proposed a method for annotation of landmark photos via learning textual tags and visual features of landmarks from geo-tagged landmark photos. Wikipedia articles are used to filter out the non-landmark tags. Cao et al. [7] have proposed a system for the task of tourism recommendation. This tourism recommendation system is built based on the representative tags and images with corresponding GPS locations. Yang et al. [8] have proposed a method to identify high quality points of interest using collections of geo-tagged and time-tagged photos. They proposed a self-tuning approach based on the cut cost similarity to eliminate the effect of parameters from spectral clustering.

Different from all these previous work, in this paper, we leverage both a high-level visual representation method and semantic city similarity to support geo-tagged image retrieval. The proposed high-level image representation method is developed by using a two-layer unsupervised learning framework, while our semantic city similarity is measured by the WordNet distance of textual tags for different cities. The diagram of the geo-tagged image retrieval system is illustrated in Fig. 2.

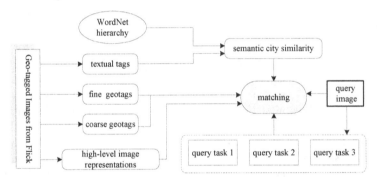

Fig. 2. The diagram of our geo-tagged image retrieval system

3   Three Types of Image Retrieval Tasks

There are a large amount of public geo-tagged images on Flickr, and there are many existing work on geo-tagged image collection, image summarization, and tourism recommendation. In this paper, for a requested image retrieval task, we first classify it into one of the three types according to the image search purpose. And then we handle each type of image retrieval task with specific method rather than a universal method to enhance the performance of the entire image retrieval system. For a given query image, the three types image retrieval tasks are defined as 1)*searching images all over the world*, 2)*searching images in the most related cities*, and 3)*searching images in a small geographic region around the query*.

Searching Images All over the World. This is the traditional content-based image retrieval problem. For this type of image retrieval task, the users are provided only the query image and there is no additional geotag information. Traditionally, the low-level visual features of the query image are first extracted and then matched with the pre-built image set to find the most similar pattern. The most popular low-level features include color histograms, GIST descriptor [18], SIFT descriptor [19], and et al. However, because the web images are usually diverse and weakly tagged, the low-level visual features are too noisy for the web image retrieval task. In this paper, a geographically informative high-level image representation method is developed leveraging the visual similarity with the representative categories.

Searching Images in the Most Related Cities. This task is very useful in tourism recommendation system. In a tourism recommendation system, once a user submits a query image with coarse geotags (i.e., the keyword geotags specially the city names), we can recommend him/her with the potential cities (and also the related images) which he/she would like. In order to achieve this task efficiently, we calculate the "city similarity" between different cities first. And then for a given query image with coarse geotags, we just search the similar photo images in the *potential cities*.

The statement "city similarity" is not very common in the computer vision community, so we make some interpretations here. It is well accepted that some cities may "look like" each other due to at least one of the following reasons: 1)they are geographical neighbors, 2)they are culturally connected for a long history, 3)they have the similar development trajectories, and et al. These *similar cities* are similar with each other in terms of architectural styles, eating patterns, or natural landscapes, so the photo images taken from these cities would be overall more visually similar than the photo images taken from other cities. For example, for three metropolises Paris, London and Beijing, it is usually that one would think London looks more like Paris than Beijing, so we can declare that Paris is more *similar* with London than Beijing in terms of visual appearance.

Based on these observations, we argue that for a given query image, it is more likely to find its visually similar images from the similar cities of which the query image belongs. And also this strategy could speed up the geo-tagged image retrieval process significantly compared with the traditional image retrieval systems. In order to measure the similarity between different cities, the WordNet hierarchy is employed to calculate the semantic city similarity using the textual tags.

Searching Images in a Small Geographic Region around the Query Image. This task is very useful for tourism guide. Imaging the situation that one is standing in front of an unknown landmark during his/her travel in a strange city, how could he/she gather as much information as possible about this landmark quickly? One of the most convenient approaches is that first capturing a photo of the landmark and then using it as a query to search the similar images in the same place to obtain the details. For this image retrieval task, the query images are specified with fine geotags, i.e., the GPS coordinates, which can be easily obtained benefiting from the widespread use of location-aware cameras and smart phones.

It is also worth noting that the obtained GPS coordinates by the location-aware cameras or smart phones are not the accurate coordinates of the landmark locations mainly because of the two issues: 1)the GPS, cellular or Wi-Fi networks are usually noised especially in the down town due to the surrounding buildings and 2)the geographic location of the photo image is measured at the location where the picture is captured rather than the location where the landmark is located. In order to overcome this problem, we should explicitly indicate a proper range of GPS coordinates in which the image retrieval process would operate. In this paper, for a given image retrieval task, we first classify it to a proper type according to the purpose and additional geographic information (i.e., non-geotags, coarse geotags, and fine geotags), and then handle it with a corresponding method.

4 The Proposed Method

4.1 Semantic City Similarity Measurement

In order to find the visually similar cities as described in section 3, we should design a city similarity measurement first. Because the photo images gathered from each city are visually diverse, it is not a good idea to use these cluttered images to measure the similarity between different cities directly. On the other hand, the abundant textual tags of the photo images could provide rich semantic information to describe a city. Compared with the low-level visual features, the semantic textual tags could be considered as the high-level city descriptors. Leveraging the semantic relationship among the textual tags of different cities, the semantic similarity between different cities could be determined reasonably.

For two given cities C_i and C_j, their semantic similarity $\gamma(C_i, C_j)$ can be determined by

$$\gamma(C_i, C_j) = \frac{1}{|N_i| |N_j|} \sum_{p \in T_i} \sum_{q \in T_j} \rho(p, q) \tag{1}$$

where T_i and T_j are the set of textual tags for C_i and C_j, respectively, N_i and N_j are the number of tags in T_i and T_j, p and q are the element of T_i and T_j. The semantic similarity between p and q is noted as $\rho(p, q)$. To calculate the value of ρ, we need to construct a hierarchy of all the textual tags.

One of the most popular semantic networks for English words is WordNet, which is a large lexical database. Nouns, verbs, adjectives and adverbs are grouped into sets of cognitive synonyms (synsets), each expressing a distinct concept. The WordNet database contains 155,287 words organized in 117,659 synsets for a total of 206,941 word-sense pairs. In this paper, the similarity computation software module from [20] is used to calculate the semantic similarities between different tags from the WordNet hierarchy.

Here we will make some intuitive explanations about our city similarity. When we say that "a city looks like the other one", we implicitly indicate that the two cities are visually similar. However, the photo images captured from different cities are too diverse to measure the city similarity directly. Fortunately, the photo images on the Internet are usually shared with rich textual tags. And we argue that the visually similar cities would share some synonymous (or even same) textual tags, so we can leverage the WordNet hierarchy to calculate the similarities among different cities.

It is worth noting that during our experiments, tags occurring twice or fewer times are ignored. This strategy could filter out the noisy tags and enhance the discrimination of the textual tags significantly. Leveraging the city similarity, the image retrieval process will be performed on only the most related cities rather than the whole image set once the coarse geotags of query images are obtained.

4.2   High-Level Image Representation Method

The low-level features are usually conducted on the salient points or patches of images, which carry little semantic meanings. Multiple types of low-level visual features are usually extracted to achieve more sufficient characterization of various visual properties of the images. Because the distributions of the images could be very sparse, the low-level visual descriptors could be very diverse in the high dimensional feature space(i.e., visual ambiguity). As a result, a high-level image representation method would be more effective to characterize the diverse and weakly tagged web images accurately.

Motivated by the pioneer work such as Object Bank and classemes, a high-level image representation method is developed in this paper. Our algorithm is built upon the idea that *a certain object category could be represented by the visual similarities with a number of other representative categories*. The details of the image representation algorithm are specified as follows.

1) Low-level visual feature extraction. We first extract the low-level visual features from raw images, including SIFT, color histogram, GIST, and et al. In this paper, the dense SIFT is employed as the low-level features due to its widespread utilization in computer vision community.
2) Visual vocabulary construction. In this paper, we perform K-means clustering method on a random subset of 10 million SIFT descriptors to construct a visual vocabulary with 1000 visual words.
3) Bag of words image representation. For each image instance, the raw SIFT descriptors are quantized into a visual word using the visual vocabulary. As a result, the 1000-bin codeword histogram is used for image content representation, so each image instance I could be noted as $I = (h_1, h_2, \cdots, h_{1000})$. This traditional BoW image representation method is widely used in computer vision community, however, the low-level visual features carry little semantic meanings.
4) Representative categories determination. Different from the image classification, there are no accurate labels for the web images which are used for image retrieval. However, we can perform some unsupervised clustering method (e.g., K-means) on the image instances which are represented by 1000-bin

histograms to obtain a number of representative cluster centers. Assuming that there are M cluster centers, which are noted as $C_1, C_2, \cdots, C_M$, and each cluster center C_i is noted by a 1000-bin histogram $C_i = (h_{i1}, h_{i2}, \ldots, h_{i1000})$. The M cluster centers can be regarded as M object categories although we do not (indeed, we need not to) know exactly the specific visual content (i.e., semantic labels) for each category.

5) High-level image representation. After the M cluster centers (i.e., visual categories) are obtained, each image instance I is represented by using the M similarity values between I and the M categories, that is

$$\widetilde{I} = (\rho(I, C_1), \rho(I, C_2), \ldots, \rho(I, C_M)) \tag{2}$$

and

$$\rho(I, C_i) = \exp(-\frac{\|I - C_i\|^2}{\sigma}) \tag{3}$$

where σ is setting to the mean values of the Euclidean distances among the image instances and the categories.

The most notable characteristic of our high-level image representation method is that an image instance is represented by its similarity with a series of categories. And these categories are learned automatically during the unsupervised clustering process rather than manually determined before the learning process. In other words, we do not know exactly what these categories are, but we know that these categories are discriminative to represent other image instances.

This high-level image representation method is largely inspired by the cognitive mechanism of human beings. For example, when we try to describe an unknown object, we usually leverage the assistance of some known categories which are visually similar with the unknown object. We often use such statements like "*it looks like ..., and it also looks like...*", and in most cases these statements are much more effective than describing the visual content of the unknown object itself directly. Based on these observations, our high-level image representation method could carry much more semantic meanings than the low-level visual features, thus this representation could also be beneficial to narrow the *semantic gap* between the visual features and the image concepts. In addition, the architecture that used for the image feature representation is very simple and it is consisted of only two unsupervised clustering layers. Benefitting from the low computational cost of the network, this image representation method is very practical for large-scale web image retrieval tasks.

5 Algorithm Evaluation and Experimental Results

In this section, we present several experiments to demonstrate the superiority of our proposed methods.

5.1   Data Preparation

We collect photo images from Flickr using Flickr APIs which are wrapped by IM2GPS scripts [5]. In our image data set, there are totally 177,158 photo images from the 120 most popular cities all over the world. The cities are listed in Table 1. In order to illustrate the geographical distribution of the images in our data set, both the global view and the enlarged views of some representative cities are shown in Fig. 3.

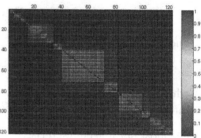

Fig. 3. The geographical distribution of our image set

Fig. 4. The semantic city similarity values between different cities

5.2   Experiments on Semantic City Similarity

According to Eq. (1), the semantic city similarity matrix is illustrated in Fig. 4. We argue that the semantic city similarity is somehow consistent with the visual similarity, although there is a semantic gap between them. So it is reasonable to leverage the semantic city similarity to recommend the potential tourist cities to the user once he/she submit a query image with geographic information for the tourism purpose. The similarity values between different cities are normalize to the range [0, 1]. For example, the most semantic similar cities with *London* are *Los Angeles*, *Madras*, *Madrid*, and *Melbourne*.

5.3   The Proposed Image Retrieval System

An image retrieval system is developed in this section, which is shown in Fig. 5. From the GUI panel of the proposed system, one can see the query image, the visual representation of the query image, the search results, and also the comments of any mouse-clicked image. The original image retrieval task is issued by this system with a specific method according to the type of the task. Some representative results on different image retrieval tasks are illustrated in Fig. 5, i.e., 1) for the general task, the image retrieval process is performed on the whole image set (as shown in Fig. 5(a)), 2)for the tourism recommendation task (i.e., searching images in the most related cities), the city name is provided with the query image, and the image search process is performed only on the images

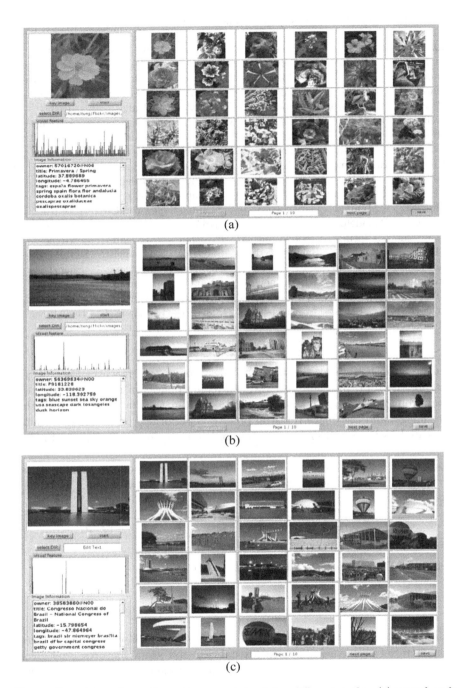

Fig. 5. Experimental results on image retrieval for different tasks: (a)general task, (b)searching only in the most semantic related cities, and (c)searching in 5km away from where the query image was captured

Table 1. The 120 most popular photographed cities on the Earth

Ahmedabad	Alexandria	Ankara	Athens	Atlanta	Baghdad
Bandung	Bangalore	Bangkok	Barcelona	Beijing	BeloHorizonte
Berlin	Bogota	Bombay	Boston	Brasilia	BuenosAires
Busan	Cairo	Calcutta	Cartagena	Chengdu	Chennai
Chicago	Chittagong	Chongqing	Cordoba	Dallas	Delhi
Detroit	Dhaka	Dongguan	Guadalajara	Guangzhou	Guiyang
Hanoi	Harbin	HoChiMinhCity	HongKong	Houston	Hyderabad
Istanbul	Jakarta	Johannesburg	Karachi	Khartoum	Kinshasa
Kolkata	KualaLumpur	Lagos	Lahore	Lima	London
LosAngeles	Madras	Madrid	Manila	Medellin	Melbourne
Mendoza	MexicoCity	Miami	Milan	Monterrey	Montreal
Moscow	Mumbai	Nagoya	Nanjing	NewYorkCity	Osaka
Paris	Philadelphia	Phoenix	PortoAlegre	Pune	PuntadelEste
Pyongyang	Recife	RiodeJaneiro	Riverside	Riyadh	Rome
Ruhr	Saigon	SaintPetersburg	Salvador	SanFrancisco	SantaMarta
Santiago	SaoPaulo	Seoul	Shanghai	Shenyang	Shenzhen
Singapore	Surat	Sydney	Taipei	Tehran	Tianjin
Tokyo	Toronto	WashingtonDC	Wuhan	Yangon	Alhambra
Capri	Caracas	EasterIsland	Florence	Galapagos	Heidelberg
Nikko	Orlando	Pisa	Pompei	Seattle	Venice

from the 5 most related cities with the query image (as shown in Fig. 5(b)), and 3)for the tourism guide task (i.e., searching images in a small geographic region around the query image), the GPS coordinates are provided with the query image, and the image retrieval process is performed on the images which are located 5km away from the location of the query image (as shown in Fig. 5(c)).

Let α_1, β_1 and α_2, β_2 be the latitudes and longitudes of two points on the Earth, then the geographical distance d_{12} between these two points can be formulated as

$$d_{12} = 2r \arcsin(\sqrt{\sin^2(\frac{\alpha_1 - \alpha_2}{2}) + \cos\beta_1 \cos\beta_2 \sin^2(\frac{\beta_1 - \beta_2}{2})}) \qquad (4)$$

where $r = 6378.137$km is the radius of the Earth.

6   Conclusion

In this work, a novel task-driven system is proposed to deal with the geo-tagged image retrieval problem using both textual tags and visual features. Any given image retrieval task will be handled by specific method due to the retrieval purpose, and this system is especially useful for tourism recommendation or tourism guide task. A WordNet hierarchy based similarity is proposed to measure the semantic (and further the visual) similarity between different cities to support effective tourism recommendation. The fine geographic information (i.e., GPS coordinates) is leveraged for tourism guide. Based on the visual category similarity, a high-level image representation method is developed to narrow the semantic gap between visual features and image concepts. The experiment on a real world image set provides very positive results.

Acknowledgements. This research is partly supported by National Natural Science Foundation of China under Grant No.(61272285, 61375023), National High-tech R&D Program of China under Grant No. 2012AA010904 and Ministry of education program for New Century Excellent Talents(NCET-11-0427).

References

1. Smeulders, A.W., Worring, M., Santini, S., Gupta, A., Jain, R.: Content-based image retrieval at the end of the early years. IEEE Transactions on Pattern Analysis and Machine Intelligence 22(12), 1349–1380 (2000)
2. Fan, J., He, X., Zhou, N., Peng, J., Jain, R.: Quantitative characterization of semantic gaps for learning complexity estimation and inference model selection. IEEE Transactions on Multimedia 14(5), 1414–1428 (2012)
3. Fan, J., Keim, D.A., Gao, Y., Luo, H., Li, Z.: Justclick: Personalized image recommendation via exploratory search from large-scale flickr images. IEEE Transactions on Circuits and Systems for Video Technology 19(2), 273–288 (2009)
4. http://www.flickr.com
5. Hays, J., Efros, A.A.: IM2GPS: estimating geographic information from a single image. In: CVPR, pp. 1–8. IEEE (2008)
6. Liang, C.-K., Hsieh, Y.-T., Chuang, T.-J., Wang, Y., Weng, M.-F., Chuang, Y.-Y.: Learning landmarks by exploiting social media. In: Boll, S., Tian, Q., Zhang, L., Zhang, Z., Chen, Y.-P.P. (eds.) MMM 2010. LNCS, vol. 5916, pp. 207–217. Springer, Heidelberg (2010)
7. Cao, L., Luo, J., Gallagher, A., Jin, X., Han, J., Huang, T.S.: A world wide tourism recommendation system based on geotagged web photos. In: ICASSP, pp. 2274–2277. IEEE (2010)
8. Yang, Y., Gong, Z., et al.: Identifying points of interest by self-tuning clustering. In: SIGIR, pp. 883–892. ACM (2011)
9. Li, J., Qian, X., Tang, Y.Y., Yang, L., Liu, C.: GPS estimation from users' photos. In: Li, S., El Saddik, A., Wang, M., Mei, T., Sebe, N., Yan, S., Hong, R., Gurrin, C. (eds.) MMM 2013, Part I. LNCS, vol. 7732, pp. 118–129. Springer, Heidelberg (2013)
10. Li, J., Qian, X., Tang, Y.Y., Yang, L., Mei, T.: GPS estimation for places of interest from social users uploaded photos. IEEE Transactions on Multimedia (2014)
11. Miller, G.: Wordnet: a lexical database for English. Communications of the ACM 38(11), 39–41 (1995)
12. Fellbaum, C.: Wordnet. Theory and Applications of Ontology: Computer Applications, pp. 231–243 (2010)
13. Li, L.J., Su, H., Fei-Fei, L., Xing, E.P.: Object bank: A high-level image representation for scene classification & semantic feature sparsification. In: NIPS, pp. 1378–1386 (2010)
14. Lampert, C.H., Nickisch, H., Harmeling, S.: Learning to detect unseen object classes by between-class attribute transfer. In: CVPR, pp. 951–958. IEEE (2009)

15. Torresani, L., Szummer, M., Fitzgibbon, A.: Efficient object category recognition using classemes. In: Daniilidis, K., Maragos, P., Paragios, N. (eds.) ECCV 2010, Part I. LNCS, vol. 6311, pp. 776–789. Springer, Heidelberg (2010)
16. Kennedy, L.S., Naaman, M.: Generating diverse and representative image search results for landmarks. In: WWW, pp. 297–306. ACM (2008)
17. Crandall, D.J., Backstrom, L., Huttenlocher, D., Kleinberg, J.: Mapping the world's photos. In: WWW, pp. 761–770. ACM (2009)
18. Oliva, A., Torralba, A.: Building the gist of a scene: The role of global image features in recognition. Progress in Brain Research 155, 23–36 (2006)
19. Lowe, D.G.: Object recognition from local scale-invariant features. In: ICCV, vol. 2, pp. 1150–1157. IEEE (1999)
20. Simpson, T., Crowe, M.: Wordnet.net (2005), http://opensource.ebswift.com/WordNet.Net

The Evolution of Research on Multimedia Travel Guide Search and Recommender Systems

Junge Shen[1], Zhiyong Cheng[2], Jialie Shen[2], Tao Mei[3], and Xinbo Gao[1]

[1] School of Electronic Engineering, Xidian University, China
{shenjunge,xbgao.xidian}@gmail.com
[2] School of Information Systems, Singapore Management University, Singapore
{zy.cheng.2011,jlshen}@smu.edu.sg
[3] Microsoft Research Asia, Beijing, China
tmei@microsoft.com

Abstract. The importance of multimedia travel guide search and recommender systems has led to a substantial amount of research spanning different computer science and information system disciplines in recent years. The five core research streams we identify here incorporate a few multimedia computing and information retrieval problems that relate to the alternative perspectives of algorithm design for optimizing search/recommendation quality and different methodological paradigms to assess system performance at large scale. They include (1) query analysis, (2) diversification based on different criteria, (3) ranking and reranking, (4) personalization and (5) evaluation. Based on a comprehensive discussion and analysis of these streams, this survey evaluates the recent major contributions to theoretical and system development, and makes some predictions about the road that lies ahead for multimedia computing and information retrieval (IR) researchers in both academia and industry world.

Keywords: Geo-Multimedia, Travel Guide systems, Recommendation, Information Retrieval, Survey.

1 Introduction

With the rapid growth of Internet, various online communication and sharing services have emerged as major communication channels for different kinds of users for different purposes. In particular, User-Generated Content (UGC) on social multimedia platforms has been significantly changing the way people understanding travel destination and the users increasingly reply on a large variety of geo-multimedia prior to finalizing the travel plan [25,41,14]. The geo-multimedia information includes not only the scenes and activities along with landmarks, but also travelers' context information, such as the number of tourist in group, the budgets, time and traveling schedules etc.

Multimedia travel guide system aims to effectively search or recommend a list of landmarks which could assist the users to plan travel and the abstract

C. Gurrin et al. (Eds.): MMM 2014, Part II, LNCS 8326, pp. 227–238, 2014.

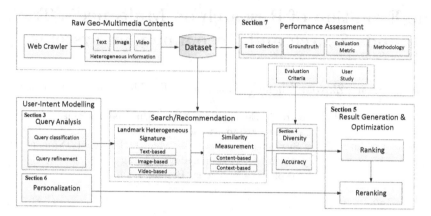

Fig. 1. The framework of travel guide systems

view of the general system architecture is illustrated in Figure 1. The system collects and analyzes information of landmarks from various online resources and ranks the landmarks based on their characteristics, as well as users' needs, preferences and current contextual information. The landmarks are characterized using heterogeneous representations to facilitate users to browse and search. In general, the system harvests geo-multimedia data online and stores the data in database. When a search query is submitted, the system calculates the relevance of each landmark with respect to the query using ranking/reranking algorithm. The query may be refined via classification or expansion. The search results are sorted in the descending order based on the relevance. In order to achieve comprehensive summary, collecting and analyzing different aspects about landmarks, such as histories, visual appearance and associated social activities or events effectively becomes a critical issue. In presentation of the retrieval results, each landmark is described by detailed textual descriptions as well as representative visual views. Further, different types of landmarks can be grouped into various clusters to facilitate fast browsing. Further, to assess the performance of a travel guide system, test collection, query set, as well as ground truth for the query set are essential. The performance of the system is evaluated by assessing the search results of all queries in the query set based on evaluation metrics, such as precision.

Developing travel guide search system is difficult. One of the key challenges is complexity and heterogeneity of online geo-multimedia. In general, they can be treated as a nonlinear composite of various kinds of contents from different sources. Applying the solutions developed for the extraction of knowledge from traditional multimedia solely might be not feasible and effective. Another problem is how to present the search and recommendation results to users so that they can quickly obtain comprehensive information about different types of landmarks. This problem is highly related to the diversification and visualization of the search results. Besides, users' queries are based on short text and thus they are not always specific and accurate. In order to capture users' intents concisely, it is important to develop an intelligent scheme to reformulate and refine the query. Moreover, users have their own travel preferences and behavior patterns. Even for the same user, the requirements may be changed dynamically under different contexts. Consequently, developing intelligent algorithm to incorporate users' context and personal preferences into design of ranking/reranking scheme becomes crucial. And a subsequential question is about how to construct users' personal profiles.

This survey reviews different aspects of multimedia travel guide search and recommendation systems in a detailed way. In Section 2, we give a detailed overview on multimedia travel guide search/recommender system and introduce a few typical systems in the domain. We present a summary of query analysis in Section 3. In Section 4, we present works done on diversity. Next, Section 5 presents the research related on ranking and reranking to list must-see attractions which can balance accuracy and diversity. In Section 6, we discuss and review the role and importance of personalization. Further, Section 7 provides overview of how to assess the system performance. In Section 8, we conclude by discussing several important issues for future study.

2   Overview

In recent years, multimedia travel guide search and recommendation system has became an important tool to assist people plan travel and understand various destinations. As a result, numerous systems have been developed, by leveraging rich online multimedia resources. This section aims to give a brief introduction to a series of representative travel guide systems. Figure 2 gives a clear overview on how the travel guide and recommendation systems evolve during the past decade.

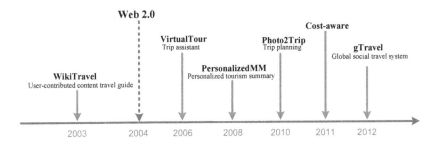

Fig. 2. The evolution of travel guide systems

To the best of our knowledge, WikiTravel [1] was the first travel guide portal, which aims to provide users timely and reliable information about landmarks. More recently, Web 2.0 sites have emerged as popular channel to allow users to interact with each other via social media. These online media sharing portals (e.g., Flickr and YouTube) typically host a large volume of accessible travel-related information, which have been widely used by travel guide systems. VirtualTour [23] is an online travel service, aiming to provide high quality images to help travelers plan their trip with the collected images from famous photo forum sites. A user interface is designed to support search based on map, location or path. By analyzing over 110,000 geo-referenced photos collected from Flickr, DiverseSearch diversifies the search results to visualize different aspects of landmarks [25]. PersonalizedMM generates personalized landmark summary with respect to a user query, by utilizing texts, geo-tagged images and videos collected from Wikipedia, Flickr, Youtube, and tourism websites [37]. W2Go recommends the top landmarks of a targeted city by ranking, with the use of photos and associated metadata from Flickr and user knowledge from Yahoo Travel Guide [13]. Representative views of the recommended landmarks are presented to users. Based on 20 million geo-tagged photos collected from Panoramio and 200,000 travelogues crawled from other websites, Photo2Trip explores the destinations and travel routes between destinations, and then makes trip planning for travelers [28]. gTravel is a global social travel system, which assists tourists in itinerary planning, tour navigation and travel knowledge sharing [43]. Besides, the system can monitor trip status and make automatic changes on itinerary, and enables users to easily explore others' experiences with its social interactive functionalities.

Table 1 gives a comprehensive summary about key characterisitics of various travel guide systems and compares the systems from different aspects. From the perspective of utilized information sources, most systems use multiple modality information. To better describe a landmark or a trip for users, the textual description and visual representation are necessary. Social information is only applied by the most recent system. We expect that more systems will leverage the social information in future, as social relationships increasingly play an important role in personalized recommendation, and attract more and more attentions in other recommendation services. Additionally, we can observe that the latest systems start to consider diversity and reranking, which will be described in subsequent sections.

Table 1. Comprehensive comparisons of current travel guide systems. "S" and "R" in the "Type" column denote "Search" and "Recommendation", respectively.

System	Type	Multi-modality	Diversity	Relevance	Reranking	Personalization
VirtualTour [23]	S*	N	N	Y	Y	Y
DiverseSearch [25]	S	Y	Y	Y	N	N
PersonalizedMM [37]	R*	Y	N	Y	N	Y
W2Go [13]	R	Y	N	N	Y	N
Photo2Trip [28]	R	Y	Y	Y	Y	Y
gTravel [43]	R	N	Y	Y	Y	N

3   Query Analysis

Query analysis is one of the most important components for modern IR systems. To enhance the effectiveness of search, various kinds of algorithms/schemes have been developed to reformulate the query in recent years. When searching for travel destinations or landmarks, users don't have a very clear description about target and thus queries are generally abstract concepts (such as "leisure place") with a set of constraints, such as *distance*(e.g., near to some place) or a *geolocation range*(e.g. a city or a district). Sanderson et al. [32] reported that one fifth of queries are geographical, in which 80% queries were associated with geographical terms. Jones et al. [24] found that users tend to search for locations which were relatively close to their current locations. And the distributions of distances between the search location and current location of different query topics varied greatly by query topics. Besides, around 10% of query reformulations related to locations involve a geocorrection. These studies reflect the importance of geo-information in users' travel related queries. Thus, many researchers analyze and utilize geo information in user queries to improve the performance of various retrieval tasks. Andrade et al. [3] applied different strategies to combine the relatedness between two geographic places with textual ranking, and found out that geographic ranking can significantly improve results for some queries. Yu et al. [42] combined the thematic and geographic relevance measures on a per-query basis. Query specificity is used as a feature to determine the weights of different sources of ranking evidence for each query. Yi et al. [41] developed a city language model to analyze users' geo intent based on large amounts of web-search logs.

Analyzing query topics can be very helpful to refine the query at the topic-level, and achieve more accurate and diverse results. He et al. [19] showed that most multiple queries include more than one topic, and reported that most users reformulate query at the topical level. Fan et al. [11] proposed a method to expand the query based on topic distributions of the input query and the candidate terms. However, little work about query topic analysis has been done in travel guide search systems. How to detect the topics and refine the query accordingly is an interesting research topic, and should attract more attentions in the future. Besides, little effort has been devoted to multi-modality query (e.g. text, image and video) in travel guide systems. Multi-modality query is more natural and able to describe user information needs about landmark search in more comprehensive way. For example, when searching a landmark, which has certain visual or ambient sound characteristics, it is difficult for users to specify information needs using only one modality. Indeed, the reformulation of multi-modality query in travel guide systems is a new research direction and there are a lot of open issues.

4   Diversification

Multimedia travel guide search and recommendation systems are becoming increasingly important to individual users and business. However, while the majority of algorithms proposed in literature have focused on improving accuracy, an essential aspect - diversity, has often been overlooked. For very specific and clearly expressed queries, a returned result is simply judged whether it is relevant to the query or not. While for abstract query concepts or the queries with multiple topics, the diversity of the results becomes important. The diverse results increase the possibility for users to get the desired results. In travel guide systems, there are two types of diversity: (1) diversified representative views of a landmark; and (2) diversified landmarks - different types of landmarks. For the query with specific landmark, the information related to the landmark is returned. However, there are many different aspects of a landmark, such as the visual appearance and history. Even only for the visual appearance, there could be different to represent views. When searching for images of a landmark, users expect to see different views of the landmark to get a whole view of the landmark. For the query with no specific targets or recommendation systems, many different landmarks may satisfy users' requirements. Clustering landmarks into categories can facilitate the users to browse and discover the landmarks they prefer. On the other hand, even for the same landmark, different users may have different perspectives or be interested in different aspects of the landmark, such as some users are interested in the activities around the landmarks and others are more interested in the scenery. To meet the diverse needs of a wide variety of users, it is necessary to present diverse results to users.

In real world, each geo-landmark can be described and modelled using various kind of media information. Currently, most researchers focus on mining knowledge from heterogeneous information, such as travelogues, geo-photos and videos, to extract multiple attributes of a landmark. In [25], Kennedy et al. proposed a method to select diverse and representative landmark images by using tags and visual features. Mei et al. [29] presented a series of methods to discover latent semantic topics from blogs and their distributions over locations. Chen et al. [6] used time sequences of photos to identify the locations. Scenery and sightseeing qualities are considered in [46] to plan a travel route. Both textual and visual metadata are used to estimate the location of Flickr videos in [12]. Hao et al.[17] generated overviews for locations via analyzing representative tags from travelogues. Different characteristics of city landmarks are extracted from the blogs by exploiting graphic models in [21]. Snavely et al. [35] generated rich representations from images taken by different people at a single location. Simon et al. [34] solved the problem of scene summarization by selecting a set of images that efficiently represented the graphic content of a given scene. Crandall et al. [10] placed images on a map with the combination of textual and visual features, using a corpus of 20 million images crawled from Flickr. Recently, Rubinac et al. [31] created landmark summarization with diversified and illustrative photos of the places. Travel search systems [30] and [40] summarized landmarks with diverse visual content of photos for travel information retrieval. As shown above, most of the recent researchers on diversity optimization focus on delivering a wide arrange of visual appearance for a targeted landmark, while little study has been done on using clustering to improve quality of end results. Further, very few existing studies consider to combine multiple modality (i.e. display results by using text, image and video) to diversify the result presentation. In addition, while achieving a good balance between search result diversification and effectiveness, no research has been carried out.

5 Ranking and Reranking

Users usually only pay attention to the top results in the returned list. Accordingly, ranking and reranking play key roles in all information search and recommendation systems. The main goal of ranking and reranking is to place the regarded most relevant results on the top, so as to maximize users' satisfaction and minimize the information load. Many reranking methods have been proposed, including the classification-based [39], clustering-based [20] and graph-based [36][47] methods. In travel guide systems, queries are usually accompanied with various constraints (e.g., distance constraints). On the other hand, there are heterogeneous information sources, such as text, image and video, associated with landmarks. Thus, in travel guide systems, ranking and reranking face different and more challenging problems, e.g., how to rank and rerank the results with respect to the constraints and these heterogeneous information.

Countless existing travel guide systems use only visual clues in ranking or reranking. By learning interested views from community photos, Ren et al. [30] created a table of content as a comprehensive summary of the landmarks via re-ranking the image search results. A DLMSearch system [40] is designed to support image query and diversify the landmark results through re-ranking to ensure the results to be highly pertinent, differ on landmark location and high visual quality. There are also systems which employ multiple information resources. Kennedy et al. [25] extracted representative views of a landmark by reranking Flickr photos using visual features based on the search results using tags. In [13], photos, user-generated reviews and ratings to Yahoo Travel Guide are combined together to generate the recommended landmarks. Reviews in Yahoo Travel Guide and image tags in Flickr are used to obtain preferences of users and determine the popularity of travel destinations. Based on user preference and travel site popularity, suitable popular places are selected for each user. Personalized travel recommendation system [9] uses locations and geo-tagged photos to rerank popular attractions.

As described, diversity is an important aspect of travel guide systems due to the intrinsic properties of travel related query. In existing systems, there are few systems considering the diversity in the design of ranking/reranking algorithms, such as [25] and [40]. More efforts should be invested into developing ranking/reranking algorithms in consideration of relevance and diversity simultaneously. This relates to the problem of how to quantitatively define the diversity of consequences, which is still an open question.

6 Personalization

With fast growth of geo-multimedia information in various social media portals, searching or recommending the contents associated with certain landmarks comprehensively becomes very tedious. To satisfy users' information needs accurately, personalization becomes one of the most essential enabling techniques to support different real life applications. For example, many users may prefer historic or cultural landmarks or travel sites, like a history museum, and some users could be interested in natural scenery. When recommending landmarks in a city, considering the preferences of users can better satisfy users' needs. Thus, intelligent personal travel guide system around the globe has created an urgent demand. As many applications have been successfully developed to enhance the quality of personal services. [37] designed a system in the multimedia view to generate personalized tourism summary in the form of text, image and video. In personal travel guide recommendation systems, users are recommended by travel routes. Xie et al. [38] proposed a method of composite recommendation of points of interest for tourists according to the tourist's budget. [28] focused on the relationships of many landmarks. Digital mobile devices are ubiquity, like mobile phone, digital cameras and tablet computer, which contain rich GPS positioning. Personalization in a mobile environment can provide more accurate and useful tourist recommendations which respect to personal preferences, usage, personal preference and other contextual information. [2] took advantages of contextual information to build prototypes of context-aware tour guide. This is one of the earliest applications to provide location-aware services in both indoor and outdoor. It also exploits user's historic locations to better assist the user. Personalized multimedia content with respect of users' preferences, location by mobile for sightseeing is studied in [33] to facilitate the individual decision. [5] presented an expert tourist guide called UbiquiTO, which adapts the provided content and user interests, as well as other context conditions like location and the device. [26] presented a system to estimate GPS for an image and [7] provided landmark identification with mobile devices. [22] can automatically generate a multimedia travelogue for mobile users. [16] utilized the compass in the mobile to give an intuitive way of information to the user in order to provide a tour guide. Averjanova et al. [4] developed a map-based mobile recommended system that provided personalized recommendations to users.

However, how to profile user's personal preference is a problem. Early personalized decision making systems asked users to input their preferences manually. [44] provides a study of exploiting online travel information to discover the interests of the tourists. Although some researchers tried to mine the personal preferences for personalized travel guide recommendation [44,15], they did not take into account contextual information in their approaches. Therefore, how to effectively mine users' preferences under different contexts from travel logs for personalized travel guide development needs to be further studied. Another interesting research direction is to explore the tourist travel graph, which contains the history of all tourists, to discover and utilize the relations or similarities between tourists for personalized travel recommendation. The upcoming field studies shall analyze the behaviours and habits of all tourists, so that the system can use the tourist graph.

7 Performance Evaluation

In the context of multimedia travel guide search and recommendation systems [1], the evaluation is to assess whether the systems can be trustworthy and easy to use and results they provide can satisfy users' information needs. A reliable and robust evaluation methodology can provide quantitative and qualitative assessments to various systems. There are several limitations of existing evaluation methods: Test collection should contain data of numerous landmarks across the world. Besides, information for each landmark should be comprehensive with heterogeneous information. The baselines used in existing works are based on simple heuristic methods (e.g. *rank-by-count* or *rank-by-frequency*) or the results of commercial travel search/recommend engines (e.g. TripAdvisor and Yahoo Travel Guide). In existing works, to evaluate the developed systems, researchers construct their own datasets which contain the necessary information to support their research results, respectively, such as [13,8,18,13,45,27]. However, the self-constructed datasets cannot support cross method evaluation. The developed travel guide systems cannot be easily compared to each other. For the travel search/recommendation system, it is hard to give universal criteria for assessment. A reliable method is to assign each query with several assessors who are very familiar with the landmarks in the queried area (and who are also the corresponding type of people if it is required in the query); and then based on the majority vote principle to decide the ground truth. As mentioned before, the used ground truth is based on either the collected data or commercial travel search/recommend engines. There are two common objective evaluation methods, one is to use the results of existing travel engines (e.g., TripAdvisor [2]) as ground truth [18]; and another is to use a set of collected data as test set [13,8]. The evaluation methodology also plays an important role in the assessment of a system. When designing it, a few issues need to be considered and they include (1) which aspects of the systems are to be evaluated (e.g., robustness, effectiveness, efficiency, etc.); (2) for the evaluation of each aspect, what kinds of data and resources are required (e.g., queries, different datasets, participants in user study, etc.); (3) the evaluation metrics for the evaluation of each aspect; and (4) when comparing the proposed systems with other competitors, the arrangement of resources and sequence of steps should not affect the fairness of the comparison, such as the used queries and data for different systems, the sequence of using different systems in user study, etc. The principle is that the methodology should comprehensively evaluate the proposed system and fairly compare with other competitors. In addition, while user study is highly important to measure the performance of IR systems, very little existing research has investigated how to evaluate travel guide search and recommendation systems based on user study.

---

[1] Travel guide search/recommendation is actually a subdomain of IR research.

[2] http://www.tripadvisor.com/Inspiration

8   Conclusion and Future Research Directions

The availability of massive online geo-multimedia information is revolutionizing the way people search and recommend landmarks and travel destinations In this survey we present a comprehensive and timely review of state-of-the-art in the domain of the multimedia travel guide search and recommendation systems. The five key aspects related to the system development and algorithm design are identified and they include (1) query analysis, (2) diversification based on different criteria, (3) ranking and reranking, (4) personalization and (5) evaluation. Further, we provide a detail discussion and analysis of the latest technical developments in the five streams. More importantly, we hope that this article provides a clear roadmap for future research. As we have discussed, there is a wide range of promising research problems in this field. For example, one distinguishing characteristic for online geo-multimedia is multi-modality. In order to generate high quality recommendation and search results, how to effectively fuse different kinds of information is very crucial. Moreover, in recent years, with an explosive growth of social network services, popular websites such as Facebook and Twitter attract millions of users. Network and related user behavior analytics are especially suited for our field. User profiles often contain geographic information that can be very helpful to understand their travel activities. Some social networks like Foursquare [3] contain a large amount of users' check-in logs, which can be used to analyze user visiting pattern and user preferences. Besides, the social relationship between users can be analyzed based on the social network and leveraged for landmark recommendations. Thus, another important but challenging problem is how to leverage these valuable information.

References

1. http://wikitravel.org
2. Abowd, G.D., Atkeson, C.G., Hong, J.: Cyber-guide: A mobile context-aware tour guide. Wireless Networks (1997)
3. Andrade, L., Silva, M.J.: Relevance ranking for geographic ir. In: ACM GIR (2006)
4. Averjanova, O., Ricci, F., Nguyen, Q.N.: Map-based interaction with a conversational mobile recommender system. In: UBICOMM (2008)
5. Cena, F., Console, L.: Integrating heterogeneous adaptation techniques to build a flexible and usable mobile tourist guide. AI Communications (2006)
6. Chen, C.-Y., Grauman, K.: Clues from the beaten path: Location estimation with bursty sequences of tourist photos. In: CVPR (2011)
7. Chen, D.M., Baatz, G., Kioser, K., Tsai, S.S., Vedantham, R., Pylv an ainen, T., Roimela, K., Chen, X., Bach, J., Pollefeys, M., Girod, B., Grzeszczuk, R.: City-scale landmark identification on mobile devices. In: CVPR (2011)
8. Cheng, J., Chen, Y., Huang, Y., Hsu, W., Liao, M.: Personalized travel recommendation by mining people attributes from community-contributed photos. ACM Multimedia (2011)
9. Clements, M., Serdyukov, P., Vries, A.P.D., Reinders, M.J.T.: Re-ranking of web image search results using a graph algorithm. In: ACM SIGIR (2010)

---

[3] https://foursquare.com/

10. Crandall, D., Backstrom, L., Huttenlocher, D., Kleinberg, J.: Mapping the worlds photos. In: WWW (2009)
11. Fan, J., Wu, H., Li, G., Zhou, L.: Suggesting topic-based query terms as you type. In: APWeb (2010)
12. Friedland, G., Choi, J., Janin, A.: Video2gps: a demo of multimodal location estimation on flickr videos. In: ACM Multimedia (2011)
13. Gao, Y., Tang, J., Hong, R., Dai, Q., Chua, T.-S., Jain, R.: W2go: a travel guidance system by automatic landmark ranking. In: ACM Multimedia (2010)
14. Gao, Y., Wang, M., Zha, Z.-J., Shen, J., Li, X., Wu, X.: Visual-textual joint relevance learning for tag-based social image search. IEEE Transactions on Image Processing (2013)
15. Ge, Y., Liu, Q., Xiong, H., Tuzhilin, A., Chen, J.: Cost-aware travel tour recommendation. In: KDD (2011)
16. Hagen, K., Modsching, M., Kramer, R.: A location aware mobile tourist guide selecting and interpreting sights and services by context matching. In: MobiQuitous (2005)
17. Hao, Q., Cai, R., Wang, X.-J., Yang, J.-M., Pang, Y.: Generating location overviews with images and tags by mining user-generated travelogues. In: ACM Multimedia (2009)
18. Hao, Q., Cai, R., Wang, C., Xiao, R., Yang, J., Pang, Y., Zhang, L.: Equip tourists with knowledge mined from travelogues. In: WWW (2010)
19. He, X., Yan, J., Ma, J., Liu, N., Chen, Z.: Query topic detection for reformulation. In: WWW (2007)
20. Hsu, W., Kennedy, L., Chang, S.: Video search reranking via information bottleneck principle. In: ACM Multimedia (2006)
21. Ji, R., Xie, X., Yao, H., Ma, W.-Y.: Mining city landmarks from blogs by graph modeling. In: ACM Multimedia (2009)
22. Jiang, S., Qian, X., Lan, K., Zhang, L., Mei, T.: Mobile multimedia travelogue generation by exploring geo-locations and image tags. In: ISCAS (2013)
23. Jing, F., Zhang, L., Ma, W.: Virtualtour: an online travel assistant based on high quality images. ACM Multimedia (2006)
24. Jones, R., Zhang, W., Rey, B., Jhala, P., Stipp, E.: Geographic intention and modification in web search. In: International Journal of Geographical Information Science (2008)
25. Kennedy, L.: Generating diverse and representative image search results for landmarks. In: WWW (2008)
26. Li, J., Qian, X., Tang, Y.Y., Yang, L., Liu, C.: GPS estimation from users' photos. In: Li, S., El Saddik, A., Wang, M., Mei, T., Sebe, N., Yan, S., Hong, R., Gurrin, C. (eds.) MMM 2013, Part I. LNCS, vol. 7732, pp. 118–129. Springer, Heidelberg (2013)
27. Li, Y., Crandall, D.J., Huttenlocher, D.P.: Landmark classification in large-scale image collections. In: ICCV (2009)
28. Lu, X., Wang, C., Zhang, L.: Photo2trip: Generating travel routes from geo-tagged photos for trip planning. In: ACM Multimedia (2010)
29. Mei, Q., Liu, C., Su, H., Zhai, C.: A probabilistic approach to spatiotemporal theme pattern mining on weblogs. In: WWW (2006)
30. Ren, Y., Yu, M.: Diversifying landmark image search results by learning interested views from community photos. In: WWW (2010)
31. Rudinac, S., Hanjalic, A., Larson, M.: Finding representative and diverse community contributed images to create visual summaries of geographic areas. In: ACM Multimedia (2011)

32. Sanderson, M., Kohler, J.: Analyzing geographic queries. In: SIGIR Workshop on Geographic Information Retrieval (2004)
33. Scherp, A., Boll, S.: Generic support for personalized mobile multimedia tourist applications. In: ACM Multimedia (2004)
34. Simon, Snavely, N., Seitz, S.M.: Scene summarization for online image collections. In: ICCV (2007)
35. Snavely, N., Seitz, S., Szeliski, R.: Photo tourism: exploring photo collections in 3d. In: SIGGRAPH (2006)
36. Tian, X., Yang, L., Wang, J., Yang, Y., Wu, X., Hua, X.-S.: Bayesian video search reranking. In: ACM Multimedia (2008)
37. Wu, X., Li, J., Neo, S.: Personalized multimedia web summarizer for tourist. In: WWW (2008)
38. Xie, M., Lakshmanan, L.V.S., Wood, P.T.: Breaking out of the box of recommendations: from items to packages. In: RecSys (2010)
39. Yan, R., Hauptmann, E., Jin, R.: Multimedia search with pseudorelevance feedback. In: Bakker, E.M., Lew, M., Huang, T.S., Sebe, N., Zhou, X.S. (eds.) CIVR 2003. LNCS, vol. 2728, pp. 38–247. Springer, Heidelberg (2003)
40. Ye, J., Chen, J.: Dlmsearch: Diversified landmark search by photo. In: ACM Multimedia (2012)
41. Yi, X., Raghavan, H., Leggetter, C.: Discovering users' specific geo intention in web search. In: WWW (2009)
42. Yu, B., Cai, G.: A query-aware document ranking method for geographic information retrieval. In: ACM GIR (2007)
43. Zhang, R., Guo, X., Sun, H., Huai, J., Liu, X.: gtravel: a global social travel system. In: ACM Multimedia (2012)
44. Zheng, V.W., Cao, B., Zheng, Y., Xie, X., Yang, Q.: Collaborative filtering meets mobile recommendation: A usercentered approach. In: AAAI (2010)
45. Zheng, Y., Xie, X.: Learning travel recommendations from user-generated gps traces. ACM Transactions on Intelligent Systems and Technology (2011)
46. Zheng, Y.-T., Yan, S., Zha, Z.-J., Zhou, X., Li, Y., Chua, T.-S., Jain, R.: Gpsview: A scenic driving route planner. In: ACM TOMCCAP (2013)
47. Zitouni, H., Sevil, S., Ozkan, D., Duygulu, P.: Re-ranking of web image search results using a graph algorithm. In: ICPR (2008)

Average Precision: Good Guide or False Friend to Multimedia Search Effectiveness?

Robin Aly[1], Dolf Trieschnigg[1], Kevin McGuinness[2],
Noel E. O'Connor[2], and Franciska de Jong[1]

[1] Human Media Interaction Group, University Twente, The Netherlands
{r.aly,d.trieschnigg,fdejong}@utwente.nl
[2] INSIGHT Research Centre for Big Data Analytics, Dublin City University, Ireland
Kevin.McGuinness@eeng.dcu.ie, Noel.OConnor@dcu.ie

Abstract. Approaches to multimedia search often evolve from existing approaches with strong average precision. However, work on search evaluation shows that average precision does not always capture effectiveness in terms of satisfying user needs because it ignores the diversity of search results. This paper investigates whether search approaches with diverse results have been neglected within the multimedia retrieval research agenda due the fact that they are overshadowed by search approaches with strong average precision. To this end, we compare 361 search approaches applied on the TrecVid benchmarks between 2005 and 2007. We motivate two criteria based on measure correlation and statistical equivalence to estimate whether search approaches with diverse results have been neglected. We show that hypothesized effect indeed occurs in the above examined collections. As a consequence, the research community would benefit from reconsidering existing approaches in the light of diversity.

1 Introduction

Progress in information retrieval is driven by improvements of performance measures, which also holds for multimedia information retrieval (MIR) [8]. Approaches with strong performance numbers are often developed further, while approaches with lower performance numbers tend to be neglected. This practice is sensible if the considered performance measure indeed captures search effectiveness, i.e. the degree of fulfilling users' information needs. Recent research in search engine evaluation suggests that average precision, which is currently the standard performance measure in MIR, does not always reflect search effectiveness well. Instead, performance measures should take into account the *diversity* of search results, which recently has been proposed for multimedia and text retrieval [9, 2].[1]

---

[1] Because there are several slightly various interpretations of diversity, we rely in this introduction on the reader's intuition for what diversity is and we will consider the different interpretations in the experiments section.

C. Gurrin et al. (Eds.): MMM 2014, Part II, LNCS 8326, pp. 239–250, 2014.
© Springer International Publishing Switzerland 2014

For example, let us consider two search approaches A and B that both only return relevant documents for the query "Ship or boat" in the top results. Suppose search approach A returns documents containing tankers while search approach B returns documents that cover multiple *aspects* of the query (tankers, sailing ships, rowing boats, etc). The two search approaches are equally effective in terms of average precision. However, search approach B is intuitively more effective because it satisfies a wider range of user needs.

Assuming that the MIR community always derives new search approaches from approaches with the strongest reported average precision, two scenarios are possible: 1) the existing search approaches have higher or similar diversity than other search approaches, in which case average precision was a good guide for the MIR community, or 2) there were other search approaches producing more diverse results that were often more effective for users that were overshadowed by approaches with higher average precision. This paper investigates the meta-research question of whether 1) or 2) is the case. This question has recently been investigated for search approaches in text retrieval [7]. For the multimedia domain this issue is perhaps even more important because here search approaches often depend on similarity to few positive training examples, which puts more weight on the role of visual characteristics instead of higher level semantic features addressed by occurrences of query terms in text retrieval.

We base our investigation on the average precision and diversity of a large number of search approaches used in TRECVid between 2005 and 2007, which is one of the standard evaluation benchmarks for video retrieval [8]. While the binary relevance judgments used to measure average precision are publicly available, we have to define the input of diversity measures. These inputs are the *aspects* of a query addressed by documents. Current methods for the definition of document aspects assume a single reason for the need of diversity in the search results, for example underspecified queries [6]. However, because little is known about the motivation for diversity required in MIR, we will develop a new method for the definition of aspects that makes fewer assumptions. Furthermore, the development of new search approaches is a creative process and clear relationships such as "approach A is derived from B" only seldom exist. It is therefore challenging to find out whether search approaches have been overlooked. To address this challenge, we define two criteria, which are likely to have played an important role in the development of MIR. Based on these criteria we judge whether search approaches with strong diversity were overshadowed in MIR.

The findings of this paper are relevant to benchmark organizers and the MIR community as a whole. It should be noted that the findings do not pertain to the quality of individual search approaches with high average precision and low diversity because many applications for such approaches exist. Furthermore, diverse search approaches can also be developed from scratch. While this is a valid and potentially fruitful strategy, we focus in this paper on deriving approaches from existing ones.

The rest of this paper is structured as follows. In Section 2 we describe related work. Section 3 presents the methodology we use to assess whether search approaches with diverse results have been overlooked. Section 4 details the experimental setup and its results. Section 5 discusses these results. Section 6 concludes this paper.

2 Background and Related Work

Goffman [5] was one of the first to propose that diversity should be taken into account in search results, which was confirmed by Xu and Yin [10] in a user study. In text retrieval, several performance measures have been proposed that assess diversity [11, 2, 1]. Diversity measures depend on the definition of query aspects and the assignment of documents to aspects that they address. Therefore, the methodology to define these aspects is important for this paper.

In the literature, there are two general methods to define the aspects of a query: top-down, based on an analysis of queries in a query log, and bottom-up, based on clustering a set of retrieved documents. Paramita et al. [6] use a top-down approach to create a test data set to assess diversity in photo retrieval. They use queries from a query log to define aspects. The query set consists of frequent queries, of which the individual query terms also occur in other queries. The aspects are defined based on the more complex queries that contain the original query as a substring. For example, for the query "boat" they consider "fast boat" and "big boat" as aspects because these terms appear in the query log. This approach assumes that queries are underspecified and search results should therefore be diverse. However, diversity can also be desirable for other reasons, which should be reflected in the aspect definition. For example, a user with the query "Ship boat" might want to create a documentary about different vessel types, and therefore desires diversity while the query is specific for his need. In this paper, we give assessors the possibility to define their own aspects for a given query set.

The bottom-up approach starts with the definition of aspects from result lists. In van Leuken et al. [9], users are asked to cluster the top-ranked search results for a query and they measure diversity by the inter-cluster agreement between users and their method. However, inter-cluster agreement does not take the ranking of documents into account, which we argue is essential. For example, a search approach that first presents documents from a non-relevant cluster should be attributed with low performance, even if the employed clustering algorithm was perfect. Therefore, although we take a similar approach to define aspects, we limit the aspects to documents that were previously judged relevant, and use state-of-the-art diversity measures from text retrieval, which consider the document ranking.

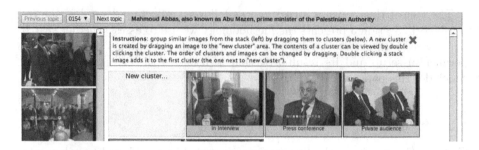

Fig. 1. User interface to define query aspects for documents that were judged relevant by TRECVid judges [8]

3   Determining Research Agendas

In order to answer our research question, whether our research agendas have overlooked search approaches that support diversity better than other with similar average precision, we need to consider a large number of search approaches over several data sets. We use the queries from TRECVid between 2005 and 2007 [8], for which many submissions by different search approaches are available. In the following, we first describe the methodology we used to measure diversity and then explain the criteria we use to answer the research question.

3.1   Aspect Definition

State-of-the-art diversity measures require the definition of aspects and the assignment of documents to them [11, 2, 1]. We asked judges to define aspects for documents that were previously judged relevant by TRECVid assessors.[2]

Figure 1 shows the interface that was used by the judges. At the top, the window displays the query text, instructions, and navigational elements. The left part of the window contains the documents that were not yet assigned to an aspect. If a judge does not assign a document to an aspect, we assume that it belongs to an aspect of its own. On the right, the interface shows the created aspects. The judges could assign documents to aspects by dragging-and-dropping them from the left to the respective aspect on the right. The aspects were visually represented by the document that was used to create them. The interface allows the judge to view the list of documents of an aspect by double-clicking. Here, the judge could enter the name of the aspect, and remove documents from the aspect. If a judge was finished with defining aspects for a query and assigning documents to them, he could use the button "next query" to move to the next query.

In order to allow judges to process as many queries as possible in the available time, the "next query" button returned the query that was the least often processed by other judges. If there were multiple queries processed by the same

---

[2] See Section 4 for a description of the data sets and the judges' demographics.

number of judges, we chose a random query from this set. Using the aspects defined by the judges, we can calculate a number of diversity measures for each search approach considered. The relevance judgements from TRECVid are used to calculate the average precision of the search approaches. The difference between the average precision measures and the diversity measures forms the basis of our further analysis.

3.2   Criteria to Analyze Research Agendas

In this paper, we investigate whether our research agendas have overlooked or neglected search approaches that support diversity. Providing an answer to this question is difficult because we cannot verify whether this has actually been the case. The available data, submitted search results to TRECVid, represent snapshots of research agendas and we cannot verify whether research approaches have been overlooked. Furthermore, the development of search approaches is not only driven by search performance, but also by other factors, such as special features or personal preference. In the following, we assume that the research community evolved purely based on the average precision of proposed search approaches. Based on this assumption, we suggest two criteria that the MIR community could have used to decide upon search approaches to further develop.

Performance Correlation. The first criterion is based on the assumption that the likelihood of further developing a search approach is motivated by the differences in average precision to other search approaches. For example, the motivation to further develop a search approach with twice the average precision as another approach would be much more likely to be further developed. However, if the differences are smaller with respect to diversity, motivation for choosing the latter approach would have been less strong. In other words, the criterion captures the quantitative differences of likelihood, if diversity is considered instead of average precision. Therefore, we investigate the relationship between the average precision and diversity measures over the available search approaches. We consider two sub-criteria: the trend, and the variability around the trend. If an increase of average precision causes a weaker increase in a diversity measure, we say that the trend is slowly increasing. In this situation, the motivation of the MIR community to further develop a search approach with high gains in average precision would have been less strong when basing their decision on diversity. The opposite holds for strong increasing trends. Furthermore, if the diversity of search approaches varies around the trend, there is an increasing risk of missing a good search approach in terms of diversity because the trend is a poor predictor for them.

Statistical Equivalence. The second criterion assumes that the research community wants to base new developments on the best search approach for a query population. However, because performance measures are only calculated on a sample from this query population, the performance differences between search approaches could originate from chance due to the sampling process. In other words, some search approaches are *statistically equivalent* to the search approach given the considered sample, and are developed further in order not to miss the search approach with the best overall performance. Note that the set of equivalent search approaches depends on the considered performance measure. Now, if the number of equivalent search approaches for a diversity measure is greater than for average precision, this means that the chance of having overlooked the best overall search approach increases.

Table 1. Dataset statistics (TVXXt: test collection of the ad-hoc search task in TREC-Vid 20XX, I: interactive, M: manual (one query reformulation), F: fully automatic)

Collection	Docs.	Queries	Used queries	Search approaches	I	M	F
TV05	45,765	14	149, 150, 153, 154, 155, 156, 158, 159, 160, 164, 165, 166, 167, 171	118	49	26	42
TV06	79,484	14	173, 175, 176, 178, 179, 180, 181, 183, 185, 187, 190, 192, 193, 194	124	36	11	76
TV07	18,142	17	197, 200, 201, 202, 203, 204, 205, 207, 208, 210, 211, 212, 214, 215, 216, 217, 219	119	32	4	82

4 Experiments

In this section, we describe the experiments we performed to investigate the answers to our research questions stated in Section 1. The results of the experiments will be discussed in Section 5.

4.1 Data Sets and Measures

Table 1 shows the statistics of the TRECVid years 2005 (TV05), 2006 (TV06), and 2007 (TV07). To reduce the amount of annotation work we filtered queries with more than 300 relevant documents. We decided to maximize the number collections given our resources on obtaining judgments.

In total, 14 persons with a computer science background ranging in age from 25 to 41 judged the aspects of documents to a query. On average, the judges defined 8.75 aspects per query with a standard deviation of 6.5, a minimum of a single aspect, and a maximum of 37 aspects. The judgments for one query roughly took 30 minutes on average. These values were similar for the three datasets. We assigned every query to two judges in order to investigate their influence on the diversity measure. We split the judgments randomly per query into two groups, creating two complete set of judgments. We describe the experiments used to measure the influence of choosing between these sets in Section 4.2.

A variety of diversity measures have received research attention recently in the text retrieval community. In this work, we considered a number of measures, which were used in the diversity sub-task of the TREC web track [3]. Zhai et al. [11] proposes the subtopic recall $SR@k$ measure that considers the number of subtopics in the top-k documents of a ranking. Subtopic recall SR rewards

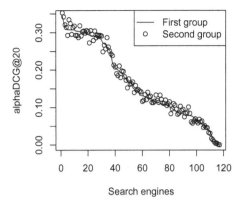

Fig. 2. Influence of the group of judges used for diversity assessment in the TV05 collection and the α-DGC measure

the first-ranked documents of an aspect and treats the remaining documents of this aspect as non-relevant. Subtopic recall assumes that a user is interested in only one document per aspect, and therefore wants variety. Clarke et al. [2] presents the $\alpha - DGC@k$ measure, which is an extension of the discounted gain measure until rank k. The measure takes the possibility of disagreement between a judge and a user into account, which is represented by the probability of disagreement α. In this paper, we set $\alpha = 0.5$, which is also done in the TREC web track. Agrawal et al. [1] present a series of intent (here: aspect) aware measures, which are based on existing measures. The intent-aware version of a measure considers each query aspect as an independent type of request, and assumes that documents on other aspects are non-relevant to requests of this type. The intent-aware measure is then the expected performance of the existing measure. Following the diversity subtask of the TREC web track [3], we use the intent-aware version of the Expected Reciprocal Rank at rank 20, IA-ERR@k20. Also following the TREC web track, we assume a uniform distribution among request types.

4.2   Influence of Selected Judges

In order to assess the influence of the considered judges, we investigated their agreement and the differences this caused in terms of diversity. Compared to the study by van Leuken et al. [9], our inter-judge agreement was relatively low according to the Folwkes-Mallows index [4]. Furthermore, Figure 2 shows the influence of the chosen group of judges on the diversity assessment in the TV05 collection for the α-DGC@20 measure. The x-axis shows the search approaches sorted by the α-DGC@20 for the first group of judges. The y-axis shows the corresponding α-DGC@20 values of the considered groups. Note that these values are actually averages over the values of the considered queries. Although there is variation between the two groups, their differences are small on average.

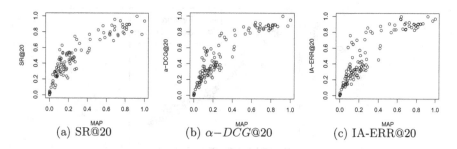

(a) SR@20   (b) $\alpha-D\tilde{C}G$@20   (c) IA-ERR@20

Fig. 3. Performance correlation on the TV05 dataset. The TV06 and TV07 dataset yielded similar results.

We plotted similar graphs for the α-DGC@20, the SR@20 and the IA-ERR@20 measures for all three collections. We do not show them because they did not contain qualitative differences. Given that the differences among the groups of judges are small, we only report results for the first group in the following section. Nevertheless, we verified that the results of the second group were roughly the same.

4.3   Diversity Assessment

In the remainder of this section, we analyze whether our research agendas have overlooked search approaches with diversity support. We use the criteria described in Section 3: the performance correlation, the statistical equivalence to the best search approaches.

Performance Correlation. Figure 3 shows the correlation between the diversity measures and average precision. Due to space limitations we only show the figures for the TV05 dataset; the figures for the other datasets look similar, however. Note that we normalized the performance values in Figure 3 to the interval [0 : 1], because it simplifies the interpretation and the units of the performances measures generally do not have a real-world explanation.

Figure 3(a) shows the correlation between average precision and SR@20. We see that the general increase in SR@20 performance is sub-linear to the increase of average precision. The variance around the trend is higher for search approaches with high average precision. Figure 3(b) shows the correlation between average precision and α-DGC@20. The trend of the curves is similar to the one of SR@20. The variance around the trend for search approaches with high average precision is, however, lower. The correlation with the intent-aware measure IA-ERR@20 is shown in Figure 3(c). The main difference to other measures is that search approaches with a relatively low average precision show a good intent-aware reciprocal rank.

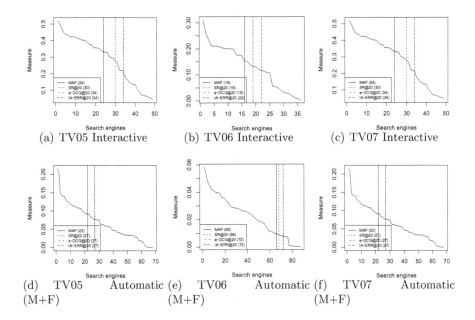

(a) TV05 Interactive   (b) TV06 Interactive   (c) TV07 Interactive

(d)   TV05   Automatic (e)   TV06   Automatic (f)   TV07   Automatic
(M+F)   (M+F)   (M+F)

Fig. 4. Statistical Equivalence: The difference in investigated search approaches when choosing engines that are not statistically different to the best search approach in terms of the respective search approach, using a two-sided student's t-test ($p < 0.05$).

Statistical Equivalence Figure 4 compares the choice in search approaches according to the statistical equivalence to the search approach that performed the best in the tested query set. We separate interactive search approaches from manual and fully automatic search approaches, because the best interactive search approach cannot reasonably be compared to fully automated search approaches. In all sub-figures, the x-axis shows the list of search approaches with decreasing average precision. The y-axis shows the average precision of the individual search approach.

Figure 4(a) shows statistical equivalence for the interactive runs of the TV05 dataset. For the subtopic recall SR@20, the number of equivalent search approaches increases from 24 to 30 (25%). For both, the α-DGC@20 and the IA-ERR@20 measure, the number of search approaches increased to 34 (41%). Figure 4(b) repeats this statistic for the TV06 dataset. Figure 4(c) shows that the top-16 search approaches are equivalent according to average precision. For SR@20 and α-DGC@20, the 19 (+18%) best search approaches are equivalent, and according to IA-ERR@20 the 22 (+37%) best search approaches are equivalent.

Figure 4(d) shows statistical equivalence for the manual and fully automated search approaches on the TV05 dataset. While the top-22 search approaches are statistically equivalent, for all diversity measures the 27 (+22%) search approaches are equivalent. Figure 4(e) shows the results for the TV06 dataset. For average precision, most search approaches (66) are statistically equivalent to the best search approach. The sub-topic recall is at 68 almost as high as for mean average precision. For the α-DGC@20 and IA-ERR@20 measure 72 search approaches are equivalent (+9%). Figure 4(f) shows the results for the TV07 dataset. For average precision, 22 search approaches are equivalent, and for each diversity measure 27 search approaches are equivalent (+22%).

5   Discussion

In this section, we discuss the results from the experiments described in Section 4.

Compared to the study by van Leuken et al. [9], the agreement between the groups of judges described in Section 4.2 is relatively low. On the one hand, low agreement could mean that the task was not clearly defined. On the other hand, low agreement can also be seen as a case for diversity, because users, in this case our judges, interpret queries and results differently. Nevertheless, the diversity measure values were comparable for the two groups of judges.

For the performance correlation criteria, we found that for search approaches with high average precision an increase of average precision did not cause an increase of diversity. Therefore, search approaches with very strong average precision do not have stronger diversity with high average precision. Furthermore, there was a large variance around this trend, which yields the suggestion that there are other factors than average precision that determine diversity.

Furthermore, we investigated the number of search approaches that were statistically equivalent with the best search approach in terms of average precision. For both interactive and fully automatic search approaches, approximately 25% more search approaches were equivalent to the best search approach when considering diversity instead of average precision. Given this high percentage, we argue that it is likely that as a community we overlooked search approaches that are potentially the best.

6   Conclusions

The evolving MIR research agenda is mainly driven by average precision. However, work in search engine evaluation suggests that the diversity of search results should also be taken into account. This paper investigated the meta-research question: "Did our research agendas possibly overlook search approaches with diversity support?" We investigated the research question in two steps: 1) the assessment of average precision and diversity of existing search results and 2) by providing criteria to decide whether it is likely that our research agendas overlooked promising results.

For the first step, we assessed the diversity of 361 search approaches for 45 queries from TRECVid between 2005 and 2007 based on average precision and three state-of-the-art diversity measures. To calculate the diversity measures, we asked two groups of judges to define aspects for each query. We found that the agreement between the groups was relatively low for individual queries. However, the judgments of both groups resulted in similar diversity measurements when averaged over all queries and therefore do not affect the results of our study.

In the second step, we motivated the following two criteria to decide whether search approaches that support diversity have been overlooked. The first criterion was focused on the *performance correlation* between average precision and diversity measures. The performance correlation allows us to identify trends between average precision and diversity measures as well as variation from these trends. A slow increasing trend means that one measure does not influence the other. A higher variation from the trend increases the risk of choosing a worse search approach according to diversity. The second criterion was the number of search approaches that were *statistically equivalent* to the best search approach according to a performance measure. If this number is higher for diversity measures than for average precision, this means that we missed search approaches, which had a fair chance to be better than the approach with the strongest performance measure on the considered query set.

Comparing the average precision and diversity of the mentioned data sets based on these criteria, indicated that there is a real danger that the research community neglected, or at least overlooked, search approaches with promising diversity support. We discovered that the trend between average precision and diversity was almost flat. This means for the MIR community that search approaches with high average precision are often indistinguishable in terms of diversity. There was also a high variation around this trend, which showed that search approaches with relatively low average precision sometimes had a much higher diversity compared to what the trend predicted. Furthermore, for the statistical equivalence criterion, we found that the MIR community could have investigated roughly 25% more search approaches based on diversity measures instead of average precision. As a result, we conclude that it is very likely that the MIR community overlooked search approaches with diversity support in the past.

This study used a sample of queries and videos from the TRECVid evaluation benchmark and other datasets, e.g. the Pascal or ImageCLEF datasets, could in principle yield different results. We therefore propose to investigate the generality of our findings in future work.

Acknowledgments. This work was co-funded by the EU FP7 Project AXES (ICT-269980). We want to thank the anonymous reviewers for their valuable input.

References

[1] Agrawal, R., Gollapudi, S., Halverson, A., Ieong, S.: Diversifying search results. In: WSDM 2009: Proceedings of the Second ACM International Conference on Web Search and Data Mining, pp. 5–14. ACM, New York (2009)

[2] Clarke, C.L.A., Kolla, M., Cormack, G.V., Vechtomova, O., Ashkan, A., Büttcher, S., MacKinnon, I.: Novelty and diversity in information retrieval evaluation. In: Proceedings of the 31st Annual International ACM SIGIR Conference on Research and Development in Information Retrieval, SIGIR 2008, pp. 659–666. ACM, New York (2008)

[3] Clarke, C.L.A., Craswell, N., Soboroff, I., Voorhees, E.: Overview of the TREC 2011 Web Track. In: Twentieth Text Retrieval Conference (TREC 2011) The Proceedings (2011)

[4] Fowlkes, E.B., Mallows, C.L.: A method for comparing two hierarchical clusterings. Journal of the American Statistical Association 78(383), 553–569 (1983)

[5] Goffman, W.: A searching procedure for information retrieval. Information Storage and Retrieval 2(2), 73–78 (1964)

[6] Paramita, M.L., Sanderson, M., Clough, P.: Developing a test collection to support diversity analysis. In: Proceedings of Redundancy, Diversity, and Interdependence Document Relevance Workshop held at ACM SIGIR, pp. 39–45 (2009)

[7] Sanderson, M., Paramita, M.L., Clough, P., Kanoulas, E.: Do user preferences and evaluation measures line up? In: SIGIR 2010: Proceedings of the 33rd International ACM SIGIR Conference on Research and Development in Information Retrieval, pp. 555–562. ACM, New York (2010) ISBN 978-1-4503-0153-4

[8] Smeaton, A.F., Over, P., Kraaij, W.: Evaluation campaigns and trecvid. In: MIR 2006: Proceedings of the 8th ACM International Workshop on Multimedia Information Retrieval, pp. 321–330. ACM Press, New York (2006)

[9] van Leuken, R.H., Garcia, L., Olivares, X., van Zwol, R.: Visual diversification of image search results. In: WWW 2009: Proceedings of the 18th International Conference on World Wide Web, pp. 341–350. ACM, New York (2009)

[10] Xu, Y., Yin, H.: Novelty and topicality in interactive information retrieval. Journal of the American Society for Information Science and Technology 59(2), 201–215 (2008)

[11] Zhai, C.X., Cohen, W.W., Lafferty, J.: Beyond independent relevance: methods and evaluation metrics for subtopic retrieval. In: SIGIR 2003: Proceedings of the 26th Annual International ACM SIGIR Conference on Research and Development in Informaion Retrieval, pp. 10–17. ACM, New York (2003)

An Investigation into Feature Effectiveness for Multimedia Hyperlinking

Shu Chen[1,2], Maria Eskevich[2], Gareth J.F. Jones[2], and Noel E. O'Connor[1]

[1] INSIGHT Centre for Data Analytics
Dublin City University, Dublin 9, Dublin, Ireland
shu.chen4@mail.dcu.ie, noel.oconnor@dcu.ie,
[2] CNGL Centre for Global Intelligent Content, School of Computing
Dublin City University, Dublin 9, Dublin, Ireland
{meskevich,gjones}@computing.dcu.ie

Abstract. The increasing amount of archival multimedia content available online is creating increasing opportunities for users who are interested in exploratory search behaviour such as browsing. The user experience with online collections could therefore be improved by enabling navigation and recommendation within multimedia archives, which can be supported by allowing a user to follow a set of hyperlinks created within or across documents. The main goal of this study is to compare the performance of different multimedia features for automatic hyperlink generation. In our work we construct multimedia hyperlinks by indexing and searching textual and visual features extracted from the blip.tv dataset. A user-driven evaluation strategy is then proposed by applying the Amazon Mechanical Turk (AMT) crowdsourcing platform, since we believe that AMT workers represent a good example of "real world" users. We conclude that textual features exhibit better performance than visual features for multimedia hyperlink construction. In general, a combination of ASR transcripts and metadata provides the best results.

Keywords: Multimedia, Hyperlinking, Crowdsourcing, Information Retrieval.

1 Introduction

Fully realizing the value of the increasing amount of multimedia archival content available online requires users to engage in exploratory search behaviour to find materials which may be of interest to them. Users are increasingly not as interested in simply re-finding information contained in known-items as in the past – they wish to explore unfamiliar archives of multimedia content. This user activity can be supported by providing a set of hyperlinks within or across documents within an archive or archives. Hyperlinks should be constructed based on the semantic information described by text or visual contents of the archive. A rich and semantically meaningful set of hyperlinks can potentially improve the user experience by enabling navigation and recommendation.

C. Gurrin et al. (Eds.): MMM 2014, Part II, LNCS 8326, pp. 251–262, 2014.

Since the requirement for hyperlinks arises from the needs and interests of users, it follows that an investigation of hyperlink generation in multimedia data collections should be user-driven. Workers engaged by crowdsourcing platforms represent a good example of real potential users of multimedia browsing applications because they fit the profile of experienced Internet users, and they are able to perform relevance assessment [1]. Thus, investigation into multimedia search and hyperlinking can be based on available research video collections, whilst workers from a crowdsourcing platform can play the role of the users that help us to define which multimedia features can contribute to effective hyperlink construction.

The main goal of this paper is to compare the performance of different multimedia features for automatic hyperlink generation. State-of-the-art multimedia retrieval techniques are used to create hyperlinks within the video collection automatically. These techniques determine the relatedness between source video segments, termed *anchors* and target video segments. Workers from the crowdsourcing platform act as real-time users of a multimedia retrieval system. They are asked to watch a query video segment (anchor) and a potentially related video segment extracted by our automatic hyperlink generation process, and provide feedback on whether those segments are indeed related.

The paper is structured as follows: Section 2 overviews related work on multimedia hyperlinking and crowdsourcing techniques. Section 3 presents the design of our hyperlinking strategy, including data description and hyperlinking algorithm description. Section 4 provides experimental results and the details of user feedback. Section 5 concludes the paper and comments on our further work plans.

2   Related Work

There are a number of examples of the utilisation of links to automatically augment textual information for research or commercial purposes. Examples of this approach include the Smart-Tag service developed by Microsoft which aims to construct links between web pages or Google AutoLink which links street addresses or ISBNs to related internet resources. However, early linking systems caused numerous controversies, since many people expressed concerns that hyperlinks were being "surreptitiously" modified for commercial purposes [16]. Hyperlinking research has gradually become oriented towards non-profit data collections, such as Wikipedia. In [15] the authors presented a link creation system "Wikify!" based on Wikpedia resources. This system combined automatic document keyword extraction and word sense disambiguation to provide a rich text annotation service. The authors in [16] presented an alternative strategy using machine learning to identify significant terms within unstructured documents and enrich them with links to the appropriate Wikipedia articles. The principle of relatedness was used to exclude the situation where links were determined by a rare sense of a word, according to the incoming and outgoing links to the current Wikipedia document. In [2], the authors presented work on linking multimedia resources for unskilled users, defined as exhibiting exploratory behaviour in [3]. Hyperlinking research has also appeared in the area of digital libraries focussing on news, multimedia and cultural heritage archives. The linking task was redefined as linking items with a rich textual representation in a news archive to items with sparse annotations in a multimedia archive, where items should be linked if they describe the same or a related event [2].

The VideoCLEF 2009 tasks included a multimedia hyperlinking task which required participants to find related resources across languages This was based on linking videos to material on the same subject in a different language [12]. The MediaEval 2012 benchmark campaign introduced the Search and Hyperlinking task as a Brave New Task. The idea of the task was to connect two activities in one framework, a video segment search task was combined with a separate sub-task which used relevant segments as anchors from which links to other video segments should be formed within the Hyperlinking sub-task [7]. The similarity between query and target anchors was determined by participants using either of both of textual information from metadata or spoken transcripts, and visual content within shot segments [6].

Evaluation of hyperlinking systems can be carried out either based on ground-truth data collections or based on human evaluation of results. The cross-lingual hyperlinking task ay NTCIR-10 in 2012 provided two evaluation instances – automatic evaluation against queries created from the Wikipedia groundtruth and manual assessment of results [20]. In our opinion, the complexity of video content means that the evaluation of multimedia hyperlinking is best served by manual evaluation based on human judgements. Crowdsourcing is a method of having people do things that we might otherwise consider assigning to a computing device to calculate automatically [9]. As such, it offers scalable pools of workers available on-demand to offer a flexible means of gathering human judgements as needed to evaluate hyperlink construction.

3   Experimental Design

This section describes the data used for our evaluation of multimedia hyperlinking, and the strategy and features used to form these links in this study.

3.1   Data Description

The dataset used for the experiment consists of semi-professional videos uploaded to the Internet video sharing platform Blip.tv[1]. These videos are gathered into the blip10000 collection [18]. Following the setup of the Search and Hyperlinking task at MediaEval 2012, for our hyperlinking experiments, we make use of the test set in the collection that contains 9,550 videos and has a runtime of 2,125 hours [7]. The dataset comprises metadata that was manually assigned to each video by the user who uploaded it. The shot boundary of each episode was automatically created by TU Berlin [10]. The number of shot segments is 42,000 with

---

[1] http://www.blip.tv/

an average duration of 30 seconds. Each shot segment has an associated keyframe extracted from the middle of the shot. To analyze spoken information, two automatic speech recognition (ASR) transcripts are provided by LIMSI/Vocapia Research[2] and LIUM Research team[3]. Spoken transcripts from LIMSI/Vocapia were created by first using a language identification detector (LID) and then running an appropriate ASRS system [11]. The LIUM system is based on the CMU Sphinx project [17]. In our investigation we use the 1-best ASR transcription hypotheses only.

We define a hyperlink as a constructed link between two video segments within the collection, one a query anchor, the other a target segment. Each anchor or segment contains the start and end time within the video, and corresponding audio and visual channels. A query anchor simulates a user's request while browsing using a hyperlinking system. All 30 query anchors used in our hyperlinking system were taken from the test set of the MediaEval 2012[4] Search and Hyperlinking task. Each query contains a corresponding filename and a duration to describe the video segment boundary of the current query. Each query is associated with a piece of text description extracted from the corresponding LIMSI or LIUM transcripts. All spoken words within the video segment boundaries are included. To represent the visual content of query anchors, a keyframe located at the middle of an anchor shot is extracted by using *ffmpeg*[5].

A target segment is a section of video within the collection which we assume to be of interest to users, and that would enrich their browsing experience. In our hyperlinking system, a target segment is based on automatically detected video shots. Since the shots vary in length, we define the length of a target segment to be between 90 and 120 seconds, based on previous crowdsourcing experience in MediaEval 2012 [5]. Thus, any shot shorter than 90 seconds is expanded by combining it with nearby shots, while any shot longer than 120 seconds is cut into a segment of 120 seconds from its start point. Each target segment is also associated with corresponding spoken transcripts and a keyframe.

3.2   Linking Algorithm

The linking algorithm uses textual and visual features to determine the similarity between query and target anchors. We use metadata descriptions and ASR transcripts (LIUM and LIMSI) to represent textual information, and describe the visual content of keyframes using both low-level and high-level features.

Text Analysis. We use the Apache Lucene 3.3.0[6] software in order to index and retrieve the segments based on textual information. ASR transcripts and

---

[2] http://www.vocapia.com/

[3] http://www-lium.univ-lemans.fr/en/content/
language-and-speech-technology-lst

[4] http://www.multimediaeval.org/mediaeval2012/

[5] http://www.ffmpeg.org/

[6] http://lucene.apache.org/core/

metadata are merged into a single field for indexing of each segment. A standard analyzer of Apache Lucene is used to convert text data into the searching format. Text data in the single field is converted into lower case. The stop words are removed using the default list provided within Lucene. The analyzer tokenizes text based on a sophisticated grammar that recognizes e-mail addresses, acronyms, and alphanumerics [19]. The searching phase chops text data within of query anchor into terms and uses a *tf-idf* measure to score retrieved documents.

Low-Level Visual Analysis. We use a colour histogram and a bag-of-visual-word model to describe the low-level features of each keyframe. The colour histogram is calculated based on the HSV space. A three-level spatial pyramid representation is applied to each keyframe, which is divided into $1{\times}1$, $2{\times}2$, and $4{\times}4$ grids. The feature vector is normalized into $[0, 255]$, then a χ^2 function is applied to compare two histograms as following:

$$d(H_1, H_2) = \sum_{1 \leq i \leq k} \frac{(H_1(i) - H_2(i))^2}{H_1(i)} \tag{1}$$

where H_1 and H_2 represent two feature histogram respectively. The length of the feature vector is k, and $H_1(i)$ means the ith point in histogram H_1.

The bag-of-visual-words model is generated by applying the SIFT descriptor [14] calculated by a total of 7,198 images randomly picked up from the video keyframe set. A K-means algorithm clusters the descriptor vectors to create visual words, where the number of cluster centres is experimentally set to 1,000. The weight vector of each keyframe is calculated based on visual words and its own SIFT descriptor. Finally, a cosine distance algorithm is applied to compute the distance between visual words.

High-Level Visual Analysis. We use two different high-level databases to extract the concepts (high level features) of each video keyframe. The first one is Object Bank[7] provided by Visual Lab, Stanford University. It contains a total of 177 high-level concepts created by a scale-invariant response map of a large number of pre-trained generic object detectors [13]. Each keyframe is described as a feature vector with the length of 44,604 which is calculated using multiple scales and different levels of a spatial pyramid. A Euclidean distance algorithm is applied to compute the distance between the high-level feature vectors.

The second high-level feature database is provided by the Vision Group at University of Oxford, specially created for the blip10000 dataset used in MediaEval 2012. It contains a set of concept detector scores for 589 concepts [4][8]. The detectors where trained by downloading positive images from Google images and learning their difference to assumed-to-be negative images in the dataset using the libLinear toolkit [8]. The distance between high-level concepts is calculated using the Euclidean distance.

---

[7] http://vision.stanford.edu/projects/objectbank/
[8] The concepts used were provided by Christoph Kofler from TU Delft.

4 Experimental Investigation

4.1 Crowdsourcing Task Design

Crowdsourcing allows us to obtain human-generated feedback about the relatedness between the video anchors, i.e. whether the hyperlinks that we create are valuable for real users. We collect feedback on whether users are interested in watching the selected video segment after having watched an initial query segment. Our investigation is carried out using the Amazon Mechanical Turk (AMT)[9] platform for crowdsourcing.

Traditionally, a task performed on the AMT platform is referred to as a Human Intelligence Task (HIT). In each HIT, our users were presented with a pair of video segments and were required to answer a number of questions to describe their opinion as to whether the two videos were related or not. Users were asked to provide details on the reason for their (un)relatedness judgement, and point out what features influenced their decision. We offered five options for the users to describe the feature selection: "Object", "Person", "Place", "Topic", and "Other" that can be the same in case of related videos or different in the case of unrelatedness. Moreover, in order to avoid spam submissions from workers and to determine reasonable answers from the workers, we also asked the workers to type in a number of meaningful words from the video segments that they had been asked to watch. The HIT reward was set at $0.11, which was found to be acceptable to the workers.

4.2 Evaluation Overview

We uploaded a total of 8 runs to AMT for human evaluation involving different multimedia features, either textual or visual – as shown in Table 1. RUN_1, RUN_2, RUN_3 and RUN_4 use textual information to create video hyperlinks and RUN_5, RUN_6, RUN_7 and RUN_8 use low-level and high-level visual features.

A total of 3,915 HITs were created by all 8 runs. We received 3,521 useful submissions that were accepted for video hyperlinking evaluation. As working with videos is an unusual task on the AMT platform, we investigated the consistency of the decisions on video segment relatedness. This was based on the condition that each HIT was supposed to be answered by two different users. As it is possible to get a disagreement on the relatedness judgement, we defined that a pair of video segments is weakly related if only one user provides a positive answer on the relatedness judgement, whereas they are strongly related if the answers of both users are positive. There were 468 HITs marked as related. Within this set, 177 HITs were regarded as strongly related.

4.3 Evaluation Results and Analysis

Table 2 shows what features influence the relatedness judgement based on user feedback. The 'Object', 'Person' and 'Place' options mean that users determined

---

[9] https://www.mturk.com/mturk/

Table 1. Overview of the Video Hyperlinking Runs

RUN_NAME	Features	Types	
RUN_1	LIUM	Textual	
RUN_2	LIUM+META		
RUN_3	LIMSI		
RUN_4	LIMSI+META		
RUN_5	Colour Histogram	Low-level	Visual
RUN_6	Bag-of-Visual-Word		
RUN_7	Visual Group (Oxford)	High-level	
RUN_8	Object Bank (Standford)		

Table 2. User Options on the Relatedness Evaluation

OPTION	Object	Person	Place	Topic	Other
No. of Selection	243	244	247	430	133

Table 3. Overview of Positive Answers on Each Run. (WR: weak related, SR: strong related).

RUN	RUN_1	RUN_2	RUN_3	RUN_4	RUN_5	RUN_6	RUN_7	RUN_8
WR	64	67	60	65	44	53	41	29
SR	70	72	60	66	71	5	13	1
Total	134	139	120	131	115	58	54	30

Table 4. Overview of MAP Values. (WR: weak related, SR: strong related, ALL: WR+SR, WV: within the videos, WC: within the collection).

RUN		RUN_1	RUN_2	RUN_3	RUN_4	RUN_5	RUN_6	RUN_7	RUN_8
WV	ALL	**0.2108**	0.2084	0.1706	0.1919	0.1934	0.0562	0.0611	0.0329
	WR	**0.0597**	0.0564	0.0482	0.0559	0.0462	0.0469	0.0443	0.0324
	SR	**0.1107**	0.1072	0.0881	0.0940	0.1070	0.0039	0.0112	0.0006
WC	ALL	0.1209	**0.1293**	0.1082	0.1277	0.0753	0.0720	0.0692	0.0393
	WR	0.0496	0.0547	0.0468	0.0591	0.0302	**0.0622**	0.0553	0.0388
	SR	0.0406	**0.0416**	0.0362	0.0387	0.0266	0.0041	0.0080	0.0006

the relatedness based on visual information, such as the same objects, location or human faces. The 'Topic' option means the users' judgement was influenced by the spoken information from video segments. Moreover, the 'Other' option was provided to allow users to express their own opinion on the relatedness judgement. Table 2 indicates that most users considered spoken information as an important aspect in evaluating hyperlinking relatedness.

Table 3 shows the number of relevant video segments retrieved by each run for both weak and strong relatedness. The query set used in the evaluation contains a total of 30 queries in each run. To evaluate the ranked list retrieved by each query, the top 10 video hyperlinking results were selected, with a total of 300 results for each run. According to table 3, the runs retrieved based on textual features achieved more positive results on the relatedness judgement. Among them, RUN_2 detects the most relevant video pairs, i.e. 139 out of 300 results. On the contrary, the performance of runs based on visual features decreases.

Average Precision (AveP) and Mean Average Precision (MAP) were used to evaluate the performance of each run. In addition to considering strong and weak relatedness, the evaluation also considers whether hyperlinks were created within the videos or within the collection. A hyperlink within a video means that a target segment exists either in the same video as the query anchor or in other different videos in the collection, while a hyperlink within the collection means a target anchor only exists in a different video. Table 4 shows an overview of MAP values for the different alternatives.

In general, the hyperlinking algorithms based on textual features perform better than those using visual features. The retrieval results using LIUM transcripts have the best score in most cases. An exception is the case of weak relatedness within the collection, where LIMSI with the corresponding video metadata achieves the best performance. MAP values and HIT feedback are consistent in the conclusion that speech data information is a bigger influence than visual data when judging the relevance of video segments. User feedback implies that they prefer to link two video segments that share the same or a similar story. The correspondence in person or object depicted is a much lower priority.

When comparing the results for visual features, both low-level feature descriptors, colour histogram and bag-of-visual words, always performs better than high-level feature descriptors. This is due to the fact that relevant video segments more easily share similar low-level visual features, such as background colour or illumination, while the performance of high-level features is seriously influenced by the Semantic Gap. This is clear when comparing the results for Visual Group (Oxford) and Object Bank (Stanford) high-level datasets. The former was specially created for the blip.tv dataset used in MediaEval 2012, whilst the latter, even if representative enough for a general image dataset, misses specific aspects within a TV dataset.

When we analyse results for within videos vs. within collection, there is a clear difference in terms of textual and visual features. Within videos, the best performance based on textual features is determined by the combination of ASR transcripts and metadata. Within the collection, LIUM+METADATA and LIMSI+METADATA show better performance than using single LIUM or LIMSI transcripts. When creating links within the same video due to the fact that metadata is always the same, the difference between spoken transcripts influences the ranked retrieval result. Within the whole collection, on the other hand, removing links locating in the same video, metadata information and ASR transcripts both exhibit differences in determining the description of video content. The performance of colour histogram features decreases significantly when linking videos within the whole collection. This is due to the fact that two video segments within the same video often share the same or similar background.

Table 5 shows an overview of AveP values of each run for a total of 6 queries. All the AveP values are calculated for linking videos within the whole collection. Both weak relatedness and strong relatedness are considered. In general,

Table 5. Overview of AveP values

RUN	RUN_1	RUN_2	RUN_3	RUN_4	RUN_5	RUN_6	RUN_7	RUN_8
Topic_1	0.205	0.243	0.230	**0.252**	0.008	0.118	0.000	0.005
Topic_2	0.305	0.228	0.339	**0.385**	0.000	0.026	0.000	0.174
Topic_3	0.330	**0.356**	0.252	0.260	0.028	0.028	0.139	0.000
Topic_4	0.240	0.240	0.146	0.156	**0.455**	0.080	0.000	0.015
Topic_13	0.025	0.040	0.000	0.020	0.000	0.000	0.050	**0.400**
Topic_14	0.000	0.014	0.034	0.010	0.000	**0.347**	0.014	0.013

Table 6. Overview of Rerank LIUM, Colour Histogram, and High-level Concept MAP values, described as MAP/Increase Rate. (WR: weak related, SR: strong related, ALL: WR+SR, LM: LIUM transcripts+metadata, CH: colour histogram, VG: Visual Group (Oxford)).

RUN	LM	LM+CH	LM+VG	LM+CH+VG
ALL	0.1293	0.1975 / +52.7%	0.1647 / +27.4%	0.2040 / +57.8%
WR	0.0547	0.1181 / +115.9%	0.0910 / +66.4%	0.1335 / +144.1%
SR	0.0416	0.0600 / +44.2%	0.0532 / +27.9%	0.0539 / +29.6%
RUN	CH	CH+LM	CH+VG	CH+LM+VG
ALL	0.0753	0.0927 / +23.1%	0.1312 / +42.6%	0.1265 / +68.0%
WR	0.0302	0.0577 / +47.7%	0.0628 / +107.9%	0.0644 / +113.2%
SR	0.0266	0.0170 / -36.1%	0.0434 / +63.2%	0.0360 / +35.5%
RUN	VG	VG+LM	VG+CH	VG+LM+OX
ALL	0.0692	0.0360 / -47.9%	0.0620 / -10.4%	0.0354 / -48.8%
WR	0.0553	0.0221 / -60.0%	0.0362 / -35.4%	0.0252 / -54.4%
SR	0.0080	0.0121 / +51.3%	0.0213 / +166.3%	0.0086 / +7.5%

AveP values are consistent with MAP evaluation. The retrieval results using the combination of LIUM/LIMSI transcripts and metadata information have better scores in the first three queries, while the scores decrease in the runs extracted by visual features. This demonstrates the conclusion that speech data has a higher priority when determining the relevance of a pair of video segments. However, in Topic_4, Topic_13, and Topic_14, the best performance is achieved by the runs using visual descriptors. In Topic_4, the run using colour histogram analysis has a score of 0.455. Figure 1 shows two groups of example keyframes associated with the retrieval results of Topic_4 in RUN_5 and Topic_14 in RUN_6. Figure 1(a) and Figure 1(b) present an introduction about certain software, with different spoken information but similar keyframes. Therefore, the analysis of visual content shows the advantage of removing the disagreement of voice messages and reflects the user's interest in the visual scene. Figure 1(c) and Figure 1(d) also suggest the same for visual high-level concepts. According to user feedback, the two video segments are regarded as relevant due to the fact that only one person gives a presentation, even if the content is quite different. Therefore, a further conclusion is that visual features can perform as a complement to textual feature analysis when constructing multimedia hyperlinks, and vice versa.

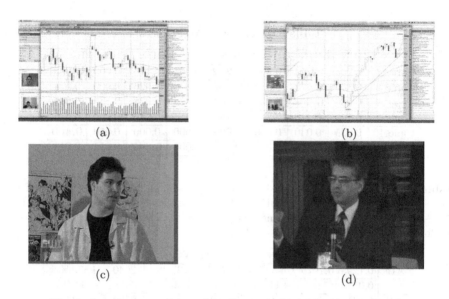

Fig. 1. Sample comparision of keyframes in Linked Video Segments

To prove this conclusion, a reranking algorithm was used to retrieve a new ranked list of the top 10 results implemented by different feature types. A total of 3 runs are selected based on LIUM ASR transcripts associated with corresponding metadata, colour histogram, and high-level concepts from Visual Group (Oxford). The top 10 results of each method were reranked by fusing normalized scores from the other two. A linear fusion algorithm was used where the weight for all scores was set to be equal. Table 6 shows the evaluation results based on the MAP measure. The reranking strategy improves most results comparing with Table 4. The improvement is clear for the linking strategies implemented by using LIUM transcripts and colour histogram. Based on these results we plan to carry out further work on multimedia hyperlinking by devising efficient fusion algorithms to utilize the advantage of textual and visual features.

5   Conclusions and Future Work

This paper describes our investigation into feature effectiveness for automatic multimedia hyperlinking. It simulates a scenario whereby the user browses a set of video data associated with existing hyperlinks across the whole collection. Our objective was to research how different multimedia features influence user performance and contribute to multimedia hyperlink generation. Automatic link construction uses both textual and visual features, including LIUM/LIMSI transcripts, metadata information, colour histogram descriptor, bag-of-visual-words extracted by SIFT descriptor, and high-level visual concepts from the Visual Group (Oxford) and the Object Bank (Stanford). The evaluation is based on using human computing techniques supported by Amazon Mechanical Turk.

Crowdsourcing evaluation concludes that textual features exhibit better performance than visual features for multimedia hyperlink construction. The textual information related to a video can be extracted from both spoken data or metadata. In general, a combination of ASR transcripts and metadata shows the best results. Moreover, the quality of hyperlinks created based on visual features is variable. However, some potential links can be determined by visual features due to the lack of spoken information or incomplete metadata.

The evaluation suggests that textual information significantly contributes to the relevance of video segments. Searching and indexing spoken words should thus consider the context information and the concept of the story described by the whole video. Moreover, it is a challenge to efficiently fuse the results from different hyperlinking frameworks based on textual and visual features. Both of these aspects will form the basis of our future work.

Acknowledgements. This work was supported by funding from the European Commission's 7th Framework Programme (FP7) under AXES ICT-269980.and Science Foundation Ireland (Grant 07/CE/I1142) as part of the Centre for Next Generation Localisation (CNGL) project at DCU.

References

1. Alonso, O., Rose, D.E., Stewart, B.: Crowdsourcing for Relevance Evaluation. SIGIR Forum 42(2), 9–15 (2008)
2. Bron, M., Huurnink, B., de Rijke, M.: Linking Archives Using Document Enrichment and Term Selection. In: Gradmann, S., Borri, F., Meghini, C., Schuldt, H. (eds.) TPDL 2011. LNCS, vol. 6966, pp. 360–371. Springer, Heidelberg (2011)
3. Bron, M., van Gorp, J., Nack, F., de Rijke, M.: Exploratory Search in an Audio-Visual Archive: Evaluating a Professional Search Tool for Non-Professional Users. In: 1st European Workshop on Human-Computer Interaction and Information Retrieval, EuroHCIR 2011 (2011)
4. Chatfield, K., Lempitsky, V., Vedaldi, A., Zisserman, A.: The devil is in the details: an evaluation of recent feature encoding methods. In: Proceedings of the British Machine Vision Conference, pp. 76.1–76.12 (2011)
5. Chen, S., Jones, G.J.F., O'Connor, N.E.: DCU Linking Runs at Mediaeval 2012 Search and Hyperlinking Task. In: MediaEval. CEUR Workshop Proceedings, vol. 927. CEUR-WS.org (2012)
6. Eskevich, M., Jones, G.J.F., Aly, R., Ordelman, R.J., Chen, S., Nadeem, D., Guinaudeau, C., Gravier, G., Sébillot, P., de Nies, T., Debevere, P., Van de Walle, R., Galuscakova, P., Pecina, P., Larson, M.: Multimedia Information Seeking Through Search and Hyperlinking. In: Proceedings of the 3rd ACM Conference on International Conference on Multimedia Retrieval, ICMR 2013, pp. 287–294 (2013)
7. Eskevich, M., Jones, G.J.F., Chen, S., Aly, R.B.N., Ordelman, R.J.F., Larson, M.: Search and Hyperlinking Task at Mediaeval 2012. In: MediaEval 2012 Multimedia Benchmark Workshop, Pisa, Italy, p. 14. CEUR-WS.org, Aachen (2012)
8. Fan, R.-E., Chang, K.-W., Hsieh, C.-J., Wang, X.-R., Lin, C.-J.: LIBLINEAR: A Library for Large Linear Classification. J. Mach. Learn. Res. 9, 1871–1874 (2008)

9. Jones, G.J.F.: An Introduction to Crowdsourcing for Language and Multimedia Technology Research. In: Agosti, M., Ferro, N., Forner, P., Müller, H., Santucci, G. (eds.) PROMISE Winter School 2012. LNCS, vol. 7757, pp. 132–154. Springer, Heidelberg (2013)

10. Kelm, P., Schmiedeke, S., Sikora, T.: Feature-based Video Key Frame Extraction for Low Quality Video Sequences. In: 10th Workshop on Image Analysis for Multimedia Interactive Services, WIAMIS 2009, London, United Kingdom, May 6-8, pp. 25–28 (2009)

11. Lamel, L., Gauvain, J.-L.: Speech Processing for Audio Indexing. In: Nordström, B., Ranta, A. (eds.) GoTAL 2008. LNCS (LNAI), vol. 5221, pp. 4–15. Springer, Heidelberg (2008)

12. Larson, M., Newman, E., Jones, G.J.F.: Overview of Videoclef 2009: New Perspectives on Speech-Based Multimedia Content Enrichment. In: Multilingual Information Access Evaluation II. Multimedia Experiments, vol. 6242, pp. 354–368 (2010)

13. Li, L.-J., Su, H., Xing, E.P., Fei-Fei, L.: Object Bank: A High-Level Image Representation for Scene Classification and Semantic Feature Sparsification. In: Neural Information Processing Systems (NIPS), Vancouver, Canada (December 2010)

14. Lowe, D.: Object Recognition from Local Scale-Invariant Features. In: The Proceedings of the Seventh IEEE International Conference on Computer Vision, vol. 2, pp. 1150–1157 (1999)

15. Mihalcea, R., Csomai, A.: Wikify!: Linking Documents to Encyclopedic Knowledge. In: Proceedings of the Sixteenth ACM Conference on Information and Knowledge Management, CIKM 2007, pp. 233–242 (2007)

16. Milne, D., Witten, I.H.: Learning to Link with Wikipedia. In: Proceedings of the 17th ACM Conference on Information and Knowledge Management, CIKM 2008, pp. 509–518 (2008)

17. Rousseau, A., Bougares, F., Delglise, P., Schwenk, H., Estv, Y.: LIUM's Systems for the IWSLT 2011 Speech Translation Tasks. In: Proceedings of IWSLT 2011I (2011)

18. Schmiedeke, S., Xu, P., Ferrané, I., Eskevich, M., Kofler, C., Larson, M.A., Estève, Y., Lamel, L., Jones, G.J.F., Sikora, T.: Blip10000: A Social Video Dataset Containing SPUG Content for Tagging and Retrieval. In: Multimedia Systems Conference 2013 (MMSys 2013), pp. 96–101 (2013)

19. Sonawane, A.: Using Apache Lucene to search text - Easily Build Search and Index Capabilities into your Applications (August 2009), http://www.ibm.com/developerworks/library/os-apache-lucenesearch/

20. Tang, L.-X., Kang, I.-S., Kimura, F., Lee, Y.-H., Trotman, A., Geva, S., Xu, Y.: Overview of the NTCIR-10 Cross-Lingual Link Discovery Task. In: Proceedings of NTCIR-10 (2012)

Mining the Web for Multimedia-Based Enriching

Mathilde Sahuguet and Benoit Huet

Eurecom, Sophia-Antipolis, France

Abstract. As the amount of social media shared on the Internet grows increasingly, it becomes possible to explore a topic with a novel, people based viewpoint. We aim at performing topic enriching using media items mined from social media sharing platforms. Nevertheless, such data collected from the Web is likely to contain noise, hence the need to further process collected documents to ensure relevance. To this end, we designed an approach to automatically propose a cleaned set of media items related to events mined from search trends. Events are described using word tags and a pool of videos is linked to each event in order to propose relevant content. This pool has previously been filtered out from non-relevant data using information retrieval techniques. We report the results of our approach by automatically illustrating the popular moments of four celebrities.

1 Introduction

Every day, millions of new documents are published on the Internet. This amounts to a huge mass of available information and it is not straightforward to retrieve relevant content. The data is out there, but a question still remains: how to make sense of it and choose which content is worth watching? Hence, we observe an important need to organize relevant data regarding a topic of interest. Indeed, organizing data amounts to choosing a way to display. Rendering of information is an integral part of the understanding: it could be made accordingly to different facets or different events. Part of this process implies strictly restricting data to relevant items, as the intrusion of some non-relevant data would alter the comprehension.

In [4], the authors discuss the issue of creating and curating digital collections by crawling and selecting media items from online repositories. They raise awareness on the possibility to use context as a cue towards understanding of a situation or a media. In a similar fashion, [9] leverages from both social media sharing and search trends as a source of knowledge to identify important events and build a timeline summarization.

We define an event an occurrence of abnormal activity or happening relative to a topic, on a limited time segment, that captured a lot of interest. In our scenario, interest can be measured using web search activity: a happening can be spotted as an event when it triggered massive web search. For a celebrity, an event could be a public event (concert), a personal event (wedding) or even a viral video.

C. Gurrin et al. (Eds.): MMM 2014, Part II, LNCS 8326, pp. 263–274, 2014.
© Springer International Publishing Switzerland 2014

In this paper, we aim at associating each of the events discovered using techniques from [9] with a filtered set of relevant media items. The process includes two stages: first, we mine multimedia content from social media sharing platforms in order to discover the semantics of events and gather a set of possibly relevant content. Then, we focus on filtering this content to discard non related items. The output of our work is a timeline of events related to a topic, each event being illustrated by a set of videos. Indeed, videos capture information in a rich and effective manner, allowing viewers to quickly grasp the whole semantic content with limited effort.

The fact that we propose to retrieve a set and not a list of media is of primary importance: we do not aim at ranking but rather at offering a pool of resources that make sense regarding the event at stake. A next step of the process (not addressed here) would be to rank items in each set to adapt each user through a personalization phase. A typical scenario is the usage of a second screen when watching television. In this scenario, the second screen device (tablet / smartphone / notepad) is used as an interface for enriching television content and achieving interaction between user and content. The additional content presented to the user is taken from the retrieved set of media, mined on social media sharing platforms . The popularity of such platforms provides access to a massive amount of multimedia documents of varying genre and quality. Therefore, filtering the dataset is a key point of our framework: we want to make sure that the content displayed is accurate and illustrative of the event.

In this paper, we address the problem of automatic multimedia content enriching on the basis of events. Important events are characterized by unusually high number of search. Using Google Trends allows to study user search behavior on a specific topic and identify key events [9]. A focused query is performed on YouTube to retrieve a set of relevant candidate videos illustrating the event. Each event is described by a tag cloud and video sets are pruned out by applying techniques inspired from pseudo relevance feedback and outlier detection, and based on textual features. We evaluate our work by illustrating events along celebrity oriented summaries.

2   Related Work

Information retrieval aims at satisfying the information need of a user. A lot of works have addressed the issue of proposing to the user a ranked list of content from a set of documents, with respect to a query. This usually implies to design a representation of the documents and the query, as well as a similarity measure between them. Typical techniques include vector space model with tf-idf weightings and cosine similarity. Probabilistic models have also been investigated, among which have been defined probabilistic language models [6]. Lucene[1] is an open source search engine that implements some of those models, such as Okapi BM25 [8] or the work of [12] that performs language model smoothing using Dirichlet prior.

---

[1] http://lucene.apache.org/

Information retrieval systems also use methods to give more accurate result to a query. Relevance feedback is a method for refining the results of a search query depending on the initial results, thus improving the retrieval effectiveness. Pseudo (or blind) relevance feedback automates this process, usually by using the top retrieved documents to refine the search query. Other methods of query expansion have been studied. For example, [3] use term classification to pick "good" expansion terms, while [11] uses the external knowledge of Wikipedia to strengthen this task.

Information retrieval and pseudo-relevance feedback methods both relate to the retrieval of accurate information to suit a user need. The difference to our work is that, while we aim at discarding non-relevant documents from an event-oriented dataset, those techniques present to the user the most relevant document concerning a query. Re-ranking the dataset amounts to using dataset knowledge in order to assign relevance scores. Overall, documents considered as non-relevant will have a low ranking, which can be a mean to filter a fixed amount of items.

We can also see this filtering task under the light of outlier detection. Indeed, outliers are data that differ from the set they are part of, in the sense that they are inconsistent with the rest of the data or deviate from a certain observed distribution. Hence, filtering out non-useful or non-relevant data could amount to discard "outlier" data. We will focus on unsupervised outlier detection techniques, as we do not have an insight on the data beforehand. Usually, this implies to make the assumption that the dataset contains mostly non-outlier data: outliers are a minority. Distance-based approaches define outlierness as a function of distance to neighborings points. Among those techniques, DB-outliers [5] considers that a datapoint is an outlier if less than p percent of the datapoints are distant by more than e, p and e being parameters of the algorithm. Other algorithms use the distance of a point to its k nearest neighbors [1], [7]. Density is another cue to mine outliers from a dataset (Local Outlier Factor, [2]).

3 Framework

Our framework (Figure 1) is composed of the following steps: we query Google Trends with the given query term in order to have an overview of its popularity through time and identify time segments of interest (moments of high popularity) and associated keywords. Then, we query social media platforms on those segments in order to get a pool of videos for each segment, that should illustrate the event at stake. Last step consists in filtering the possible videos to display: we prune out non relevant videos from each pool by analyzing the datasets. For each event, the user (or a personalization step) can then choose from each pool of videos which content to watch in order to have an overview of the event.

3.1 Time Segment Extraction

Time segment extracting is performed using previous work in [9]. Basically, we leverage search trends (week-based time series representing the popularity of a

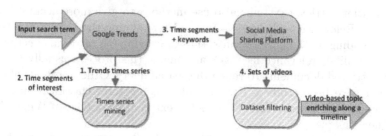

Fig. 1. Proposed framework

search term) in order to identify peaks in the popularity of a term. The corresponding time segments or *bursts* are associated a *burst value*. During a burst, high people interest with respect to the topic is characteristic of an event.

The output of this operation is a set of week dates and associated keywords that we want to link to some events in the real world. The next step is then to query online social sharing platforms (using their provided API) in order to give context to those events.

3.2 Video Focused Search

We perform multiple queries on the YouTube API on the relevant time intervals and their associated terms. For each time segment, we obtain a set of videos (issued from different queries with diverse search terms) that were uploaded during the queried week and are supposed to be related to the event at stake. Users of such storing platforms are aware that in some cases, retrieved documents may not fit the query perfectly. Therefore, it is necessary to filter out the returned videos in order to keep the most relevant videos only.

3.3 Candidate Set Filtering

Candidate set filtering is performed based on the semantics of the user-generated text that surrounds each video (title and description). First, we discard non English-language content using [10], so it is possible to compare textual features on their semantic meaning. Next, we extract textual features in order to be able to use natural language processing techniques for the analysis. Each video is associated with a textual document corresponding to its title and description. We index each of these documents into the Lucene search engine. Last, we extract terms frequency across all documents for each set of videos, i.e. for each event.

Candidate set filtering is made using documents scorings. A number n documents will be pruned out from the dataset based on this ranking. The value of n will be discussed in section 4.3.

Pseudo-relevance Feedback Methods. First, we propose to perform dataset filtering by using an approach based on pseudo-relevance feedback technique.

We perform query expansion based on the top terms of the retrieved dataset. Then, the system ranks this same dataset based on the automatically formulated query. We prune out the n documents with the lowest scores.

Pseudo relevance feedback techniques are not directly applicable, because they imply to have an ordered dataset, which is not the case of the document set we obtained in the previous steps: it is the result of multiple automatic queries. We do not use the query terms of the initial queries, but rather formulate a new query using the most frequent terms of each set. Indeed, we assume that the top terms associated to each video set are representative of the event (see 4.1).

We compare the results when ranking using three different methods: the default Lucene scoring function based on a TF-IDF model; the probabilistic relevance model with Okapi BM25 ranking [8]; and the language model that we note "LMDirichlet" from [12]. As we rely on user-generated textual content, we do not trust this source for releasing a good ordering our dataset, but we rather consider that we can discard content that have the lowest relevancy scores.

Outlier Detection Method. We can also see non relevant items as outlier data in the dataset: we expect "outlier videos" to be videos that do not depict the event at stake, contrarily to other videos. The open-source software ELKI[2] defines DB outlier scores as a generalization of the DB-outlier algorithm [5] to a ranking method: the outlier score of an item is the fraction of other items that lie further than a distance d to the item. The highest scores are assigned to items more likely to be outliers. Scores are computed using the cosine distance on TF-IDF vectors. As for the previous set of algorithms, we discard items that have the highest outlier score.

4   Experiments

Our goal is to illustrate what captured people's attention regarding a certain topic: we define events and we search related multimedia content for hyperlinking. We will focus on the person scenario. One requirement of our framework is that the topic, here a person, should have raised enough queries in the past to have results in Google Trends. Hence, in this paper we will generate a multimedia biography of popular moments of a celebrity, but this work could apply to many different concepts. We populated timelines with suggestions of relevant videos, for the following persons: Oscar Pistorius (O.P), Beyonce Knowles (B.K), Mark Zuckerberg (M.Z.) and Batman (B.). The timeline is drawn from January 2004 (start date for Google Trends data) to present.

4.1   Popular Event Extraction

First, we look at the performance of our event extraction framework. For each query, we want to compare the extracted time segments or burst weeks to a

---

[2] http://elki.dbs.ifi.lmu.de/

manually created ground truth (set of events with date and description). This ground truth was constructed based on expert biographies[3] and Wikipedia data, although for the "Batman" query the motivation was different: as it is a fictional character, we created ground truth by listing movies and video game releases that are the most generally popular associated events.

The k top terms of each dataset illustrate what those documents have in common. The choice of the number of terms k is crucial: too small, it does not give enough information nor describes the event; too big, k includes terms that are too specific of a subset of documents. After looking at various values of k, we found that using $k=12$ was a good compromise. For the first event in the query "Beyonce", we illustrated the top terms on a tag cloud in figure 2. Hence, given those terms and the date, we matched this event to Beyonce's performance during the half-time of the Super Bowl on February 3, 2013.

beyonce bowl child destiny facebook fan fortune full go half halftime hq illuminati live love new perform performance show audience super superbowl video time tour twitter video watch world xlvii

Fig. 2. Tag cloud associated with the documents relative to the first burst when querying Beyonce

We compared top terms (using $k=12$) with description of the events in our ground truth to reveal matches or misses. For each person, a manual evaluation showed that the top terms were good cues for description of the event. Table 1 displays the results in term of: true positive events (TP), false positive events (FP), false negative events (FN) and discovered events (DE) which are events not described in the ground truth but we could find trace of on the web.

Table 1. We report the number of bursts along with the number of true positives (TP), false positives (FP), false negatives (FN) and discovered events (DE) for each topic

person	# burst	TP	FP	FN	DE
O.P.	1	1	0	8	0
B.K.	9	7	1	21	1
M.Z.	6	2	4	25	0
B	6	2	2	7	2

As the timeline is based on popular moments which do not exactly match official biographies, evaluation of such results is neither straightforward nor trivial. On the one hand, it does not return all highlights of a biography, but only unforeseen events that caught public attention. In this sense, true negative is

---

[3] http://www.biography.com/

hard to assess: how can one classify an event as worth appearing on the time-line? If all happenings of a lifetime are displayed, we are loosing the point in the summarization. We generated the ground truth by exhaustively taking every date and event mentioned in the expert biography and Wikipedia page, with no consideration of the importance of the event. Hence, false negatives are not very representative of the capacity of the algorithm to capture "important" moments.

On the other hand, it may reveal events that are not part of a classic biography, hence not part of the ground truth, but that could be linked to actual events that were discussed a lot: they are the ones we call *discovered events* (DE). For example, Beyonce falling during a live show in Orlando was not part of any descriptive biography, but we could discover this happening with our system.

Also, a dissimilarity of granularity between our framework (week unit) and Google Trends (month unit) made it hard to extract focused search terms when several events happened during the same month. While our algorithm has se-lected the week from the 9th to 15th of January 2011 as a peak week for the M.Z. query, the top words did not reveal a unified event; external knowledge lead us to correlate the peak in the search to rumors of Facebook shutting down.

4.2   Evaluation Dataset

In order to evaluate our filtering methods, we created a ground truth on events taken from the extracted timeline. It was done as follows: for each of the four celebrity, we took the first burst and manually associated it with an event, rely-ing on our knowledge of what happened (see table 2). An annotator assessed, for each video, if it was relevant to the event or not, not taking into account personal interest in this video. The criteria for relevancy was: if the event de-scribed, discussed or depicted in an informative manner? For M.Z., the event was a private event (his wedding) that had limited coverage (only a few pictures were disclosed) and happened very close to another event that had more video coverage (Facebook's introduction on the stock market). Hence, this dataset has not been evaluated; the ground truth was therefore made on three events.

During this process were marked as not relevant videos that were:

- clearly out of topic
- relevant to the celebrity but not to the actual event
- personal reaction or discussion about the event, not part of an aired television show (we deemed this kind of live reaction or personal feeling relevant only to very few individuals). The type of video pollutes the dataset to a great extent.
- only partly relevant (e.g., the video is a news report covering different topic, so the user would still have to choose part of the video)
- a television screen capture
- not English-speaking

We were assuming that most of the videos would depict the event at stake, because the query was focused on a very limited time segment and on specific keywords extracted from the search trends. Nevertheless, we soon realized that

Table 2. Presentation of the dataset corresponding to three events

person	event description	week	#videos	TP	FP	burst
O.P.	murder of his girl-friend	2013/02/10-16	97	74 (76.29%)	23 (23.71%)	100
B.K.	superbowl halftime performance	2013/02/3-9	165	66 (40%)	99 (60%)	79
B	shooting at the first of Dark Knight Rises	2012/07/15-21	167	55 (32.93%)	112 (67.07%)	56

this hypothesis did not hold: for some datasets, less than half of the videos were relevant to the subject. We matched this figure with the burst value of the event: the higher the burst, the cleaner the dataset (see table 2). This finding highlighted the need to prune out irrelevant media from the dataset.

4.3   Candidate Set Filtering

As seen earlier, document scoring (either degree of outlierness or ranking given a search query) will be base of the decision to prune out videos from the dataset.

Filtering Out n Videos Per Datasets. We consider non-relevant data as false positives and relevant documents as true positives. We plot the number of false positives (FP), the number of true positives (TP), the false positive rate (FPR) and the number of true positive rate (TPR) in the n documents pruned out of the set. The results are given on figure 3.

The results should be interpreted as follows: the last 15 items of the dataset with the default Lucene similarity contain:

- for O.P., 8 non-relevant items out of 23 (34.8% are detected) and 7 relevant documents out of 74
- for B.K. and B., 15 non-relevant items out of 99 (15.2% are detected) and no relevant document (out of 66)

We can see on figure 3 that the four algorithms have very similar performances, so we will work with the default similarity for the remaining of this paper. For a given number of items, the false positive rate (percentage of non-relevant items) is above the true positive rate for all events. This means that by pruning out the last n videos, we discard more non-relevant content than relevant content relatively to the number in the initial set. Discarding some true positive content is a drawback that we cannot avoid.

Choice of the Parameter n. We need to choose the number of items pruned out. This parameter cannot have a fixed value : it should first depend on the size of the dataset (e.g., pruning out 20% of the dataset). Also, different dataset contains more or less non-relevant videos. As said in 4.2, we use the burst value of the event as a cue towards the composition of the dataset.

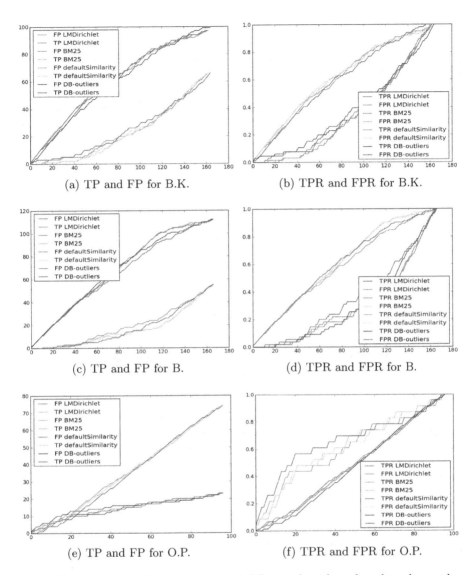

(a) TP and FP for B.K.

(b) TPR and FPR for B.K.

(c) TP and FP for B.

(d) TPR and FPR for B.

(e) TP and FP for O.P.

(f) TPR and FPR for O.P.

Fig. 3. Results for the different queries and different algorithms, based on the number of filtered documents

We defined a threshold that is adaptive to the dataset by taking into account its size and the burst value. The number of videos pruned out should increase with the number of videos in the dataset and decrease, but less than linearly, with the value of the burst. Hence, we choose to use:

$$\left(n = \frac{\#videos}{\sqrt{burstvalue}} * \alpha\right) \tag{1}$$

where α is a parameter that controls the relative size of the dataset. We performed the filtering using this formula for α ranging from 1 to 7 and measured the false positive rate in the final dataset (see figure 4). We aim to have a low percentage of non-relevant data in the final dataset, so this figure suggests 6 as a suitable value for this parameter: there is a relatively small error rate across all three queries.

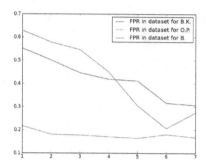

Fig. 4. We choose the alpha parameter by plotting the false positive rate in the final dataset for the three events. Given this observation, alpha should be chosen equal to 6 to have the lower false positive rate across all dataset.

Final Evaluation. We ran our dataset filtering technique on the three test sets with the default similarity function and reported the performances in table 3. The percentage of non-relevant items discarded ranges from 69.57% to 93.75%, at the cost of discarding around 50% of relevant videos in the dataset. Those numbers should be examined in the light of the resulting datasets. The results suggest that, while we do not obtain perfectly clean sets, there is a significant improvement in term of false positive rate, between the initial dataset and the final one. The improvement is minor for O.P., while it is very important for B. and B.K. An illustration of results is shown in figure 5.

Table 3. Results of our methods. FP and TP are the number of false positives and true positives in the content that we pulled out of the dataset, and the associated percentage is the rate of false (true) positive among all false (true) positive. The last two columns compare the false positive rate in the initial and final datasets.

celebrity	# videos pruned out	FP	TP	FPR (final)	FPR (initial)
O.P.	58	16 (69.57%)	42 (56.76%)	17.95%	23.71%
B.K.	111	82 (82.83%)	29 (43.94%)	31.48%	60%
B	133	105 (93.75 %)	28 (50.91%)	20.59%	67.07%

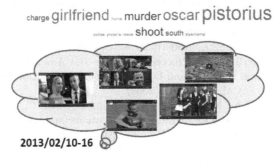

Fig. 5. Results for the first event mined from the query 'Oscar Pistorius'

5   Conclusion

In this paper, we tackled the issue of topic enriching by linking events to media items. We focused on events that were revealed by their existence on both people's interest and by the existence of related content in social media platform such as YouTube. We designed a framework that automatically proposes a pool of media items corresponding to each event and that removes non-relevant content. We describe events using words that can be visualized on a tag cloud. Textual features are used to automatically refine the dataset: items are ranked using different measures and a varying number of items (that is adaptive to the dataset and to the event) are pruned out. We compared different techniques that perform similarly. When filtering the dataset, our priority is to get a set that is as clean as possible from non-relevant items, at the cost of discarding some relevant data. Results suggest a significant decrease of the rate of non-relevant items between the base dataset and the final one.

Our approach is based on textual features which are user-generated; in order to have an insight on the actual video content, future work will perform video content analysis based on visual and audio information. We will also attempt to discover long-term events whose atomic unit will be more than a week by time-series mining.

Acknowledgments. This work was supported by the European Commission under contracts FP7-287911 LinkedTV and FP7-318101 MediaMixer.

References

1. Angiulli, F., Pizzuti, C.: Fast Outlier Detection in High Dimensional Spaces. In: Elomaa, T., Mannila, H., Toivonen, H. (eds.) PKDD 2002. LNCS (LNAI), vol. 2431, pp. 15–26. Springer, Heidelberg (2002)

2. Breunig, M., Kriegel, H.-P., Ng, R.T., Sander, J.: LOF: Identifying Density-Based Local Outliers. In: Proceedings of the 2000 ACM SIGMOD International Confrence on Management of Data, pp. 93–104. ACM (2000)
3. Cao, G., Nie, J.-Y., Gao, J., Robertson, S.: Selecting good expansion terms for pseudo-relevance feedback. In: Proceedings of the 31st Annual International ACM SIGIR Conference on Research and Development in Information Retrieval, SIGIR 2008, pp. 243–250. ACM, New York (2008)
4. Capra, R.G., Lee, C.A., Marchionini, G., Russell, T., Shah, C., Stutzman, F.: Selection and context scoping for digital video collections: an investigation of youtube and blogs. In: Proceedings of the 8th ACM/IEEE-CS Joint Conference on Digital Libraries, JCDL 2008, pp. 211–220. ACM, New York (2008)
5. Knorr, E.M., Ng, R.T.: Algorithms for Mining Distance-Based Outliers in Large Datasets, pp. 392–403 (1998)
6. Ponte, J.M., Croft, W.B.: A language modeling approach to information retrieval. In: Proceedings of the 21st Annual International ACM SIGIR Conference on Research and Development in Information Retrieval, SIGIR 1998, pp. 275–281. ACM, New York (1998)
7. Ramaswamy, S., Rastogi, R., Shim, K.: Efficient algorithms for mining outliers from large data sets. SIGMOD Rec. 29(2), 427–438 (2000)
8. Robertson, S., Walker, S., Jones, S., Hancock-Beaulieu, M., Gatford, M.: Okapi at TREC-3, pp. 109–126 (1996)
9. Sahuguet, M., Huet, B.: Socially Motivated Multimedia Topic Timeline Summarization. In: Proceedings of the 2013 International Workshop on Socially-aware Multimedia, SAM 2013. ACM (2013)
10. Shuyo, N.: Language Detection Library for Java (2010)
11. Xu, Y., Jones, G.J., Wang, B.: Query dependent pseudo-relevance feedback based on wikipedia. In: Proceedings of the 32nd International ACM SIGIR Conference on Research and Development in Information Retrieval, SIGIR 2009, pp. 59–66. ACM, New York (2009)
12. Zhai, C., Lafferty, J.: A study of smoothing methods for language models applied to Ad Hoc information retrieval. In: Proceedings of the 24th Annual International ACM SIGIR Conference on Research and Development in Information Retrieval, SIGIR 2001, pp. 334–342. ACM, New York (2001)

Spatial Similarity Measure of Visual Phrases for Image Retrieval

Jiansong Chen*, Bailan Feng, and Bo Xu

Institute of Automation, Chinese Academy of Sciences
Zhongguancun East Road No.95, Haidian District, Beijing, China
{jiansong.chen,bailan.feng,xubo}@ia.ac.cn

Abstract. Spatial information plays an essential role in accurate matching of local features in applications, e.g., image retrieval. Despite of previous work, it remains a challenging problem to extract appropriate spatial information. We propose an image retrieval framework based on visual phrase. By encoding the spatial information into the similarity measure of visual phrases, our approach is able to capture accurate spatial information between visual words. Furthermore, the image-specific visual phrase selection process helps to reduce large number of redundant visual phrases. We have conducted experiments on two datasets: UKbench and TRECVID, which shows that our ideas significantly improve the performance in image retrieval application.

Keywords: Image retrieval, visual phrase, spatial similarity measure.

1 Introduction

Most state-of-the-art image retrieval approaches are based on the bag-of-visual-word (BoW) model[1], due to its compactness and efficiency. Although the model works generally well, it loses much spatial information, leading to large number of false matches. An important way to improve BoW model is capturing the co-occurrence patterns among visual words[10,3,9,4]. The idea behind the method is that some visual words have strong co-occurrence relations with each other, thus combining them together should be more reasonable.

One of the key issues of the method is to define a co-occurrence pattern and then mine all the visual phrase(VP)s fitting in with the pattern. Simple co-occurrence pattern without space division has been proposed by researchers[10,3]. However, these approaches neglect the relative spatial layouts among visual words. Pattern with rough space division either in local space[9,4] or in entire image space[12] can only capture weak spatial information because dense space division could also bring much dislocation error. In the case, identical VPs are possibly identified as different ones due to slightly different spatial layouts of them. However, intuitively, exploiting accurate spatial information in visual

---

* The work was supported by National Nature Science Foundation of China(grant No.61202326 and No.61303175).

C. Gurrin et al. (Eds.): MMM 2014, Part II, LNCS 8326, pp. 275–282, 2014.

phrase method would benefit the application of image retrieval. Additionally, since the total number of VPs increases exponentially with the number of visual words, selecting descriptive VPs for an image is another problem should be considered.

Based on the above analysis, this paper proposes a novel VP-based image retrieval approach. After mapping local invariant features to visual words, we combine pairs of visual words within local space into VPs. For each VP, we calculate its internal spatial information and save them into inverted file index. In the retrieval process, we encode the spatial information into the similarity measure of matching VPs, and treat it as the voting weight of the corresponding VPs. In addition, for each image, whether query or database image, an image-specific VP selection strategy is adopted to reduce the redundancy of VPs.

The rest of the paper is organized as follows. In the following section, we describe our algorithm, including VP model, VP selection strategy and its similarity measure. The extensive experiments and discussions are presented in section 3. Finally, we conclude this paper in section 4.

2   Algorithm

2.1   VP Model

The VP model is built upon visual words. We exploit K-Means algorithm to train visual word vocabulary by clustering a large number of SIFT descriptors (DoG detector is adopted). Afterwards, we map each local interest feature to corresponding visual word by nearest neighbor search. The vocabulary is organized by k-d tree structure for efficiency during quantization.

In this paper, a VP is defined as a commonly co-occurring visual word pair in particular spatial constraint. According to the definition, to construct a VP, the corresponding visual words should satisfy two conditions: (1) their locations should satisfy a certain spatial constraint, and (2) they should be frequently co-occurring together. This subsection presents the VP structure satisfying the first condition, and the second condition will be satisfied in the VP selection process (see next subsection).

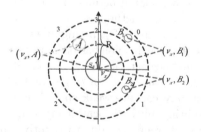

Fig. 1. Illustration of VP model

As shown in Fig.1, for a visual word v_x, we define its co-occurrence space as a circular region with radius of r_x. r_x is computed by $r_x = s_x * \gamma$, where s_x is the scale of local interest feature corresponding to visual word v_x, and γ is a parameter controlling the scope of co-occurrence space. Because r_x is in proportion to the scale of word v_x, it is not sensitive to scale change. Each word v_y occurring in the co-occurrence space is paired with v_x constituting a VP candidate, signified as $vpc = (v_x, v_y)$, in which v_x is called center word and v_y is called co-occurrence word. In Fig.1, taking v_x as center word, there are totally three VP candidates: (v_x, A), (v_x, B_1), and (v_x, B_2). Note that, B_1 and B_2 are the same visual words with different positions, however we consider (v_x, B_1) and (v_x, B_2) to be identical VP ignoring their spatial layout difference. So that, the model is much robust to object variations among different images.

2.2   Image-Specific VP Selection Strategy

The number of VP candidates in an image could be much larger than the number of visual words. For example, in our experiments, the average number of visual words in an image is about 1k, while the number of VP candidates is approximately 10k (γ is set to 8). Much of the VP candidates are redundant as well as increasing the memory cost and reducing the retrieval efficiency. In order to reduce such redundancy, we present our VP selection strategy as follows.

The selection process is performed on each image independently. In an image I_j, we calculate a significance score $L(i, j)$ for each VP candidate vpc_i in it. According to the score $L(i, j)$, we rank and select the top K VP candidates with the highest scores as the final reserved VPs for the image. In this paper, we set K as:

$$K = min(K, M_j) \tag{1}$$

where M_j is number of VP candidates in image I_j.

The significance measure used in this paper for VP selection contains two components: the database significance score and the image-specific significance score. Formally, the significance score $L(i, j)$ is defined as:

$$L(i, j) = L(vpc_i, I_j) = \Theta_i \cdot \Phi_{i,j}, \tag{2}$$

where Θ_i represents the significance of vpc_i in overall training database, while the term $\Phi_{i,j}$ indicates the significance of vpc_i associating to the specific image I_j. Following [4], we measure the database significance score as:

$$\Theta_i = freq(vpc_i) \cdot \frac{P(vpc_i|D_I)}{1 + \prod_{n=1}^{2} P(v_{i,n}|D_I)}, \tag{3}$$

where $freq(vpc_i)$ is the total occurrence number of vpc_i in training corpus, $P(vpc_i|D_I)$ is the probability of vpc_i occurring in document D_I and $P(v_{i,n}|D_I)$ is the probability of the component words of vp_i occurring in document D_I. $P(vpc_i|D_I)$ and $P(v_{i,n}|D_I)$ can be approximated by:

$$P(vpc_i|D_I) \approx \frac{docfreq(vpc_i)}{T} \tag{4}$$

$$P(v_{i,n}|D_I) \approx \frac{docfreq(v_{i,n})}{T}, \tag{5}$$

where $docfreq(x)$ is the number of documents with VP candidate or visual word x in training corpus, and T is the total number of documents in training corpus. The intuition behind formula (3) is to weight the VP whose component words are much more highly correlated than independent.

The image-specific significance function $\Phi_{i,j}$ is defined as the normalized term frequency of vpc_i in image I_j:

$$\Phi_{i,j} = \frac{n_{ij}}{n_j}, \tag{6}$$

where n_{ij} and n_j respectively signify the number of vpc_i and the total number of VP candidates in I_j. The formula gives high importance of a VP candidate, if it occurs more often in a specific image.

The image-specific selection strategy has two benefits. First, the numbers of selected VPs for different images are basically the same, avoiding the feature imbalance problem. Moreover, the strategy allows to consider both database significance and image-specific significance in selecting VPs. Therefore, the selected VPs would be more suitable for images.

2.3   Spatial Similarity Measure (SSM)

Although the spatial information is neglected in VP model, it is still helpful for discrimination of VPs. To quantitatively describe the spatial information, the co-occurrence space of each visual word is equally divided into $N \times N$ (4×4 is illustrated in Fig.1 grids based on the orientation and scale of corresponding local feature. Then, we obtain the positions of the co-occurrence words, relative to the center word, in the coordinate system. Specifically, for a VP $vp = (v_x, v_y)$, its spatial layout information is quantitatively described by $(s_{x,y}, \theta_{x,y})$, in which $s_{x,y}$ and $\theta_{x,y}$ are the grid coordinates of v_y in the co-occurrence space of v_x in scale and orientation axis respectively.

Given two matching VPs: (v_x^q, v_y^q) from query image I_q and (v_x^d, v_y^d) from database image I_d, their corresponding spatial layouts are signified as $(s_{x,y}^q, \theta_{x,y}^q)$ and $(s_{x,y}^d, \theta_{x,y}^d)$ respectively. We measure the inconsistencies of their spatial layouts and transform them to spatial similarity terms. Specifically, the inconsistency in scale axis $\Delta s^{q,d}$ is defined as the absolute difference of their scale coordinates:

$$\Delta s^{q,d} = |s_{x,y}^q - s_{x,y}^d|. \tag{7}$$

Since given two orientation coordinates $\theta_{x,y}^q$ and $\theta_{x,y}^d$, their distances can be calculated in two opposite directions. We define the orientation inconsistency $\Delta \theta^{q,d}$ as the minimum absolute distance of orientation coordinates:

$$\Delta \theta^{q,d} = min(|\theta_{x,y}^q - \theta_{x,y}^d|, N - |\theta_{x,y}^q - \theta_{x,y}^d|). \tag{8}$$

Afterwards, the spatial inconsistencies are transformed to spatial similarity terms by exponential function:

$$\omega_s^{q,d} = e^{-\delta \Delta s^{q,d}}, \tag{9}$$

$$\omega_\theta^{q,d} = e^{-2\delta \Delta \theta^{q,d}}, \tag{10}$$

where δ is a positive parameter controlling the influence of inconsistencies. The ranges of $\omega_s^{q,d}$ and $\omega_\theta^{q,d}$ are both from 0 to 1. Thus, the spatial similarity measure (SSM) $\omega^{q,d}$ of the matching VPs is given by:

$$\omega^{q,d} = \omega_s^{q,d} \cdot \omega_\theta^{q,d} = e^{-\delta(\Delta s^{q,d} + 2\Delta \theta^{q,d})}. \tag{11}$$

Image retrieval based on BoW of the cosine similarity can be interpreted as a voting mechanism. In this paper, we use $\omega^{q,d}$ as a weighting term to update the standard idf (inverse document frequency) weighting:

$$\omega(vp) = \omega^{q,d} \cdot idf(vp). \tag{12}$$

Based on formula (12), the VPs with the consistent spatial layouts across two images probably score highly and contribute more weighting to the similarity of images. The proposed method quantitatively describes the spatial layout of VP instead of using it to differentiate VPs directly. Therefore, it has more discriminative power. In real experiments, the local co-occurrence space for a visual word is divided into 256×256 grids, which is hardly possible in the traditional methods.

To implement the SSM for large-scale retrieval, we use an inverted file index to store spatial layout information of VPs. Fig.2 shows the structure of our index. We keep one entry for each VP occurrence. In each entry, we use 4 bytes for the ID of the image in which the VP occurs, and 2 bytes for quantized $s_{x,y}^d$, $\theta_{x,y}^d$ respectively. In the process of retrieval, $s_{x,y}^d$ and $\theta_{x,y}^d$ are obtained from index directly and then used to compute the corresponding SSM, which saves a lot of computation.

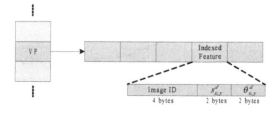

Fig. 2. Inverted file structure

3 Experiments

The image retrieval system was experimented on two dataset: University of Kentucky (Ukbench) dataset[2] and an image corpus collected from TRECVID dataset[7]. Ukbench dataset contains 10200 images of 2550 objects. Each object has four visually similar images, which are taken under four different viewpoints of the objects. For TRECVID image corpus, we collected 211 images with 20 categories from keyframes of TRECVID 2005 video dataset. In each category, we select 3 images in each category as queries comprising 60 query images in total. To construct a real image retrieval environment, we distract the images by adding another 21,532 keyframes from the TRECVID 2007 video dataset.

The common settings for all the experiments are summarized here. N, γ, δ are set to 256, 8 and 0.02 experimentally. For training, the visual word vocabulary and database significance score are both trained on Ukbench dataset. For evaluation, the mean average precision (mAP) is adopted as our evaluation metric, which is generated as follows: for each query, we compute its average precision and then take the mean value over all queries. The N_s score[2] is also reported on Ukbench dataset, as it is the performance measure usually reported for this dataset. It is computed as the average number of recalled images at the first four returned images.

3.1 Influence of Parameter K

In the framework, the parameter K controls the number of preserved VPs for an image. We conduct experiments on the Ukbench dataset to examine its effect on retrieval performance. In the experiments, 1k word vocabulary is used and the results are shown in Fig.3. The curve shows a clear tendency that the performance of our system improves steadily along with the decrease of K when $K > 3000$. It validate our idea that the image-specific selection strategy could reduce redundant VPs and preserve descriptive VPs for the images. Note that it makes a bad effect on the performance when K is too small due to the fact that some helpful VPs are also discarded in this case. Finally, we choose to set $K = 3000$ in all the following experiments.

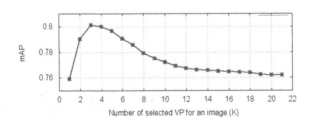

Fig. 3. System performance with different values of K

Table 1. Comparison of different methods

Method	Vocab	Selection	Ukbench $(mAP(N_s))$	TRECVID (mAP)
BoW[1]	1k	no	0.527(1.83)	0.499
VP	1k	no	0.687(2.43)	0.543
VP	1k	yes	0.727(2.68)	0.581
SVP(2x2)	1k	yes	0.749(2.74)	0.632
SVP(4x4)	1k	yes	0.756(2.76)	0.654
SVP(6x6)	1k	yes	0.741(2.72)	0.625
VP+SSM	1k	no	0.762(2.81)	0.657
VP+SSM	1k	yes	0.801(2.96)	0.669
VP+SSM	2k	yes	0.810(3.08)	0.683
VP+SSM	4k	yes	0.815(3.13)	0.678
Wu[13]	1M	no	n/a(3.11)	n/a
Zhang[8]	122.5k	no	n/a(3.19)	n/a

3.2   Comparison of Retrieval Performances

To certificate the effectiveness of the proposed method, we compare it with other algorithms of different settings both on Ukbench and TRECVID dataset. The results are shown in Table.1. In the table, VP means the simple co-occurrence visual phrase model described in section 2.1. SVP refers to [9], in which the same co-occurrence words falling into different grids of the co-occurrence space are considered to constitute different VPs with the center word. The content within parentheses after SVP represents the space division parameters.

Apparently, combining visual words into VPs improves the performance a lot than using dependent visual words (BoW). For VP method, the image-specific VP selection strategy improves it from further which is also demonstrated in previous experiment. Incorporating spatial information by using SVP model brings some benefits. However, excessive division of the model also hurts the performance, for example SVP(6x6) performs worse that SVP(4x4). This might be because the SVP model is very rough and slightly excessive division would bring more dislocation error than discrimination ability. In contrast, the proposed SSM shows much better performance than the algorithm without SSM and algorithms using SVP model. This demonstrates the effectiveness of SSM, which softly incorporates spatial information in similarity measure for image retrieval application. Since the image-specific selection and SSM enhance the discrimination ability of VP in different aspects, combing them together (VP+SSM with selection) obtains further improvement.

We also experiment our method on different sizes of visual word vocabulary. The results reveal that increasing the vocabulary size promotes the performance of our method analogous to classic BoW method. However, it is noticeable that the improvement is not considerable when the vocabulary is large. Our explanation is that the quantization error of VP is also magnified with the combination of visual word. When the vocabulary is large, it brings much quantization error and counteracts the discrimination ability. Finally, comparing with [13] and [8], which encode spatial contexts into BoW method, our method is almost comparable with them. However, considering that the vocabulary size of our method is much smaller than theirs, our method shows an advantage of efficiency in training visual word vocabulary.

4   Conclusion

In this paper, a novel image retrieval approach based on visual phrase is reported. In the approach, a spatial similarity measure of VP is proposed. The measure has more descriptive power than traditional spatial VP by imbedding accurate spatial information. Furthermore, we make use of an image-specific VP selection strategy to select sufficient and suitable VPs for different images. The experimental results both on Ukbench and TRECVID dataset shows significant performance improvement.

References

1. Sivic, J., Zisserman, A.: Video Google: A text retrieval approach to object matching in videos. In: CVPR 2003, pp. 1470–1477. IEEE (October 2003)
2. Nister, D., Stewenius, H.: Scalable recognition with a vocabulary tree. In: CVPR 2006, vol. 2, pp. 2161–2168. IEEE (2006)
3. Zheng, Q.F., Gao, W.: Constructing visual phrases for effective and efficient object-based image retrieval. In: ACM TOMCCAP 2008, vol. 5(1) (2008)
4. Zheng, Y.T., Zhao, M., Neo, S.Y., Chua, T.S., Tian, Q.: Visual synset: towards a higher-level visual representation. In: CVPR 2008, pp. 1–8. IEEE (June 2008)
5. Yuan, J., Wu, Y., Yang, M.: Discovery of collocation patterns: from visual words to visual phrases. In: CVPR 2007, pp. 1–8. IEEE (June 2007)
6. Hu, Y., Cheng, X., Chia, L.T., Xie, X., Rajan, D., Tan, A.H.: Coherent phrase model for efficient image near-duplicate retrieval. IEEE Trans. Multimedia 11(8), 1434–1445 (2009)
7. http://trecvid.nist.gov/
8. Zhang, S., Huang, Q., Hua, G., Jiang, S., Gao, W., Tian, Q.: Building contextual visual vocabulary for large-scale image applications. In: ACM Multimedia, pp. 501–510 (2010)
9. Liu, D., Hua, G., Viola, P., Chen, T.: Integrated feature selection and higher-order spatial feature extraction for object categorization. In: CVPR 2008, pp. 1–8. IEEE (June 2008)
10. Zhang, S., Tian, Q., Hua, G., Huang, Q., Li, S.: Descriptive visual words and visual phrases for image applications. In: ACM Multimedia, pp. 75–84. ACM (October 2009)
11. Jégou, H., Douze, M., Schmid, C.: Improving bag-of-features for large scale image search. IJCV 87(3), 316–336 (2010)
12. Zhang, Y., Jia, Z., Chen, T.: Image retrieval with geometry-preserving visual phrases. In: CVPR 2011, pp. 809–816. IEEE (June 2011)
13. Wu, Z., Ke, Q., Isard, M., Sun, J.: Bundling features for large scale partial-duplicate web image search. In: CVPR 2009, pp. 25–32. IEEE (June 2009)

Semantic Based Background Music Recommendation for Home Videos

Yin-Tzu Lin[1], Tsung-Hung Tsai[2], Min-Chun Hu[2,3], Wen-Huang Cheng[2], and Ja-Ling Wu[1]

[1] National Taiwan University, Taipei, Taiwan, R.O.C.
{known,wjl}@cmlab.csie.ntu.edu.tw
[2] Academia Sinica, Taipei, Taiwan, R.O.C.
{thtsai925,whcheng}@citi.sinica.edu.tw
[3] National Cheng Kung University, Tainan, Taiwan, R.O.C.
anita_hu@mail.ncku.edu.tw

Abstract. In this paper, we propose a new background music recommendation scheme for home videos and two new features describing the short-term motion/tempo distribution in visual/aural content. Unlike previous researches that merely matched the visual and aural contents through a perceptual way, we incorporate the textual semantics and content semantics while determining the matching degree of a video and a song. The key idea is that the recommended music should contain semantics that relate to the ones in the input video and that the rhythm of the music and the visual motion of the video should be harmonious enough. As a result, a few user-given tags and automatically annotated tags are used to compute their relation to the lyrics of the songs for selecting candidate musics. Then, we use the proposed motion-direction histogram (MDH) and pitch tempo pattern (PTP) to do the second-run selection. The user preference to the music genre is also taken into account as a filtering mechanism at the beginning. The primitive user evaluation shows that the proposed scheme is promising.

1 Introduction

With handy video recorder, many people like to record adorable moments in their lives and share the videos on social medium. A suitable soundtrack usually adds interest to the shared video just like dotting the dragon's eye on the painting. However, it is time-consuming and labour-intensive to find such music in a large music collection containing all kinds of songs nowadays. Despite this demand, it is surprising that few researches have focused on the recommendation of matching soundtracks specific to home videos. Most of the existing works focused on editing the video to better align the motion to a given music's tempo [14,3,11] or finding matching music for just photo slideshows [2,4,5], but few efforts are made to find a proper soundtrack for a given home video since it is challenging to define a good matching between the visual and aural content. The quality of the matching is subjective and highly related to personal preference to the music.

C. Gurrin et al. (Eds.): MMM 2014, Part II, LNCS 8326, pp. 283–290, 2014.

Feng and Ni [6] used movie clips as the training data to model the relationship between visual and aural contents. However, their method does not take the semantics in the video and audio into account. Besides, we doubt weather the relation of video and audio in movie clips can be applied to home videos since home videos often contains chaotic scenes with almost disordered motion and color information whereas movies often contain designed scenes and background musics. Furthermore, the matched soundtrack will not be the original music but a cut-and-paste version of several short music segments which may lead to audible artifacts. Chen *et al.*[2] and Dunker *et al.*[4,5] associated photos and music through a perceptual way, e.g. emotion recognition on the both. But for home videos, emotion is not the only thing to convey. Typical home videos are usually recorded for specific events such as wedding ceremony, birthday party, etc. If we just match the video according to the emotion criterion, we may pick a happy music for a wedding video but the topic of the music is nothing about love. Instead, we proposed to match the video and music through a semi-semantic way. That is, the recommended music should contain semantics that relate[1] to the ones in the input video, e.g. a love song to a wedding video.

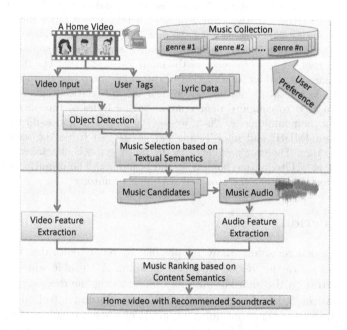

Fig. 1. System Overview

Figure 1 illustrates the proposed framework of background music recommendation for a given home video. The video semantics come from few tags given

---

[1] We employ the word "relation" not word "equality", and that's why we call it semi-semantic.

by the user and object detection results obtained from video keyframes, and the audio semantics come from the lyrics of the music. Although music is more often used to express emotion than to convey ideas, the topics described in lyrics will somewhat illustrate what the music itself recalls people when listening to it. We believe that people would think the soundtrack and the video are matching if the topic lyrics of the music is related to the topic event of the video. However, music depicting the same topic may be composed in various styles and with different tempi, and a textual-semantically matched song may not be content-semantically matched to the video. For example, the tempo of the music and visual motion of the video may be inharmonious. Many previous "music video" and "photo slideshow" generation works just edited the video's motion and the photos' show up timings to make them synchronize with the rhythms of the music. Instead, we "find out" songs whose aural tempi best match the visual motions throughout the video from candidate musics picked according to textual-semantics (c.f. Figure 1). If the motions in video and the rhythms in music matched, the joint video will conform to the principle of media aesthetics[13] and is believed to be entertaining. Another issue that should be considered is the user preference. As we mention before, the quality of the matching result may highly depend on user's personal preference to the music's style. A classical music fan may not enjoy rock music even though the topics of the music is suitable for the input video. As a result, we let the users choose the genre of the candidate music at the beginning (c.f. Figure 1). We believe acceptably few user intervention will improve the result a lot.

2   Music Selection Based on Textual Semantics

In this section, we will describe details about how we match the textual semantics between video and audio (c.f. the upper part of Figure 1).

2.1   Textual Semantics in Video

The textual semantics of the given video can be extracted from two kinds of sources, i.e. the tags given by the user and the tags automatically annotated by object detectors. Users are asked to provide related tags when they upload videos to existing social medium. However, the number of tags given by users is usually too small to properly describe the corresponding video. As a result, we further analyze the video contents and automatically annotate the video with the involved objects by the technique of object detection. We use the "object bank" proposed by Li et al.[8] to detect 177 types of objects in the key frames of the given video. The detected object names are then regarded as extra tags to annotate the video. Unlike tags given by the user, tags annotated by the "object bank" are not trusty enough since the appearance of an object does not imply that this object is an important semantics in the current video. For example, "lamp" will be detected in many home videos, but "lamp" is usually less related to the event that the video wants to record. To better reflect the importance of the detected objects, we compute the well-known TF*IDF^2 weighting ρ_{ik} for each object i in video k.

2.2   Textual Semantics in Music

For gathering semantics in music, we use the lyric data because lyrics are the most reachable metadata for music. Similar to the video tags annotated by "object bank", the occurrence of a specific word in the lyrics does not imply that the corresponding song is actually about that word. Therefore, the technique of TF*IDF is again utilized to compute the weighting ρ_{jl} for word j in song l. We use the lyric data provided by the musiXmatch lyrics dataset (c.f. Section 4), which is in bag-of-words format. That is, for each song, we have a histogram of the appearing times of 5000 words (the top 5000 frequent words in the whole dataset). The words have been run a stemming process to link the words like "vitori" to "victory" by the data provider. We further remove stop words from the data and remain 2085 words for the further analyses.

2.3   Textual Semantics Matching

Given the video tag set $\mathcal{V}_k$ of video k and the lyrics word set $\mathcal{M}_l$ of song l, we construct the relation between the video and the music according to conceptual-semantic and lexical relation network provided by WordNet [9]. For each pair of words in $\mathcal{V}_k$ and $\mathcal{M}_l$, we compute the corresponding textual semantic similarity $S_N(w_i, w_j)$ based on WordNet distance. Then, the similarity between video k and song l can be calculated as follows,

$$S(k,l) = \sum_{i \in \mathcal{V}_k, j \in \mathcal{M}_l} \frac{2 \cdot \rho_{ik} \cdot \rho_{jl}}{\rho_{ik} + \rho_{jl}} \cdot S_N(w_i, w_j). \tag{1}$$

We rank the musics according to $S(k,l)$ and choose the top L songs (L=20 in our work) as candidates for content matching.

3   Content Semantics Matching

After selecting songs based on textual semantics, we further consider the content semantics to align the video with each selected song in the temporal axis. A corresponding matching score will be obtained and we choose the song with the largest matching score as the background soundtrack for the given video. Generally, a music piece and a video segment are perceptually matched if the short-term rhythm of the music is consistent with the short-term motion in the video. For instance, if the video has high motion at the first half and low motion at the last half, the matched music's tempi should be in the same manner, i.e. "fast tempo → slow tempo". Therefore, we extract short-term feature with a sliding window (says 6 seconds in our work) for both video and music audio.

---

^{2} http://en.wikipedia.org/wiki/Tf*idf

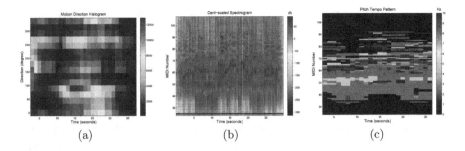

Fig. 2. (a) The proposed motion direction histogram of one of the home video 'Ronin the crazy cat" in our dataset. (b) Cent-scaled Spectrogram (c) The proposed pitch tempo pattern feature of the 30-second preview audio in the MSD[1], track_id: TRAQLBA12903CD1135.

3.1   Video Representation

We apply optical flow[3] to the whole video and obtain local motion vectors for each frame. The short-term motion feature of the tth time instance in the video is calculated based on all local motion vectors of the frames inside the current sliding window. To be precise, for each time instance t, we quantize the directions of all local motion vectors in the following 6 seconds and construct a 16-bins motion-direction histogram (MDH). For each local motion vector v, if its direction is quantized to the histogram bin b_i, the motion magnitude of v is added to the bin b_i. Figure 2(a) shows an example of the MDH of all time instances in a given video.

3.2   Music Audio Representation

Corresponding to motion-direction histogram, which depicts the short-term motion directions/magnitudes in the video, we proposed pitch tempo pattern (PTP) to model the short-term tempo of each pitch at time t. The PTP is an extension of the fluctuation pattern (FP) [10], which extracts a matrix $\mathbf{P}_t \in \mathbb{R}^{M \times N}$ for each t, describing the amplitude modulation of the loudness per pitch. M is the number of notes, and N is the number of FFT coefficients picked (in our work M=84, N is determine using the same scheme as [10]). We calculate the PTP of the pitch s at time t as

$$C(s,t) = \Gamma(\arg\max_{\tau} P_t(s,\tau)), s = 1 \ldots M, \tau = 1 \ldots N,$$

where $\Gamma(\cdot)$ is a function that map the FFT coefficient indices to real modulation frequency values. That is, for each semitone band, we find the index of the coefficient that has the maximum response in the FFT and use its corresponding modulation frequency to represent the tempo in that pitch. Figure 2(b) shows the cent-scaled spectrogram (CSS) value [7] at each pitch along the temporal axis of

---

[3] http://grauonline.de/wordpress/?p=14

a song in our dataset. Based on the CSS values in Figure 2(b), the corresponding PTP values are shown in Figure 2(c).

3.3   Content Matching

The aforementioned MDH and PTP features describe the short-term moving speed in the video and audio content, respectively. Generally, a video and a song are regarded as perceptually matched if the trend of the motion features in the video are related to those in the music. However the MDH at time t_i in the video and the PTP features at time t_j in the music can not be directly matched since the dimension of these two feature vectors are different. Therefore, we extract a 2-D speed descriptor from the MDH and PTP features, respectively, to further illustrate the short-term speed in the video and the music. For each time instance t_i in the video, the 2-D speed descriptor is obtained by computing the average magnitude among all motion directions and the entropy with respect to the motion direction, which imply the total degree of motion and the degree of chaos of motions, respectively. For each time instance t_j in the music, the 2-D speed descriptor is obtained by computing the average and the entropy with respect to the modulation frequency values, which imply the degree of total tempo and the degree of asynchronous tempo in pitch. Given two sequences of 2-D descriptors extracted from a video and a music respectively, we apply approximate string matching (ASM) [12] to compute a fuzzy aligning score,which allows subtle time shift between the two sequences. Given a home video and the musics retained by the textual semantics selection process, we use the corresponding aligning scores computed by ASM to rank musics and recommend the one with the highest score to be the background soundtrack.

4   Evaluation

Our experiment was conducted on a subset of Columbia Consumer Video (CCV) Database [4]. The CCV database contains 9317 youtube video IDs belonging to 20 different categories. We further crawled the videos to obtain its title and tags via youtube API. For music collection, we used the intersection of MSD [1] genre dataset[5] and MSD musiXmatch lyrics dataset [1]. Since we need custom audio features that are not provided in the dataset, we downloaded the preview audio (around 30 seconds) via 7digital API[6] for audio feature extraction. After removing non-downlowdable musics, our music collection is composed of 10,492 songs.

In the user evaluation, we compared three recommendation schemes: recommendation using textual semantics only, the proposed scheme, and randomly recommendation. To reduce the cognitive load of the participants, we chose 10

---

[4] http://www.ee.columbia.edu/ln/dvmm/CCV/

[5] http://labrosa.ee.columbia.edu/millionsong/blog/
   11-2-28-deriving-genre-dataset

[6] http://developer.7digital.net/

videos from the category "dog" and the musics in the "folk" genre (2441 songs) to run the three recommendation schemes. For each video, we created three sets of recommend music pieces: the top 5 songs derived according to textual-only scheme, the top 5 songs recommended according to the proposed scheme, and 5 random selected songs. The recommended music has composed with the original video and display to the participants. Five users aged around 20-30 participated in the evaluation. At each round, we ask the users to give a five-point Likert scale regarding the degree of matching on each set of the music videos. The results are illustrated in Figure 3. The proposed scheme better recommends proper background music for the home videos. Despite the evaluation is primitive, we can still see the trend that most users are satisfied with the recommended soundtracks.

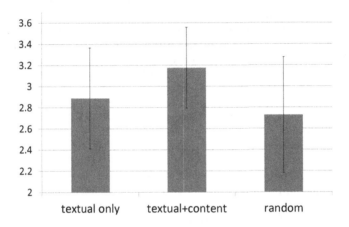

Fig. 3. Five-point Likert scale evaluated on the results of textual only recommendation scheme, proposed recommendation scheme and random selection

5   Conclusions and Future Works

We propose a new background music recommendation scheme for home videos which incorporates the textual semantics and content semantics to determine the matching degree. The key idea is to recommend the music containing semantics that relate to the ones in the input video, and the rhythm of the music should be harmonious enough with respect to the visual motion of the video. Two new features, i.e. MDH and PTP, are proposed to describe the short-term speed in the visual/aural content and the technique of ASM is applied to align the music with the video. Based on the aligning scores, we recommend the music with the highest score to the user. Subjective evaluation shows that overall the users think the proposed scheme better recommends proper background music for the home videos. However, some issues are worthy to be investigated . For example, the emotion of the visual/aural content can be taken into account to select the musics. Moreover, for musics without lyrics, we can apply the technique audio auto-tagging so that the proposed music selection mechanism based on the textual semantics can achieve better performance.

References

1. Bertin-Mahieux, T., Ellis, D.P., Whitman, B., Lamere, P.: The Million Song Dataset. In: ISMIR (2011)
2. Chen, C.-H., Weng, M.-F., Jeng, S.-K., Chuang, Y.-Y.: Emotion-based music visualization using photos. In: Satoh, S., Nack, F., Etoh, M. (eds.) MMM 2008. LNCS, vol. 4903, pp. 358–368. Springer, Heidelberg (2008)
3. Chu, W.-T., Tsai, S.-Y.: Rhythm of Motion Extraction and Rhythm-Based Cross-Media Alignment for Dance Videos. IEEE Trans. MM 14(1), 129–141 (2012)
4. Dunker, P., Dittmar, C., Begau, A., Nowak, S., Gruhne, M.: Semantic High-Level Features for Automated Cross-Modal Slideshow Generation. In: Int. Workshop on Content-Based Multimedia Indexing, pp. 144–149 (June 2009)
5. Dunker, P., Popp, P., Cook, R.: Content-aware Auto-soundtracks for Personal Photo Music Slideshows. In: IEEE ICME (2011)
6. Feng, J., Ni, B.: Auto-Generation of Professional Background Music for Home-made Videos. In: Int. Conf. on Internet Multi. Comput. and Serv. (2010)
7. Goto, M.: A chorus section detection method for musical audio signals and its application to a music listening station. IEEE Trans. Audio, Speech, Lang. Process. 14(5), 1783–1794 (2006)
8. Li, L.-J., Su, H., Xing, E.P., Li, F.-F.: Object Bank: A High-Level Image Representation for Scene Classification & Semantic Feature Sparsification. In: NIPS, pp. 1378–1386 (2010)
9. Miller, G.A.: WordNet: A Lexical Database for English. Commun. ACM 38(11), 39–41 (1995)
10. Pampalk, E.: Computational Models of Music Similarity and Their Application in Music Information Retrieval. Ph.d., Vienna University of Technology (2006)
11. Wang, J., Chng, E., Xu, C.: Fully and Semi-automatic Music Sports Video Composition. In: IEEE ICME, pp. 1897–1900 (2006)
12. Yeh, M.-C., Cheng, K.-T.: A String Matching Approach for Visual Retrieval and Classification. In: ACM MIR, p. 52. ACM Press, New York (2008)
13. Zettl, H.: In: Dorai, C., Venkatesh, S. (eds.) Media Computing: Computational Media Aesthetics, pp. 11–38. Springer, US
14. Zhang, W., Xing, L., Huang, Q., Gao, W.: A System for Automatic Generation of Music Sports-Video. In: IEEE ICME, pp. 1286–1289 (2005)

Smoke Detection Based on a Semi-supervised Clustering Model

Haiqian He, Liqun Peng, Deshun Yang, and Xiaoou Chen

Institute of Computer Science and Technology, Peking University, China
hehaiqian@pku.edu.cn

Abstract. Video-based smoke detection is regarded as an effective way for fire detection in open spaces. In this paper, a classification model based on a semi-supervised clustering method is introduced to improve the performance of smoke detection. In our model, we present a novel method to automatically determine the number of clusters K. Considering the randomness of the initial centers in K-means++, a voting strategy is proposed to maintain a stable clustering performance. Besides, the scene-related information is added to our clustering data to obtain a self-adaptive model. Finally, the experimental results show that our classification model outperforms other state-of-the-art methods and has great improvement in terms of generalization (i.e. can adapt to the unknown scenes).

Keywords: Smoke detection, Voting strategy, Self-adaptive, Semi-supervised.

1 Introduction

Forest fires represent a constant threat to ecological systems, infrastructure and human lives. In most cases, it is difficult to control a forest fire beyond 15 minutes following ignition [1]. Due to the transport delay of smoke to the sensor, traditional smoke sensors cannot be operated in the forest fire detection. Video-based detection system is a good option for forest fire detection. It does not have the delay of smoke transportation, and it can provide additional information about fire circumstances, such as size, location and propagation [2]. Generally, methods for detecting fire in video can be categorized as flame detection and smoke detection [3]. For most cases, we can see smoke before flame in a forest fire, so smoke is a good indicator of forest fire.

Toreyin et al. [4] proposed a smoke detection method based on temporal and spatial wavelet transformation. They checked features including motion, edge-blurring, flickering, color and shape of the smoke regions based on background subtraction. The key component developed for smoke detection of Xiong etal.[5] includes background subtraction, flickering extraction, contour initialization, and contour classification using both heuristic and empirical knowledge about smoke. Gubbi et al.[6] used discrete wavelet transform along with support vector machines for smoke detection. The six statistical features extracted

C. Gurrin et al. (Eds.): MMM 2014, Part II, LNCS 8326, pp. 291–298, 2014.

from an image block were arithmetic mean, geometric mean, standard deviation, skewness, kurtosis and entropy. Gonzalez et al.[7] proposed an outdoor smoke detection algorithm using a stationary wavelet transform(SWT) to remove high frequencies on horizontal, vertical, and diagonal details. In order to determine the possible regions of interest, the inverse SWT is implemented and finally the image is compared to a non-smoke scene.

Usually, smoke detection can be phrased as a classification problem. A model is built on a training dataset with a learning algorithm (e.g. SVM). Then, the trained model is used to identify which categories a new observation belongs. However, the model performance relies heavily on the training data. It usually performs poorly in an unseen scene where too many instances are not included in the training dataset. The generalization ability of the smoke detection model is severely limited by the train-test mode. Unfortunately, the previous work mentioned above did not address this problem. In this paper, contrast to the train-test mode, a semi-supervised clustering model is introduced to enhance the adaptability greatly. And the experimental results show that our model outperforms other state-of-the-art methods.

In this paper, K-means++ [8] is adopted as the clustering model. And, we present a novel effective method to automatically determine K. Considering that the randomly selected initial centers are prone to local minima, a voting strategy is proposed to maintain a stable clustering performance. With the class labels for some items, the semi-supervised clustering model [9,10] can use this knowledge for the external validation of the results of clustering and further letting it guide and adjust the clustering process. In our paper, labeled smoke blocks are used to identify which clusters belong to smoke clusters.

2 Dataset and Features

2.1 Dataset

In our model, every frame in a video stream is divided into small blocks of 32×32 pixels. Every block is checked for the presence or absence of smoke. Two different datasets are used in our experiments. The first one is publicly available at [11]. It contains a variety of scenes, such as c-h in Fig(1). Another dataset is obtained from a local forestry bureau, which records the real smoke and cloud scenes in high altitude, such as a,b,i,j in Fig(1).

Ten videos, including nine smoke scenes and one cloud scene, are selected from the two datasets for the purpose of smoke detection. Of the ten videos, six videos (about forty-five minutes, Fig. 1(a-f)) are designated as the training data. The other four videos (about twenty-five minutes, Fig. 1(g-j)) act as the testing data. For the training data, we first randomly select 230 smoke image blocks (denoted as SBS, a fixed smoke block set, independent of any test videos) from the six training videos as positive examples, and then randomly select 230 non-smoke image blocks as negative examples. To evaluate the effectiveness and generalization of our model, 300 positive examples and 400 negative examples are obtained from the four testing videos. In our experimental design, three scenes

a.forest0526   b.forest0602   c.ridge   d.town   e.the fence1

f.Manavgat   g.PelcoColakli   h.the fence2   i.forest0517   j.forest0809

Fig. 1. Different scenes in our dataset

in the testing videos don't appear in the training videos, so the adaptability of our model can be reflected from the performance on the three scenes.

2.2 Features

Representation of instances plays a very important role in smoke detection problem (also in other classification problems). As the scene blurring is an important signal when smoke appears, wavelet transformation, mainly capturing frequency information, is usually applied to detect smoke. According to [8], we extract from an image block five statistical features, including arithmetic mean, standard deviation, skewness, krutosis and entropy, from each coefficient matrix of three-level discrete wavelet transform (DWT). Besides, the same operation is applied to the image block in RGB color space. These sixty five features are used to represent an image block.

3 Methodology

3.1 The Basic Semi-supervised Clustering Model

In this section, we present our semi-supervised clustering model in details. As Fig (2) shows, there are three steps to construct our model. The basic idea is to combine the SBS and a set of blocks from a testing video, and partition the blocks in the resulted union into K (specified later) clusters. With the help of the SBS, each cluster can be labeled as smoke cluster or non-smoke cluster. A new image block from a testing video can be assigned to a cluster whose center is closest to it. If the closest cluster is smoke cluster, the test image block is labeled as smoke, and vice versa. Considering that the randomly selected initial centers are prone to local minima, a voting strategy is proposed to maintain a stable clustering performance. Each step is discussed in details as follows.

Collect Clustering Data. The clustering data consists of the SBS and scene-related information from a testing video. The SBS is a fixed labeled smoke block set and independent of any test videos. Scene-related information is a collection of image blocks from the first N frames of the testing video.

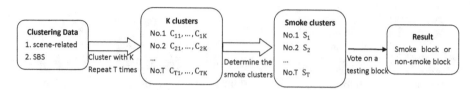

Fig. 2. An overview of the semi-supervised clustering model

Determine the Smoke Clusters. We partition the clustering data into K clusters (denoted as $C_i(i = 1, 2, \ldots, K)$) with the K-means++ [8]. In our model, each cluster will be labeled as smoke or non-smoke tag. With the help of the SBS, the model can automatically identify the smoke clusters from the K clusters. Here, we denoted the smoke clusters as S. In the clustering process, every smoke block in SBS is assigned to a cluster whose center is closest to it. For a cluster C_i, the number of smoke blocks coming from SBS is denoted as N_i. Sort the K clusters in decreasing order of N_i. The number of smoke blocks in the i-th cluster is denoted as $SortN_i$. The top M clusters are treated as smoke clusters. To determine the value of M, a threshold is introduced in the following inequality:

$$\sum_{i=1}^{M} SortN_i \geq \alpha \times \|SBS\| \quad \alpha \in [0,1] \tag{1}$$

Where $\|SBS\|$ means the number of smoke blocks in SBS. We choose the minimum M that can satisfy inequality (1). The value of α will be studied in section(4.2). In our experiment, $\alpha=0.85$.

A Voting Strategy. After two steps mentioned above, a test image block can be assigned to a cluster whose center is closest to it. The label of the test image block is determined by the label of the closest cluster. However, in the process of clustering, the initial centers are chosen randomly and K-means++ offers no global optimum guarantees. To avoid an unstable result, the clustering process is repeated T times and the final result is determined by a voting strategy. Considering the computational cost and statistical effectiveness, we set $T=100$. Let $S_t(t = 1, 2, \ldots, T)$ denote the set of smoke block clusters of the t-th clustering operation. For a testing image block x, the cluster whose center is closest to x is denoted as C_x. The voting strategy is defined as follows:

$$vote_t(x) = \begin{cases} 1 & C_x \in S_t \\ -1 & C_x \notin S_t \end{cases}$$
$$votes = \sum_{t=1}^{T} vote_t(x) \quad t = 1, 2, \ldots, T \tag{2}$$

If the value of votes is positive, the testing image block is smoke block. Otherwise, it is a non-smoke block. According to our experiment, the voting strategy always performs well. To explain the stable performance, we make the following two assumptions: (1) the T clustering processes are independent of each other; (2)

In each clustering process, the probability of correct classification p is greater than 0.5.

The first assumption is reasonable, as the initial centers are chosen randomly in each clustering process and a clustering process has no influence on another. As far as the second assumption is concerned, it is based on the fact that ideally, the probability of correct classification for a random binary classification is equal to 0.5. Classification based on the clustering model should be better than the random strategy.

Now, we get a probability distribution problem, which takes value 1 with success probability p and value -1 with failure probability $q = 1 - p$. When the value of p is not big, e.g. 0.7, it is undependable enough for once classification. If the process is repeated T times, the probability of correct classification P_{repeat} is defined as follows:

$$P_{repeat} = \sum_{k=\frac{T}{2}}^{T} \binom{T}{k} p^k (1-p)^{T-k} \tag{3}$$

As we can see, for $T=100$, $P_{repeat} \approx 0.9832$ with $p = 0.6$ and $P_{repeat} \approx 1$ with $p = 0.7$. The voting strategy can be quite stable even with an unsatisfied probability p.

3.2   The Weighted Semi-supervised Clustering Model

In the basic semi-supervised clustering model, all votes are treated equally and have the same weight. However, it may be more reasonable to discriminate different votes. Therefore, for an image block, we consider not only its nearest cluster but also its second nearest one, to determine the weight of its vote.

We still let S_t be the smoke block clusters for the t-th clustering. When both the closest cluster and the second-closest cluster belong to S_t, the vote will be given the heaviest weight. When the closest cluster belongs to S_t and the second-closest cluster does not, the weight of vote depends on the ratio of the distance from the closest cluster center to that of the second-closest cluster center. The voting strategy of the weighted semi-supervised clustering model is defined by the following formula:

$$vote_t(x) = \begin{cases} 2 & C_x^1 \in S_t \text{ and } C_x^2 \in S_t \\ 1.5 & C_x^1 \in S_t \text{ and } C_x^2 \notin S_t \text{ and } d_1 \leq 0.8d_2 \\ 1 & C_x^1 \in S_t \text{ and } C_x^2 \notin S_t \text{ and } d_1 > 0.8d_2 \\ -1 & \text{Otherwise} \end{cases} \tag{4}$$

$$votes = \sum_{t=1}^{T} vote_t(x) \quad t = 1, 2, \ldots, T$$

Where C_x^1 is the closest cluster and d_1 is the corresponding distance, C_x^2 is the second-closest cluster and d_2 is the corresponding distance.

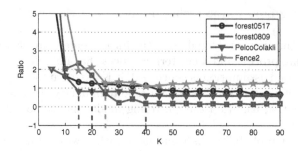

Fig. 3. The ratio of N_1' to N_2' with different K

4   Performance Study

4.1   The Number of Clusters

The number of clusters K is scene-related. Different videos contain different scenes and need a different values of K. A model that can automatically determine K can be self-adaptive to different scenes. Let K_{opt} denote the optimal K, under which the classification performs stable and best.

A set of blocks is randomly selected from the first N frames (a random block set, denoted as RBS). Every block in RBS can be viewed as a testing block and can be labeled as smoke block or non-smoke block by our semi-supervised model. After classification, there are N_1' smoke blocks and N_2' non-smoke blocks in the RBS. N_1' and N_2' are predicted values. According to our experiments on training dataset, there are three reasonable assumptions:

1. When $K \ll K_{opt}$, clusters become mixture of positive and negative instances, and as K increases, the classification performance is unstable.
2. When the value of K approximates K_{opt}, the classification performance approaches best, and as K increases further, it will remain stable, i.e. no longer change significantly.
3. the ratio of N_1' to N_2' will reach a stable value when K reaches K_{opt}

The first two assumptions are consistent with our common sense, and we don't do additional analysis. As far as the third assumption is concerned, clearly when a model reaches stable, the classification result of a random block set will no longer change significantly. The K_{opt} can be discovered from the change of the ratio of N_1' to N_2'. When the ratio reach a stable value, the corresponding value of K is required. In other words, our model can automatically determine the number of clusters without manual participation. As Fig (3) shows, the values of K of the four testing videos are 20, 40, 15 and 25 respectively.

4.2   The First N Frames and the Threshold α

According to our experiment, the parameters, N and α, have little effect on the final result. We determine the two parameters in advance and do not change them with the different testing videos. Finally, we take N=9 and α=0.85.

5   Experimental Result

In this section we report our experiment results on the testing dataset (mentioned in 2.1). We compare the overall performance of our model with the traditional classification model, named SVM+Wavelet [6]. Since the color of the smoke is a notable feature, features from RGB color space are added into the SVM+Wavelet model. So, we get SVM+Wavelet+RGB model. Table (1) shows the performance of the various models on the testing dataset. Precision (P), recall(R) and F_1-measure (F) [12] were used to evaluate these different models.

Table 1. An overview of the experimental result

Video Name		SVM + Wavelet		Semi-supervised	
		basic	+RGB	basic	weighted
Forest 0517 k=20	P	0.841	0.945	0.873	0.896
	R	0.986	0.986	0.986	0.986
	F	0.908	**0.965**	0.926	0.939
Forest 0809 k=50	P	1	1	1	1
	R	0.540	0.330	0.850	0.860
	F	0.701	0.496	0.919	**0.925**
Pelco Colakli k=15	P	0.707	0.714	0.949	0.949
	R	0.933	1	1	1
	F	0.805	0.833	**0.974**	**0.974**
Behind Fence k=25	P	0.741	0.706	0.977	1
	R	0.819	0.962	0.810	0.838
	F	0.778	0.815	0.885	**0.912**
Average	P	0.656	0.631	0.881	0.899
	R	0.900	0.980	0.916	0.928
	F	0.759	0.768	0.898	**0.913**

The SVM+Wavelet+RGB model is slightly better than the SVM+Wavelet model except on Forest0809. In most cases, color plays a positive role in distinguishing smoke and non-smoke blocks. However, the cloud is very similar to the smoke in color, so the SVM+Wavelet+RGB model performs poorly on Forest0809, which describe the movement of cloud in high altitude. Thanks to scene-related information, the optional K_{opt} and the voting strategy, the semi-supervised clustering model greatly outperforms the SVM+Wavlet model and the SVM+Wavelet+RGB model, expecially on the cloud scene (Forest0809). The results show that the self-adaptability of our model is significantly improved, and the weighted voting strategy is effective.

6   Conclusion

In this paper, a semi-supervised clustering model is introduced to solve the problem of smoke detection in open spaces. To improve the model's self-adaptability, a method that automatically determines the number of clusters K is proposed. Through the voting strategy, we obtain a stable clustering performance. The experiment results show that our model greatly outperforms the traditional model. In the future, our work will pay more attention on the movement of the smoke plumes and add the time information into our model.

This work was supported by the Natural Science Foundation of China (Multimodal Music Emotion Recognition technology research No.61170167) and the National Development and Reform Commission High-tech Program of China under Grant No. [2010]3044.

References

1. Vicente, J., Guillemant, P.: An image processing technique for automatically detecting forest fire. International Journal of Thermal Sciences 41, 1113–1120 (2002)
2. Verstockt, S., Lambert, P., Van de Walle, R., Merci, B., Sette, B.: State of the art in vision-based fire and smoke dectection (2009)
3. Kim, D., Wang, Y.: Smoke detection in video. In: 2009 WRI World Congress on Computer Science and Information Engineering, vol. 5, pp. 759–763. IEEE (2009)
4. Toreyin, B., Dedeoglu, Y., Cetin, A.: Wavelet based real-time smoke detection in video. In: 13th European Signal Processing Conference, EUSIPCO (2005)
5. Xiong, Z., Caballero, R., Wang, H., Finn, A., Lelic, M., Peng, P.: Video-based smoke detection: possibilities, techniques, and challenges. In: Suppression and Detection Research and Applications, A Technical Working Conference (SUPDET 2007) (2007)
6. Gubbi, J., Marusic, S., Palaniswami, M.: Smoke detection in video using wavelets and support vector machines. Fire Safety Journal 44(8), 1110–1115 (2009)
7. Gonzalez-Gonzalez, R., Alarcon-Aquino, V., Rosas-Romero, R., Starostenko, O., Rodriguez-Asomoza, J., Ramirez-Cortes, J.: Wavelet-based smoke detection in outdoor video sequences. In: 2010 53rd IEEE International Midwest Symposium on Circuits and Systems (MWSCAS), pp. 383–387. IEEE (2010)
8. Arthur, D., Vassilvitskii, S.: k-means++: The advantages of careful seeding. In: Proceedings of the Eighteenth Annual ACM-SIAM Symposium on Discrete Algorithms, pp. 1027–1035. Society for Industrial and Applied Mathematics (2007)
9. Grira, N., Crucianu, M., Boujemaa, N.: Unsupervised and semi-supervised clustering: a brief survey. A review of machine learning techniques for processing multimedia content, Report of the MUSCLE European Network of Excellence, FP6 (2004)
10. Basu, S., Banerjee, A., Mooney, R.J.: Semi-supervised clustering by seeding. In: ICML, vol. 2, pp. 27–34 (2002)
11. http://signal.ee.bilkent.edu.tr/VisiFire/Demo/SmokeClips/
12. http://en.wikipedia.org/wiki/F1_score

Empirical Exploration
of Extreme SVM-RBF Parameter Values
for Visual Object Classification

Rami Albatal and Suzanne Little

INSIGHT Centre for Data Analytics,
Dublin City University, Dublin, Ireland
rami.albatal@dcu.ie
http://www.insight-centre.org/

Abstract. This paper presents a preliminary exploration showing the surprising effect of extreme parameter values used by Support Vector Machine (SVM) classifiers for identifying objects in images. The Radial Basis Function (RBF) kernel used with SVM classifiers is considered to be a state-of-the-art approach in visual object classification. Standard tuning approaches apply a relative narrow window of values when determining the main parameters for kernel size. We evaluated the effect of setting an extremely small kernel size and discovered that, contrary to expectations, in the context of visual object classification for some object and feature combinations these small kernels can demonstrate good classification performance. The evaluation is based on experiments on the TRECVid 2013 Semantic INdexing (SIN) training dataset and provides initial indications that can be used to better understand the optimisation of RBF kernel parameters.

Keywords: Visual Object Classification, SVM, RBF, optimisation, extreme parameter values.

1 Introduction

Optimising machine learning algorithm parameters is a crucial task for building reliable classifiers. In the context of automatic image indexing and visual object classification, the Support Vector Machine SVM [1] is the most popular machine learning algorithm and is frequently adopted as the state-of-the-art algorithm relied on in manny applications. The SVM is a supervised learning algorithm that analyses data to recognise patterns for classification and regression analysis. In the case of visual descriptor classification, the complex feature space is often non-linearly separable, a fact that leads to the usage of the kernel trick [2] in order to implicitly map the features into high-dimensional spaces where it is easier to find hyperplanes that separate the positive from the negative examples.

Visual object classification aims to identify objects of interest in image or video keyframes based on low-level features generally using trained classifiers such as SVMs. Exhaustive parameter tuning is often infeasible due to the very

C. Gurrin et al. (Eds.): MMM 2014, Part II, LNCS 8326, pp. 299–306, 2014.
© Springer International Publishing Switzerland 2014

large, heterogeneous datasets, sparse, high-dimensional feature space and need to manage computational complexity. However, as a technical report by Lin, Hau & Chang [3] illustrates, even small efforts toward parameter tuning can yield good performance gains.

While there are many configurable options throughout the visual object classification pipeline, including data normalisation, distance metrics, kernel type etc., we focus in this paper on using the *de facto* settings for visual descriptor classification and choose to examine one parameter of the most popular SVM kernel used for visual feature classification (and for many other classification tasks) the Radial Basis Function (RBF) kernel. In [4] we can find the following definition of the RBF kernel that has two feature vectors x and x':

$$K(x, x') = exp \left(-\frac{\|x - x'\|_2^2}{2\sigma^2} \right) \qquad (1)$$

where $\|x - x'\|_2^2$ is the squared Euclidean distance between the vectors x and x', and σ is a parameter to control the size of the kernel. The σ parameter is often optimised through another representation called *gamma* represented as follows:

$$\gamma = -\frac{1}{2\sigma^2} \qquad (2)$$

As we can see in the formula 2, σ^2 and γ are inversely proportional. The larger the value of γ the smaller the value of σ and the smaller is the kernel size. From a very simplified point-of-view, a large γ value leads to narrow influence zone for each single training example in the new feature space. When a new example is evaluated, and if *gamma* is well tuned, this new example will fall into the influence zone of positive example(s) if it is a positive data sample, or it will fall into the influence zone of negative example(s) in the opposite case.

In this article, we explore the values of the γ parameter in the RBF kernel with a fixed *cost* value in the context of visual object classification. Most of the works in the domain explore some restricted range of γ values and avoid to go to extreme limits in order to avoid over-fitting, where large ratio of training examples will be considered as support vector. Our initial observations show that it is not always the case, and for some Object/Feature combinations, choosing large γ values can lead into good classification performance. These observations will be the subject of a future in-depth analysis in order to determine when performance may benefit from extreme gamma values.

The rest of this article is organised as follows: First, in section 2 we will present the state-of-the-art works on optimising the SVM parameter for visual object classification, then in section 3 we will present our results obtained from experiments on TRECVid 2013 SIN training set [5], and finally we will discuss our results in section 4 where we will present future work based on our observations.

2   Related Works

Optimisation efforts are commonly concerned with the problem of choosing the best features (see a recent review [6]) however the problem of parameter optimisation, both in general and for SVMs with RBF kernels in particular, has been tackled in a variety of ways ranging from heuristic methods to machine learning approaches designed to reduce the problem space. Some have applied genetic algorithms [7] or other machine learning techniques (e.g., particle swarm optimisation [8]) to identify optimal settings. Others perform grid search using a tuning dataset to find the best settings without over-tuning or apply optimisation strategies. For instance, Duan, Keerthi & Poo [9] examined the computational cost of various optimisation approaches. Past efforts have focussed on domains such as bioinformatics, genomics and finance. The standard approach in visual object classification is generally to follow the configurations suggested in [10]. The main challenge is to balance the computational requirements with the risk of over-fitting to a particular dataset and the potential performance gains from careful parameter tuning.

Many of the state-of-the-art works use cross-validation functions and optimisation tools provided by SVM libraries like LibSVM [11]. These libraries often sequentially test a set of gamma and cost values chosen *a priori* and do not rely on the feature dataset. Most of the works focus on the feature space itself by normalising the descriptors using L1 or L2, widely used for histogram-like descriptors [12,13], or Min_max normalisation like the one used in LibSVM, or Zero-Mean and unit-variance, or the Power transformation proposed by [14]. After normalising the descriptors, SVM parameters will be optimised to build classifiers; and as far as we know not no state-of-the-art focused its work on studying the ranges of SVM parameters and more precisely on the γ values of the RBF kernel.

In [10] we can find a good indication about how to choose the *gamma* value of RBF:

> "We first subsample the training data (if the training data set is not large, use the whole training data), then compute the distance between the points and find the distances at 0.9 and 0.1 quantile of all the distances, the average distance of these two distances is set to be the initial σ_0. This is to guarantee that the kernel parameter is neither too big or too small. Other values of σ to be selected in the experiments (via cross-validation) are $[10^{-4}\sigma_0, ..., \sigma_0, ..., 10^3\sigma_0, 10^4\sigma_0]$".

This heuristic is adopted and adjusted by Safadi and Quenot [14] for the domain of visual descriptors learning and classification, they calculate the average Euclidean distance between a subset of the descriptors then fix *gamma* as follows:

$$\gamma = \frac{2^i}{meanDist^2} \tag{3}$$

with $meanDist$ is the average distance between the descriptors in the dataset, and i is a positive integer parameter, fixed as 1 or 2 in the case of descriptors with large dimensionality, and 3 or 4 for descriptors with small dimensionality (up to few hundreds). The reasons why the *gamma* validation does not cover more i values is the computational cost (the larger the i the slower the learning and the test), and to avoid the over-fitting caused by very large *gamma* values. In this paper, we adopt the *gamma* optimisation formula mentioned above in 3 and perform an empirical evaluation to test the maximum boundary for *gamma* values in the case of low dimensionality descriptors (chosen for computational reasons).

3   Experiments

The experiments were applied on the the development set of the large scale TRECVid 2013 Semantic INdexing task collection (SIN). TRECVid is an evaluation campaign organised on a yearly basis by the US National Institute of Standards and Technology (NIST)[1], focusing on a set of different information retrieval (IR) research areas in content-based retrieval and exploitation of digital video. Started in 2010, the SIN task allow the research community to evaluate and compare methods for automatic assignment of semantic tags representing visual or multimodal concepts (previously "high-level features") to video segments. SIN is attracting more and more research interest and in 2013 there are more than 100 submissions from over 20 research projects and teams. The development collection (used in this exploration) contains 545,923 video shots, split into two subsets for cross validation: 2013x that contains 268,986 shots, and 2013y that contains 276,937 shots, and 60 concepts to be classified. Three small dimensionality descriptors were evaluated:

- gab40: normalized Gabor transform, 8 orientations x 5 scales → 40 dimensions.
- h3d64: normalized RGB Histogram 4x4x4 → 64 dimensions.
- hg104: early fusion of h3d64 and gab40 → 104 dimensions.

These descriptors have been produced and shared by various partners of the IRIM (Indexation et Recherche d'Information Multimédia) project of GDR-ISIS research network from CNRS-France.

The Formula 3 was applied with $i = [0, ..., 10]$ which allow *gamma* to have very large values that are not usually evaluated in these kinds of tasks. The optimisation was performed by 2-fold cross validation (trai on 2013x then test on 2013y, train on 2013y then test on 2013x), and the goal is maximising the Mean extended inferred average precision [15] used as performance measure in TRECVid 2013 SIN task. Figures 1 and 2 show examples of the classification performances on four different objects using the 10 values of *gamma* calculated using Formula 3 and $i = [0, ..., 10]$ for gab40 and hg104 descriptors respectively[2].

These figures are merely samples of results to illustrate that descriptors differs in term of dimensionality, quality of classification results and in γ values tuning profiles — indications that favour per-class classification optimisation rather than multi-class optimisation.

---

[1] http://trecvid.nist.gov/

[2] Note that the scale of the y axis is not the same in all the charts, because it is adjusted to include the minimum and the maximum results.

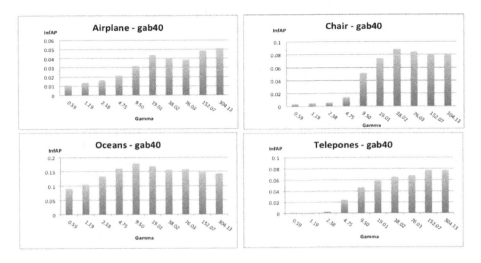

Fig. 1. Validation results for gab40 feature

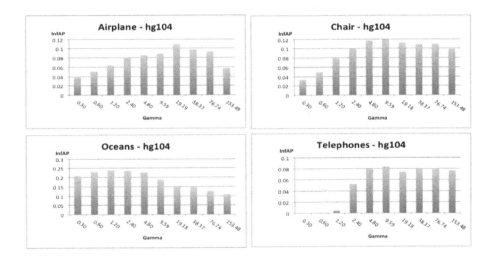

Fig. 2. Validation results for gab40 feature

Table 1 presents the optimal γ value that gives the best results for each of the 60 visual concepts in TrecVid 2013. We did evaluate setting small values of γ $(2^{-15}, 2^{-12}, 2^{-9})$ but results had lower infAP than those presented in the table, thus are not included here.

Table 1. Optimal γ values for all TrecVid 2013 SIN training dataset, value are found by cross validation on 2013x and 2013y subsets, large values (> 30) are in bold

	Optimal γ Values				Optimal γ Values		
	gab40	h3d64	Hg104		gab40	h3d64	Hg104
Airplane	**304.54**	19.34	19.19	Anchorperson	19.01	2.42	1.20
Animal	**76.03**	19.37	19.18	Baby	**76.03**	2.42	1.20
Basketball	**76.03**	1.21	0.60	Beach	**38.02**	1.21	1.20
Bicycling	**304.13**	**38.73**	9.59	Boat_Ship	**152.27**	**309.84**	4.80
Boy	**304.13**	**38.73**	**76.74**	Bridges	**304.13**	**154.92**	9.59
Bus	**304.13**	**309.84**	**76.74**	Car_Racing	**152.07**	**38.73**	**38.37**
Chair	**38.02**	19.37	9.59	Cheering	9.50	9.68	2.40
Classroom	**304.13**	**154.92**	**38.37**	Computers	**38.02**	**38.73**	4.80
Dancing	**304.13**	4.84	4.80	Protest	4.75	4.84	4.80
Door_Opening	9.50	4.84	2.40	Explosion_Fire	**304.13**	19.37	2.40
Female_Face	**152.07**	9.68	9.59	Fields	**76.03**	0.61	0.60
Flags	2.38	2.42	1.20	Flowers	**38.02**	2.42	1.20
Forest	**76.03**	4.84	2.40	George_Bush	**38.02**	**38.73**	4.80
Girl	**76.03**	19.37	9.59	Government_Leader	19.01	9.68	4.80
Greeting	4.75	4.84	1.20	Hand	**38.02**	**38.73**	4.80
Highway	**38.02**	**38.73**	4.80	Hill	**76.03**	2.42	1.20
Instr_Musician	**38.02**	19.37	4.80	Kitchen	**304.13**	**38.73**	9.59
Lakes	**38.02**	4.84	4.80	Meeting	9.52	19.34	0.60
Military_Airplane	**76.03**	**309.84**	**76.74**	Motorcycle	1.19	**38.73**	**38.37**
News_Studio	19.01	2.42	2.40	Nighttime	**152.07**	**309.84**	9.59
Oceans	9.50	9.68	1.20	Office	**38.02**	9.68	2.40
Old_People	**38.02**	9.68	4.80	People_Marching	9.50	9.68	4.80
Press_Conference	**304.54**	**309.44**	4.80	Quadruped	**152.07**	**77.46**	9.59
Reporters	**38.02**	0.61	1.20	Roadway_Junction	**76.03**	19.37	9.59
Running	**76.03**	4.84	2.40	Singing	**76.03**	19.37	4.80
Sitting_Down	**38.02**	9.68	4.80	Skating	19.01	4.84	2.40
Skier	**76.03**	**38.73**	19.18	Soldiers	**38.07**	**154.72**	**76.74**
Stadium	**76.03**	**154.92**	**38.37**	Studio_AP	19.01	2.42	1.20
Swimming	9.50	0.61	0.30	Telephones	**304.13**	**154.92**	9.59
Throwing	**152.07**	4.84	1.20	Traffic	**152.07**	**77.46**	4.80

4   Discussion and Future Work

From Figures 1,2 and Table 1 we give the following remarks:

- Optimal γ values can be very large in the context of visual descriptor classification.
- The smaller the dimensionality of the visual descriptor, the more likely the optimum value is large (75% of γ values of gab40 are high, 37% in the case of h3d64, and 13% in the case of hg104). This observation should be consolidated by testing other low dimensionality descriptors as well as large descriptors such as those based on the Bag-of-Visual-Words model [16].

– There is insufficient indication about why and when we should set and evaluate large values of γ, and how large they should be.

Contrary to generally accepted practise that a small kernel size is undesirable, the preliminary results obtained from the experiments presented in this paper show that large values of RBF γ parameter does not necessarily appear to lead to over-fitting. These results are clearly preliminary. However in order to better understand our results, we are planning to perform further analysis on:

– The descriptors in the dataset (distances between the descriptors in the original feature space as well as in the high RBF dimensional space).
– The relations between the positive/negative examples in the dataset and the γ values.
– The relation between the dimensionality of the descriptors and the γ values.

In conclusion, the motivation behind this work was to improve the performance of our classifiers in the TRECVid2013 SIN task while minimising computational effort. The results from this will be available shortly and we hope to conduct further analysis of the impact of extreme parameter values based on this work. We hope to produce some guidelines on the dataset characteristics where such large values are worthy of investigation.

Acknowledgements. Part of the research leading to these results has received funding from the European Union Seventh Framework Programme (FP7/2007-2013) under grant agreement number 285621, project titled SAVASA.

References

1. Cortes, C., Vapnik, V.: Support-vector networks. Machine Learning, 273–297 (1995)
2. Aizerman, M.A., Braverman, E.A., Rozonoer, L.: Theoretical foundations of the potential function method in pattern recognition learning. Automation and Remote Control (25), 821–837 (1964)
3. Lin, C.J., Hsu, C.W., Chang, C.C.: A practical guide to support vector classification (2003)
4. Vert, J.P., Tsuda, K., Scholkopf, B.: A primer on kernel methods. In: Kernel Methods in Computational Biology, pp. 35–70
5. Smeaton, A.F., Over, P., Kraaij, W.: Evaluation campaigns and trecvid. In: MIR 2006: Proceedings of the 8th ACM International Workshop on Multimedia Information Retrieval, pp. 321–330. ACM Press, New York (2006)
6. Bolón-Canedo, V., Sánchez-Maroño, N., Alonso-Betanzos, A.: A review of feature selection methods on synthetic data. Knowledge and Information Systems 34(3), 483–519 (2013)
7. Huang, C.-L., Wang, C.-J.: A ga-based feature selection and parameters optimization for support vector machines. Expert Systems with Applications 31(2), 231–240 (2006)

8. Lin, S.-W., Ying, K.-C., Chen, S.-C., Lee, Z.-J.: Particle swarm optimization for parameter determination and feature selection of support vector machines. Expert Systems with Applications 35(4), 1817–1824 (2008)

9. Duan, K., Sathiya Keerthi, S., Poo, A.N.: Evaluation of simple performance measures for tuning svm hyperparameters. Neurocomputing 51, 41–59 (2003)

10. Takeuchi, I., Le, Q.V., Sears, T.D., Smola, A.J., Williams, C.: Nonparametric quantile estimation. Journal of Machine Learning Research 7, 7–1231 (2006)

11. Chang, C.-C., Lin, C.-J.: LIBSVM: A library for support vector machines. ACM Transactions on Intelligent Systems and Technology 2, 27:1–27:27 (2011), Software available at http://www.csie.ntu.edu.tw/~cjlin/libsvm

12. Sivic, J., Zisserman, A.: Video Google: A text retrieval approach to object matching in videos. In: Proceedings of the International Conference on Computer Vision, vol. 2, pp. 1470–1477 (October 2003)

13. van de Sande, K.E.A., Gevers, T., Snoek, C.G.M.: Evaluating color descriptors for object and scene recognition. IEEE Transactions on Pattern Analysis and Machine Intelligence 32(9), 1582–1596 (2010)

14. Safadi, B., Quénot, G.: Descriptor optimization for multimedia indexing and retrieval. In: CBMI, pp. 1–6 (2013)

15. Yilmaz, E., Kanoulas, E., Aslam, J.A.: A simple and efficient sampling method for estimating ap and ndcg. In: Proceedings of the 31st Annual International ACM SIGIR Conference on Research and Development in Information Retrieval, SIGIR 2008, pp. 603–610. ACM, New York (2008)

16. Csurka, G., Dance, C.R., Fan, L., Willamowski, J., Bray, C.: Visual categorization with bags of keypoints. In: Workshop on Statistical Learning in Computer Vision, ECCV, pp. 1–22 (2004)

Real-World Event Detection Using Flickr Images

Naoko Nitta, Yusuke Kumihashi, Tomochika Kato, and Noboru Babaguchi

Graduate School of Engineering, Osaka University
2-1 Yamada-oka, Suita, Osaka, 565-0871 Japan
{naoko,kumihashi,kato,babaguchi}@nanase.comm.eng.osaka-u.ac.jp

Abstract. This paper proposes a real-world event detection method by using the time and location information and text tags attached to the images in Flickr. Events can generally be detected by extracting images captured at the events which are annotated with text tags frequently used only in specific times and locations. However, such approach can not detect events where only a small number of images were captured. We focus on the fact that semantically related events often occur around the same time at different locations. Considering a group of these events as an *event class*, the proposed method firstly detects event classes from all images in Flickr based on their similarity of the captured time and text tags. Then, from the images consisting each event class, events are detected based on their similarity of the captured locations. Such two-step approach enables us to detect events where a small number of images were captured.

1 Introduction

In the real-world, various events occur at different times at different locations. Since people all over the world constantly observe the real world, they can be used as citizen sensors[4] to detect many types of events around the world. Especially, since people often post their observations with the observed time and location information to social media sites such as Flickr, Twitter, and Facebook recently, such observations are open to the general public and can be obtained with low cost.

Some existing researches have tried to detect real-world events by using social media to obtain the observations by citizen sensors. Sakaki et al.[3] have proposed a method for detecting and tracking real-world events experienced by many people such as earthquakes by using Twitter. By setting a specific target event such as earthquakes, a set of event-related keywords can be determined in advance and be used to extract the tweets posted on Twitter which mention the occurrence of the target event. Rattenbury et al.[2] have proposed a method for extracting event-related tags from Flickr without specifying the target events based on the intuition that event-related tags are used frequently over only small segments of time. Further, Chen et al.[1] have proposed to detect real-world events by using Flickr by extracting event-related tags which are frequently used over only small segments of time and place. This approach can detect various events as long as their images are uploaded by many people and the spatial and temporal usage distributions of their related tags exhibit strong burst patterns.

C. Gurrin et al. (Eds.): MMM 2014, Part II, LNCS 8326, pp. 307–314, 2014.

In order to detect events whose images are uploaded by a small number of people, this paper proposes a two-step event detection approach. Our intuition is that semantically related events often occur around the same time at different locations. For example, many events related to cherry blossom, marine sports, autumn leaves, and Christmas occur at various locations in spring, summer, autumn, and winter, respectively. Even when only a small number of images of each event are shared on Flickr and the spatial and temporal usage distributions of their related tags do not exhibit burst patterns, the images of these semantically related events can be represented with common tags, whose temporal usage distribution should exhibit burst patterns. Considering a group of these semantically related events as an *event class*, the proposed method firstly detects event classes based on the similarity of the captured time and tags of all images posted on Flickr. Then, events are detected based on the similarity of the captured locations of the images of each event class.

2   Two-Step Event Detection

Given an image set P consisting of the N images $I_n (n = 1, \cdots, N)$ uploaded to Flickr, the objective of the work is to detect events $e_i (i = 1, \cdots)$ and event classes $ec_j = \{e_{j1}, e_{j2} \cdots \}$ $(j = 1, \cdots)$ consisting of semantically related events.

Since an event e_i should occur at a specific time and location, event e_i can be defined by the time e_i^t, location e_i^l, and a set of representative keywords e_i^W. Semantically related events such as New Year ceremonies, cherry blossom viewings, commencements, fireworks displays, Halloween, and Christmas parties often occur around the same time at different locations. Therefore, an event class ec_j can be defined by the time e_i^t and a set of representative keywords e_i^W. In this paper, event and event class detections are to determine $e_i = (e_i^t, e_i^l, e_i^W)$ and $ec_j = (ec_j^t, ec_j^W)$, respectively.

Each image I_n uploaded to Flickr is associated with its captured time t_n, captured location $l_n = (lat_n, lon_n)$, which consists of latitude and longitude coordinates, a set of text tags W_n, and user u_n. The images of semantically related events uploaded by different users are expected to have similar captured time and common text tags. Thus, the proposed method firstly extracts an image set $P_{ec_j} (j = 1, \cdots)$ which represents the event class ec_j from P based on their similarity of t_n and W_n, and determines ec_j^t and ec_j^W from t_n and W_n of the images in P_{ec_j}. Then, an image set $P_{e_i} (i = 1, \cdots)$ which represents the event e_i is extracted from P_{ec_j} based on their similarity of l_n, and e_i^t, e_i^l, and e_i^W are determined from t_n, l_n, and W_n of the images in P_{e_i}.

As shown in Figure 1, the proposed method is composed of 3 steps, whose details are described in the following sections.

2.1   Preprocessing

This step removes redundant images and text tags which would interfere the event detection from the images on Flickr to construct the image set P. Firstly, in order to avoid false event class detection, the images with similar captured

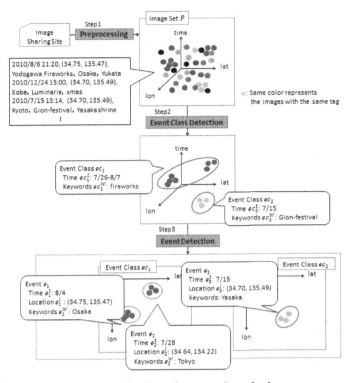

Fig. 1. Outline of proposed method

time, location and text tags taken by the same user need to be removed. Two images I_α and I_β that meet the following conditions are merged into an image I_n:

$$u_\alpha = u_\beta, \quad |t_\alpha - t_\beta| < TH_t, \quad |l_\alpha - l_\beta| < TH_l, \quad \frac{|W_\alpha \cap W_\beta|}{\min(|W_\alpha|, |W_\beta|)} > TH_W.$$

t_n, l_n, and W_n of the image I_n are determined as the average of t_α and t_β, the average of l_α and l_β, and the union of W_n. In addition, in order to improve the detection accuracy and reduce the computational time simultaneously, tags shared by too many images, more than $r\%$ of all images, are removed as general terms. The remaining N images $I_n(n = 1, \cdots, N)$ attached with the remaining tags are used as the image set P.

2.2 Event Class Detection

This step extracts the image set P_{ec_j} based on the similarity of the captured time and text tags of the images in P to determine $ec_j = (ec_j^t, ec_j^W)$. Since many images in P are not captured at specific events and the number of events captured in P is unknown, a graph clustering method called SCAN[5], which can identify an unknown number of clusters and outliers isolated as noise in the data, is applied to P. In order to apply SCAN, P needs to be represented as a graph, with I_n as a node and the similarity between images as the weight of the edge between the corresponding nodes. Text tags attached to the images of semantically related events should vary depending on the images such as the names of the captured persons and the type of the cameras; however, some text tags, presumably the terms which represent the event class, are expected to be shared among the images. Further, since the events in an event class should occur around the same time, their captured time should be close to one another. Therefore, $E(I_n, I_m)$, the similarity of the images I_n and I_m, is determined as follows so that it would be high when the images share the same text tag and their captured time is fairly close.

$$E(I_n, I_m) = \begin{cases} \exp(-\frac{|t_n - t_m|}{\frac{1}{N^2}\sum_{k=1}^{N}\sum_{l=1}^{N}|t_k - t_l|}) & (W_n \cap W_m \neq \emptyset) \\ 0 & (otherwise) \end{cases}.$$

Since SCAN is a graph clustering method for an unweighted graph, it needs to be adapted to be used for an weighted graph. SCAN uses the structural similarity between nodes a and b defined as follows.

$$\sigma_{unweighted}(a, b) = \frac{|\Gamma(a) \cap \Gamma(b)|}{\sqrt{|\Gamma(a)||\Gamma(b)|}}, \tag{1}$$

where $\Gamma(a)$ is the neighborhood of the node a with the node a itself and $\sigma_{unweighted}(a, b)$ represents the normalized number of common neighbors between the nodes a and b. SCAN classifies the nodes which share common neighbors in the same cluster. Now, in order to apply SCAN to an weighted graph, the structural similarity between the nodes a and b is determined based on the weights of their edges as follows:

$$\sigma_{weighted}(a, b) = \frac{2 \times E(a, b) + \sum\limits_{c \in \{\Gamma(a) \cap \Gamma(b)\}} \min(E(a, c), E(b, c))}{\sqrt{\sum_{x \in \Gamma(a)} E(a, x) \sum_{y \in \Gamma(b)} E(b, y)}}, \tag{2}$$

where $E(a, b)$ is the weight of the edge between the nodes a and b. If $E(a, b)$ can only take 0 or 1, Eq. (2) is equivalent to Eq. (1).

SCAN is applied to the graph of P by using $\sigma_{weighted}(I_n, I_m)$ to obtain the candidate clusters. Since the clusters containing the images posted by only a single user is unlikely to represent an event, such clusters are removed from the candidate clusters to obtain P_{ec_j} $(j = 1, \cdots)$. ec_j^t is determined as the time interval defined by the minimum and maximum times of the images in P_{ec_j} and ec_j^W is determined as a set of text tags attached to more than $s\%$ of the images in P_{ec_j}.

2.3   Event Detection

Finally, this step extracts the image set P_{e_i} based on the similarity of the captured location of the images in P_{ec_j} to determine $e_i = (e_i^t, e_i^l, e_i^W)$. Some events(e.g. Christmas, Halloween) are semantically related and occur around the same time, while some events(e.g. earthquake, Rio's Carnival) occur independently. Therefore, the detected event classes can consist of an independent event which occurs at a specific location or of several semantically related events which occur at different locations. Further, among the semantically related events, some are captured by many people while some are captured by a small number of people. Thus, the three types of events are detected based on the similarity of the captured locations of the images in P_{ec_j}.

P_{ec_j} is represented as a graph in the same way as discussed in Section 2.2. If images are captured by many people at an event, their captured locations should be close to one another. Therefore, for event detection, the similarity between the image I_n and I_m is determined as follows:

$$E(I_n, I_m) = \exp\left(-\frac{\sqrt{(\text{lat}_n - \text{lat}_m)^2 + (\text{lon}_n - \text{lon}_m)^2}}{\frac{1}{N^2}\sum_{k=1}^{N}\sum_{l=1}^{N}\sqrt{(\text{lat}_k - \text{lat}_l)^2 + (\text{lon}_k - \text{lon}_l)^2}}\right).$$

SCAN is again applied to the graph of P_{ec_j} by using $\sigma_{weighted}(I_n, I_m)$. When a single cluster is obtained, it is determined as P_{e_i}. When more than two clusters are obtained, each cluster is determined as P_{e_i}. When outliers are obtained, each outlier is determined as P_{e_i}. Then, e_i^t and e_i^l are determined as the time interval and the region defined by the minimum and maximum times and locations of the images in P_{e_i} and e_i^W is determined as a set of text tags attached to more than $s\%$ of the images in P_{e_i}. Note that the text tags in ec_j^W are excluded since they are considered to be commonly attached to all events in ec_j. When $P_{e_i} = P_{ec_j}$, e_i^t and e_i^W are determined as ec_j^t and ec_j^W, respectively.

3   Evaluation

Images captured in the area ranging from 33.5 to 35.5 northern latitude and from 134.5 to 136.5 eastern longitude (the Kansai region of Japan) during the year of 2011 and attached with at least one text tag were collected from Flickr. A total of 108,477 images uploaded by 3,287 users were collected. They are annotated with 808,586 text tags, among which 35,521 are unique. Setting $TH_t = 3\ days$, $TH_l = 0.1\ degree$, $TH_W = 0.5$, and $r = 2\%$ in the preprocessing removed 86% images and text tags and obtained 14,663 images annotated with 113,418 text tags, among which 35,488 are unique. From the 14,663 images, we created a small dataset of images captured at various types of 22 predetermined events, which construct 8 event classes, so that the ground truth for the event class and event detections can be defined. As Table 1 shows, 4 event classes are composed of 2 to 7 semantically related events, while each of the other 4 event classes is composed of a single independent event. The time spans of the events vary from 1 day to 1 month. From the 14,663 images, I) the images captured at the times and locations of the events, and II) the images annotated with manually specified event-related text tags, were extracted. As a result, 1,406 images annotated with 18,085 tags, among which 9,535 are unique, were extracted and are used as the small dataset P in the experiments. The 1,406 images should either be a) the images captured at the events annotated with various text tags, b) the images that were captured at the times and locations of the events but are not related to the events, or c) the images that are annotated with the event-related tags but were not captured at the times and locations of the events, where b) and c)

Table 1. Event classes and detection results. 'sakura', 'hanabi', 'momiji', 'gionmatsuri', and 'jidaimatsuri' are cherry blossom, fireworks, autumn leaves, gion festival, jidai festival in Japanese. 'arashiyama' is one of the popular places for leaf-viewing and also the location of Hanatoro.

event class	♯ of events	actual time	estimated time ec_j^t	determined keywords ec_j^W ($s = 50$)	♯ of detected images
cherry blossom	7	4/1-4/20	4/1-4/19	sakura	138
viewing			4/2-4/12	cherryblossom	10
fireworks displays	5	7/25-8/17	7/25-8/17	fireworks, hanabi	13
leaf-viewing	4	11/1-12/4	11/12-12/9	autumn, momiji	114
			11/18-12/3	arashiyama	9
Christmas	2	12/1-12/25	12/9-12/21	illumination	7
Gion Festival	1	7/1-7/31	7/4-7/25	gionmatsuri	42
Jidai Festival	1	10/23	10/18-10/25	gion, jidaimatsuri	8
Hanatoro	1	12/9-12/18	12/10-12/18	arashiyama	8
Plum Festival	1	2/16-3/10	–	–	–

are considered as the noises which can interfere the event class detection. Table 2 and Figure 2 show the times and locations of the 5 events constructing an event class *fireworks displays* as an example. The manually specified event-related text tags are very basic ones as shown in Table 2. The number of the images which were annotated with these event-related tags vary for each event, ranging from 1 to 33, totaling 202 among the 1,406 images.

The results of the event class detection is evaluated with the precision rate $= \frac{N_{r1}}{N_a}$ and the recall rate $= \frac{N_{r2}}{N_c}$, where N_a is the number of detected event classes, N_c is the number of predetermined event classes, and N_{r1} and N_{r2} are the numbers of correctly detected event classes. If each event class is correctly detected is checked manually based on the detected time, location, and keywords. Some event classes can be detected as more than two clusters or outliers, all of which can be considered to be correct. Since the precision rate evaluates how many of the detected event classes are correct, N_{r1} is the number of correctly detected event classes allowing such duplicates. The recall rate evaluates how many of the predetermined event classes were correctly detected; therefore, N_{r2} is the number of correctly detected event classes excluding such duplicates. The results of the event detection are evaluated in the same manner.

Applying SCAN to P and removing the clusters with the images posted only by a single user obtained the precision rate of 37.5%(9/24) and the recall rate of 87.5%(7/8). Table 1 shows the detection results. As expected, *Plum Festival* was not detected since this event occurred independently from other events and the number of its images was too few. From this result, it can be inferred that one-step detection by considering the time, location, and text tags simultaneously can fail in detecting events with a small number of images. For other event classes, the estimated times mostly coincide with their actual occurrence times and the determined keywords represent their semantics. The duplicates were detected due to different event-related tags such as 'sakura' and 'cherryblossom' for the same event class *cherry blossom viewing*. Most of the falsely detected event

Table 2. Predetermined and detected events for *fireworks displays*. 'text tags' represents manually specified event-related text tags and '♯ of images' represents the number of images annotated with the event-related text tags.

Predetermined Events				Detected Events	
event name	time	text tags	♯ of images	estimated time e_i^t	determined keywords e_i^W ($s = 50$)
Tenjin Fireworks	7/25	'fireworks' & 'tenjin'	1	7/25	yodogawa
Yodogawa Fireworks	8/6	'fireworks' & 'yodogawa'	4	-8/6	
Kobe Fireworks	8/6	'fireworks' & 'kobe'	2	8/6	kobe, minatokobe, ...
Lake Biwa Fireworks	8/8	'fireworks' & 'biwako'	1	8/8	shiga, biwako, ...
Kumano Fireworks	8/17	'fireworks' & 'kumano'	1	8/17	kumano, mie, ...

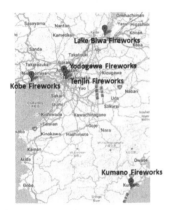

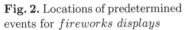

Fig. 2. Locations of predetermined events for *fireworks displays*

Fig. 3. Locations of detected events for *fireworks displays*

classes were due to the text tags automatically attached by services for posting images to Flickr such as 'hefe', 'brannan', 'amaro', 'hudson', 'earlybird', 'rise', 'eyefi' and 'uploaded: by=flicksquare'. Such text tags exhibited burst patterns at the peaks of the image uploading, and as a result, were falsely detected as the event classes. Although the images taken at the events in the same event class can capture different subjects such as christmas trees, santa clause, or family gathering for event class *Christmas*, the visual variations of the images of an event class should be smaller than those of the falsely detected event. Thus, filtering out the falsely detected event classes based on the visual similarity among the images would be our future work.

Detecting events from the correctly detected 9 event classes obtained the precision rate of 88.9%(48/54) and the recall rate of 81.8%(18/22). Table 2 and Figure 3 show an example of the detected events for an event class *fireworks displays*.

As can be seen, the detected times and locations of the events mostly coincide with their actual times and locations. *Tenjin Fireworks* and *Yodogawa Fireworks* were detected as a single event since they were located too closely to each other. Thus, the time and location of the event were determined as a time interval and a region containing both events. Other fireworks displays were correctly detected even though the images of each event were shared by only a small number of people, which verifies the effectiveness of our two-step approach.

When using 14,663 images as P as a more realistic dataset, the recall rate for the event class detection was 75%(6/8). Similar to the results for the small dataset, *Jidai Festival* and *Hanatoro* were not detected since these events occurred independently from other events and the number of their images became too few in the large dataset. On the contrary, *Plum Festival* was correctly detected as an event class since the large dataset contained other coinciding plum-related events. This result has also verified that, while one-step approach is likely to fail in detecting the events with a small number of images, the two-step approach can detect such events if they coincide with other semantically related events. The recall rate for the event detection was 72.7%(16/22). As the predetermined event classes consist of more events in the large dataset, more of the closely located events were detected as an event; however, much more events coinciding at various locations were detected successfully. Aside from the predetermined event classes, other more irregular event classes which consist of several semantically related events such as the events related to *The Great East Japan Earthquake* and *snows* were also detected.

4   Conclusions

This paper proposed a two-step method for detecting real-world events by using the time and location information and text tags attached to the images in Flickr. Experiments using Flickr images captured in Kansai region of Japan during the year of 2011 have verified that the proposed method can detect events where only a small number of images were captured when the events coincide with other semantically related events. Filtering out the falsely detected event classes, more accurately detecting closely located events, and comparison with the one-step approach by using a larger set of images which are captured over a wider range of time and location would be our future work.

References

1. Chen, L., Roy, A.: Event Detection from Flickr Data through Wavelet-based Spatial Analysis. In: Proc. CIKM, pp. 523–532 (2009)
2. Rattenbury, T., Good, N., Naaman, M.: Towards Automatic Extraction of Event and Place Semantics from Flickr Tags. In: Proc. SIGIR, pp. 103–110 (2007)
3. Sakaki, T., Okazaki, M., Matsuo, Y.: Earthquake Shakes Twitter Users: Real-time Event Detection by Social Sensors. In: Proc. WWW 2010, pp. 851–860 (2010)
4. Sheth, A.: Citizen Sensing, Social Signals, and Enriching Human Experience. IEEE Internet Computing 13(4), 87–92 (2009)
5. Xu, X., Yuruk, N., Schweiger, T.A.J.: SCAN: A Structure Clustering Algorithm for Networks. In: Proc. KDD, pp. 824–833 (2007)

Spectral Classification of 3D Articulated Shapes

Zhenbao Liu, Feng Zhang, and Shuhui Bu*

Northwestern Polytechnical University, Xi'an, 710072, China
{liuzhenbao,bushuhui}@nwpu.edu.cn

Abstract. A large number of 3D models distributed on internet has created the demand for automatic shape classification. This paper presents a novel classification method for 3D mesh shapes. Each shape is represented by the eigenvalues of an appropriately defined affinity matrix, forming a spectral embedding which achieves invariance against rigid-body transformations, uniform scaling, and shape articulation. And then, Adaboost algorithm is applied to classify the 3D models in the spectral space according to its immunity to overfitting. We evaluate the approach on the McGill 3D shape benchmark and compare the results with previous classification method, and it achieves higher classification accuracy. This method is suitable for automatic classification of 3D articulated shapes.

Keywords: 3D Shape, Spectral classification, Boosting.

1 Introduction

Automatic classification of 3D models has a very wide range of applications in different domains such as multimedia organization. Large databases of 3D models available in the public domain have produced the demand for shape analysis such as search, retrieval, and classification. Many techniques are inspired by image or video search and classification. In addition, 3D shape classification can also help to solve graphics problem such as shape registration, texture mapping, shape segmentation, shape deformation, and shape modeling. For example, recent popular segmentation methods [1] greatly rely on 3D shape classification, because a category of 3D shapes are required to train these segmentation methods.

In the past ten years, many feature description methods of 3D models [2] emerged. But few of them are remarkable when it comes to shape classification. Barutcuoglu and Decoro [3] proposed hierarchical shape classification using Bayesian aggregation and spherical harmonics descriptor. However, this method has a limitation that it is only applicable to hierarchical shape classification. A semi-supervised semantic clustering and retrieval method based on Support Vector Machines (SVM) is put forword by Hou et al. [4], and its query is labeled with its semantic concept to conduct content-based search from the resulting semantic clusters. Liu et al. [5] proposed a 3D shape classifier with neural network classifier. But this method has a low precision of classification.

---

* Corresponding author.

C. Gurrin et al. (Eds.): MMM 2014, Part II, LNCS 8326, pp. 315–322, 2014.

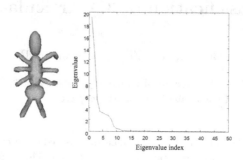

Fig. 1. An example of extracted feature of an ant model

In order to overcome the limitations of above methods, we propose a method of 3D shape classification to improve classification accuracy. The feature extraction method presented in this paper utilizes the spectral embedding and calculates eigenvalues of the affinity matrix of 3D model. Then we train several main eigenvalues in the spectral space using an adaptive boosting algorithm to learn spectral distribution of examples.

To evaluate the proposed method, we adopt shapes with articulating parts in the McGill 3D Shape Benchmark [6] as training and testing set. There are 10 classes of shapes and total 200 shapes with articulating parts in the database. In the experiments, 3D models are equally divided into two sets, the training set and the testing set. The experiment results show that our method achieves a high averaged classification accuracy above 95%. In addition, the method has several advantages, e.g., robust against non-rigid deformation, insensitive to model scale, and high processing speed.

2   The Proposed Method

The proposed method include the following steps. First, each shape is represented by the eigenvalues of a geodesic matrix, which is applied as classification feature. And then we adopt χ^2 distance to measure 3D shape similarity to overcome the bias caused by the absolute size of eigenvalues. Finally, an adaptive boosting algorithm of classifier is trained to classify 3D shapes.

2.1   Feature Extraction

The feature extraction method used in this paper is based on the spectral embeddings of 3D mesh. Spectral techniques first aim at the problem of point correspondence between two images or sets of extracted image features [7] [8]. This technique is introduced into 3D shape analysis by Jain and Zhang [9] [10]. Zhang et al. [10] surveyed spectral methods for mesh processing relying on the eigenvalues, eigenvectors, or eigen-space projections derived from appropriately defined mesh operators to carry out desired tasks. Spectral analysis makes use of shape features based on the nature of connection graph, so that an affinity matrix can be directly constructed in the triangular mesh surface.

For a given 3D model with n vertices, we can construct an $n \times n$ affinity matrix $\mathbf{A}$, and the entry A_{ij} means the affine relationship between the i-th and the j-th vertex in the 3D model. The calculation of A_{ij} is defined as follows.

$$A_{ij} = \exp(-d_{ij}^2/\sigma^2), \tag{1}$$

where d_{ij} is the approximate geodesic distance between the vertex i and vertex j, and σ is the Gaussian width. We know from the definition that the affine relationship between two points is represented with their approximate geodesic distance. The use of the Gaussian affinity can effectively reduce the influence of further points, and it also makes the eigenvalues constant to the uniform scaling. Here, we define σ is the maximum approximate geodesic distance for all vertices,

$$\sigma = \max(d_{ij}), \ i, j \in \{1, 2, \cdots n\}. \tag{2}$$

Then, the matrix $\mathbf{A}$ is is eigen-decomposed as $\mathbf{A} = \mathbf{V}\mathbf{\Lambda}\mathbf{V}^T$, $\mathbf{\Lambda}$ is a diagonal matrix where the diagonal elements $(\lambda_1 \geq \cdots \lambda_n)$ are the eigenvalues of $\mathbf{A}$ sorted in descending order. $\mathbf{V} = [v_1 | \cdots v_n]$ is a $n \times n$ matrix and $v_1, \cdots v_n$ are eigenvectors corresponding to the eigenvalues. Figure 1 illustrates an example of eigenvalues of an ant model.

For constructing the affinity matrix of a 3D model, approximate geodesic distances between any two points in 3D model are computed. For a given source vertex in the graph, Dijkstra algorithm is applied to find the path with lowest cost between the vertex and other vertices. Therefore, we first calculate the Euclidean distance between two adjacent vertices as the weight of this edge, and then the Dijkstra algorithm is used to approximate the geodesic distance.

The time complexity of construction and eigen-decomposition of the $n \times n$ affinity matrix is $O(n^2 \log n + n^3)$. Because a 3D model usually contains thousands of vertices, to decrease time complexity, we utilize Nystrom approximation [11] to compute the eigenvalues of the sub-sampling matrix and these eigenvalues approximate the eigenvalues of original affinity matrix. The sampling points are obtained by farthest points sampling. At each sampling time, we choose a point with the maximum geodesic distance to the former sampling point. The advantage is that we make the sampling points distributed on the cusp of the model. The Nystrom approximation reduces the time complexity to $O(n \log n + l^3)$, and l is the number of samples, generally $l << n$.

2.2 Feature Distance

To eliminate the bias caused by the absolute size of eigenvalues, we calculate the feature distance χ^2 used to measure feature similarity. For two given 3D models P, Q and their eigenvalues $\lambda_i^P, \lambda_i^P (i = 1, 2, \cdots, 20)$, χ^2 distance is defined as follows.

$$Dist(P, Q) = \frac{1}{2} \sum_{i=1}^{20} \frac{(|\lambda_i^P|^{\frac{1}{2}} - |\lambda_i^Q|^{\frac{1}{2}})^2}{|\lambda_i^P|^{\frac{1}{2}} + |\lambda_i^Q|^{\frac{1}{2}}}. \tag{3}$$

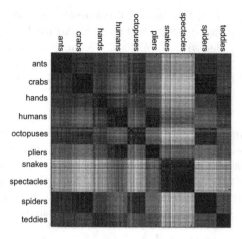

Fig. 2. The similarity matrix of 3D models in the McGill 3D shape benchmark. The deeper color identifies the smaller feature distance, which means two shapes belong to the same category.

Figure 2 shows the similarity matrix of ten categories of 3D models in the McGill 3D Shape Benchmark. Deeper color represents smaller χ^2 distances which infers that they are relevant shapes.

2.3   Classification

We adopt a robust classifier, Adaptive Boosting, to classify 3D objects with rich geometrical variation. AdaBoost classifier does not require classification parameters except the number of cycles to control the training accuracy. In addition, an important property is its ability of relative immunity to over-fitting. Therefore, we decide to apply AdaBoost algorithm to classify complex 3D models. AdaBoost algorithm first trains basic weak classifiers for different training sets, and then put these weak classifiers together to form a strong classifier. It is proved that as long as classification capability of each weak classifier is better than random guess, when the number of weak classifiers tends to infinity, training error rate of strong classifier will approach zero.

In the classification algorithm, the training set is extrinsically changed by adjusting the weight of each sample. At the beginning the weights are kept same, and a basic weak classifier $h_1(x)$ is trained on this same distribution. For the samples that $h_1(x)$ misclassified the corresponding weights increases, and for the samples correctly classified, the corresponding weights will be reduced. This makes the misclassified samples attached more attention, and generates a new sample distribution. Meanwhile, according to the degree of wrong classification, we update the weight α_t (t is the sequence number corresponding to the weak classifier $h_t(x)$) to indicate the importance of this weak classifier in the final strong classifier and α_t is defined as Equation 4. ϵ_t is the classification error rate

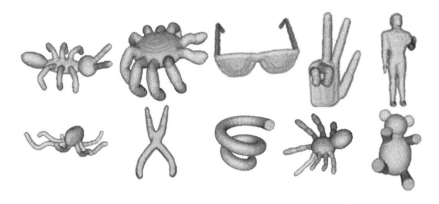

Fig. 3. The McGill 3D shape benchmark

of weak classifier $h_t(x)$. If fewer samples are misclassified, the weight of the weak classifier becomes larger.

$$\alpha_t = \frac{1}{2} \ln[(\frac{1 - \epsilon_t}{\epsilon_t})]. \tag{4}$$

The sample weights are changed as described in Equation 5. $D_t(i)$ is previous sample distribution function and the $D_{t+1}(i)$ is the new sample distribution function. Z_t is a normalization factor to ensure $\sum_i D_{t+1}(i) = 1$.

$$D_{t+1}(i) = \frac{D_t(i)}{Z_t} \times \begin{cases} e^{-\alpha_t}, if h_t(x_i) = y_i \\ e^{\alpha_t}, if h_t(x_i) \neq y_i \end{cases} \tag{5}$$

Therefore, in the new sample distribution all the samples are retrained to obtain the weak classifier $h_2(x)$ and its weight. After T train times the last weak classifier $h_T(x)$ obtains a least classification error rate. We get T weak classifiers and their corresponding weights. Finally, these T weak classifiers are summed together according to its weights to produce the final strong classifier $H(x)$ defined as Equation 6. $H(x)$ has stronger classification ability than any of T weak classifiers. Through the strong classifier $H(x)$ we classify the samples in the testing set.

$$H(x) = sign \left(\sum_{t=1}^{T} \alpha_t h_t(x) \right). \tag{6}$$

3   Experiments

In this paper, we adopt 10 categories of 3D models with articulating parts in the McGill 3D Shape Benchmark [6] to evaluate the proposed method and compare our results with previous classification method [4]. The dataset consist of ants, crabs, hands, humans, octopuses, pliers, snakes, spectacles, spiders, teddies as shown in Fig. 3. There are in total 200 3D models in the benchmark. These 3D models are equally divided into two parts. In the training and testing steps, we adopt $1 : n$ multi-classification for all categories of models, then average total classification results.

Table 1. The numerical results of previous 3D model classification [4]

Category	Number of Misclassified Model	Error Rate	Accuracy
ants	10	10%	90%
crabs	10	10%	90%
hands	10	10%	90%
human	10	10%	90%
octopuses	10	10%	90%
pliers	10	10%	90%
snakes	10	10%	90%
spectacles	10	10%	90%
spiders	12	12%	88%
teddies	13	13%	87%
average	10.5	10.5%	89.5%

Table 2. The numerical results of the proposed classification algorithm

Category	Number of misclassified models	Error rate	Accuracy
ants	4	4%	96%
crabs	3	3%	97%
hands	6	6%	94%
human	8	8%	92%
octopuses	3	3%	97%
pliers	1	1%	99%
snakes	0	0%	100%
spectacles	3	3%	97%
spiders	7	7%	93%
teddies	6	6%	94%
average	4.1	4.1%	95.9%

The numerical results of the proposed method and Hou's method [4] are reported in Table 1, and 2, respectively. The error rate listed in the two tables is the percentage of the misclassified models to all the models, and the accuracy is the percentage of correctly classified models in the dataset. Figure 4 illustrates the error rates of two classification methods on different categories. Figure 5 shows a classification example of our method on pliers, and only one shape is mis-classified. From the results of 3D shape classification, we see that our method achieved an average accuracy of 95.9%, which is clearly higher than Hou's 90.5%.

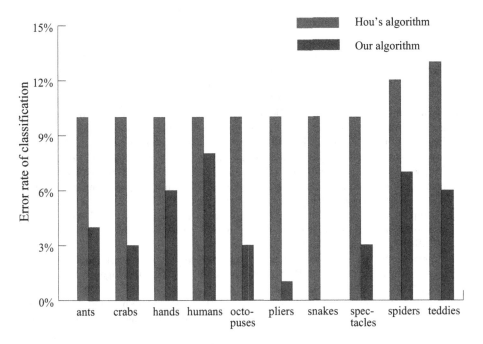

Fig. 4. Comparison of classification error rates

Fig. 5. A classification example with one error for pliers

4   Conclusion

This paper presented a novel classifier based spectral feature extraction and Adaboost for three-dimensional articulated models. Compared with the experimental results generated by previous classification algorithm, we conclude that the accuracy of classification has been improved, and it can be used to online classification of 3D shapes.

Acknowledgments. The work was supported by NSFC (61003137, 61202185), NPU-FER(JC201202, JC201220), Shaanxi NSF(2012JQ8037), and Open Fund from State Key Lab of CAD&CG of Zhejiang University.

References

1. Liu, Z., Tang, S., Bu, S., Zhang, H.: New evaluation metrics for mesh segmentation. Computers and Graphics (SMI) 37(6), 553–564 (2013)
2. Liu, Z., Bu, S., Zhou, K., Gao, S., Han, J., Wu, J.: A survey on partial retrieval of 3D shapes. Journal of Computer Science and Technology 28(5), 836–851 (2013)
3. Barutcuoglu, Z., DeCoro, C.: Hierarchical shape classification using bayesian aggregation. In: Proceedings of IEEE Shape Modeling International (2006)
4. Hou, S., Lou, K., Ramani, K.: Svm-based semantic clustering and retrieval of a 3d model database. Computer Aided Design and Application 2, 155–164 (2005)
5. Liu, Z., Mitani, J., Fukui, Y., Nishihara, S.: A 3d shape classifier with neural network supervision. International Journal of Computer Applications in Technology 38(1-3), 134–143 (2010)
6. Siddiqi, K., Zhang, J., Macrini, D., Shokoufandeh, A., Bouix, S., Dickinson, S.: Retrieving articulated 3D models using medial surfaces. Machine Vision and Applications 19(4), 261–274 (2008)
7. Umeyama, S.: An eigen decomposition approach to weighted graph matching problem. IEEE Transactions on Pattern Analysis and Machine Intelligence 10(5), 695–703 (1988)
8. Scalaroff, S., Pentland, A.: Modal matching for correspondence and recognition. IEEE Transactions on Pattern Analysis and Machine Intelligence 17(6), 545–561 (1995)
9. Jain, V., Zhang, H.: A spectral approach to shape-based retrieval of articulated 3d models. Computer-Aided Design 39(5), 398–407 (2007)
10. Zhang, H., Kaick, O.V., Dyer, R.: Spectral mesh processing. Computer Graphics Forum 29(6), 1865–1894 (2010)
11. Fowlkes, C., Belongie, S., Chung, F., Malik, J.: Spectral grouping method using the nystrom intelligence. IEEE Transactions on Pattern Analysis and Machine Intelligence 26(2), 214–225 (2004)

Improving Scene Detection Algorithms Using New Similarity Measures

Stefan Zwicklbauer, Britta Meixner, and Harald Kosch

Chair of Distributed Information Systems, University of Passau
{stefan.zwicklbauer,britta.meixner,harald.kosch}@uni-passau.de
http://www.fim.uni-passau.de/index.php?id=483&L=1

Abstract. The creation process of interactive non-linear videos affords the definition of scenes which are connected in a scene graph. These might be available in form of raw material (shots) or need to be extracted from existing films. In the latter case, the scenes have to be defined by the author in a time-consuming process. A semi-automated scene extraction helps the user perform this task. Detected shots and scenes may be corrected in a graphical user interface which is integrated in our authoring tool. To provide satisfying results in the automated scene detection process our main goal was to improve an existing algorithm. Different shot comparison functions were evaluated regarding the overall performance index. The commonly used color histogram intersection was outperformed by a combination of χ^2-distance and complexity comparison, which unexpectedly attained the best results.

Keywords: Scene Detection, Shot Similarity, User Correction GUI.

1 Introduction

The use of different consecutive algorithms is necessary to successfully extract scenes from a video. The beginning and the end of a scene is a transition between shots. According to that, the detection of shot boundaries is the first step. A keyframe extraction is the second step before executing the scene detection as the third step. Concrete criteria have to be determined whose characteristics exist within and between scenes to be able to extract scenes. The following characteristics can be noted according to [16]: (1) Shots of a scene show similar color properties. (2) Audio tracks between different scenes differ significantly. In case of no cut at a shot boundary it is a scene transition. (3) If two shots A and B show similarities regarding the used metrics and are part of the same scene, all shots between A and B are part of the scene, too. Basically we use the word 'scene' as a division of an act presenting continuous action in one place in our work. Speaking of action within one place is not an exact statement because we can distinguish between different radii of an area. Therefore we make the term 'in one place' more specific by graduating scenes into different granularity levels. Nearly all available algorithms use color histogram intersection as the main shot comparison function due to its simplicity and computability. Faster and

C. Gurrin et al. (Eds.): MMM 2014, Part II, LNCS 8326, pp. 323–330, 2014.
© Springer International Publishing Switzerland 2014

parallel processing units allow the usage of more complex features. Overall, only a well-chosen combination of algorithms (shot detection, keyframe extraction) and functions (color differences, spatial differences) provide acceptable results. In this work we focus on the improvement of an existing scene detection algorithm instead of already convincing pre-processing steps like shot detection (Section 3). For this purpose we tried to spot the best visual shot similarity function which is mainly responsible for the later results. A combination of this function with our new complexity feature improves the algorithm in terms of error susceptibility and detection rate. An evaluation shows a comparison of shot similarity functions (Section 4).

2   Related Work

A literature review of approaches on scene detection basically revealed three main types of algorithms: visual-based segmentation, audio-based full segmentation and audio-visual full segmentation. Each category offers subcategories to classify algorithms in more detail. Visual-based segmentation algorithms constitute the fundamentals of more complex approaches. A graph-based approach is described by Ngo et al. [9]. Shots are arranged to a graph whereby the similarity of shots is represented by the weight of edges (based on color similarity). In the following, shots are grouped into scenes. Another algorithm focuses on shot clustering with the help of a sequence alignment algorithm [1]. The authors cluster shots into visually similar groups which are labeled. After that, "a sequence alignment algorithm is applied to detect when the pattern of shot labels changes, providing the final scene segmentation result" [1]. Lu et al. in [7] present a video segmentation algorithm which uses the audio component of videos. Key audio elements are identified by dividing the audio stream into audio segments with the help of low-level features. In [10] the authors introduce semantic audio textures. These are semantically consistent chunks of audio data which are trained by identifying basic audio classes to the former and merging the audio classes according to predefined audio textures based on genre-specific heuristics to the latter. The third algorithm category is a combination of both visual and audio-based segmentation approaches. The approach presented in [6] primarily uses audio information for the scene segmentation by applying a principal component analysis to detect background noise. In addition, a shot boundary detection is used to reduce the number of false positive results. A similar work to ours from Sidiropoulos et al. uses audio-based and visual-based graphs [14]. Performing a probabilistic graph merging process results in identifying scene boundaries with promising result values.

Unfortunately, it is not possible to make accurate quantitative comparisons of the approaches, as most authors performed their evaluation on different data sets [2]. The main difference in scene detection algorithms can be affiliated to shot similarity calculation. Many approaches neglect complex shot similarity functions due to performance problems.

3   Scene Detection

In this work we focus on improving scene detection results with optimizing and defining new similarity functions. Basically we use the graph-based algorithm from Ngo [9] which models scenes by means of a generated temporal graph.

The first step in our algorithm will determine the difference between all shot pairs. Our goal was to calculate shot similarities in a better way than most available algorithms. We tried to combine a function which delivers the best color differences between shots with a function that describes the complexity differences. χ^2-distance [12] is defined as:

$$D_\chi(kf_g^a, kf_h^b) = \sum_{k=0}^{J-1} \frac{(H_{kf_g^a}(k) - f'(k))^2}{f'(k)} \tag{1}$$

$$f'(k) = \frac{H_{kf_g^a}(k) + H_{kf_h^b}(k)}{2}$$

where kf_g^a denotes an arbitrary keyframe g within a shot a. Additionally $H(k)$ specifies the color histogram with bin k, with J being the number of bins in a frame. The major drawback that χ^2-distance accounts only for corresponding bins with the same index, does not affect results in a bad way. Contrary to χ^2-distance the quadratic form distance, which is described in [4], regards not only corresponding bins (cross-bin). This measurement is defined as:

$$D_{QF}(kf_g^a, kf_h^b) = \sqrt{(H_{kf_g^a} - H_{kf_h^b})^T A (H_{kf_g^a} - H_{kf_h^b})} \tag{2}$$

where $A = [a_{i,j}]$ denotes a matrix with the weights $a_{i,j}$ denoting the similarity between color bins i and j. Again H denotes the color histogram of the respective keyframe. A more detailed overview of this approach can be found in [4].

The most used image comparison function color histogram intersection [15] is also implemented for evaluation purpose. After computing the visual similarity between two keyframes we analyze them concerning the image complexity. Therefore, the amount of existing edges will be compared. We categorize the following edges: strong, middle and small:

$$D_{Comp}(kf_g^a, kf_h^b) = \frac{\frac{1}{2} \cdot min(X_{strong}^{kf_g^a}, X_{strong}^{kf_h^b})}{max(X_{strong}^{kf_g^a}, X_{strong}^{kf_h^b})}$$

$$+ \frac{\frac{1}{3} \cdot min(X_{mid}^{kf_g^a}, X_{mid}^{kf_h^b})}{max(X_{mid}^{kf_g^a}, X_{mid}^{kf_h^b})} + \frac{\frac{1}{6} \cdot min(X_{small}^{kf_g^a}, X_{small}^{kf_h^b})}{max(X_{small}^{kf_g^a}, X_{small}^{kf_h^b})} \tag{3}$$

where $X_{strong}^{kf_g^a}$ denotes for instance the amount of strong edges in keyframe g which belongs to shot a. Color similarity, complexity similarity, and the time distance in frames will be combined to an overall similarity value ([9]):

$$w(S_i, S_j) = exp\left(\frac{-k \cdot |f_{S_i} - f_{S_j}|}{T} \cdot sim(S_i, S_j)\right) \tag{4}$$

The temporal distance $|f_{S_i} - f_{S_j}|$ between two shots is measured in frames. The associated parameter k describes the importance of the temporal distance. The authors of [9] suggest $k = 8$ as satisfying. T constitutes the amount of frames in the video. The highest distance value of two keyframes constitutes the shot similarity $sim(S_i, S_j)$ which depicts a linear combination $\Theta^T x$ with Θ being the weight vector of the similarity features x of two frames. We cast the problem to a ranking problem and train the weights with Sofia-ML[1] machine learning framework. For training purpose we manually selected and rated a few shots of our evaluation data set and mark a shot pair as true if they belong to the same scene and false if not. Due to a needed comparison of every shot pair we create a shot similarity matrix. The matrix will be divided into shot groups where shots with a high degree of similarities belong together. Additionally different shots should be classified into different groups. For this purpose we use the Normalized Cut approach (NCut) from [13]. Since belonging to NP-complete problems the authors suggest to reduce NCut to a standard-eigenvalue-problem. Fortunately we only need the eigenvector corresponding to the second smallest eigenvalue, which helps us split our shot set recursively until a given threshold is exceeded. Eigensolver methods like the Lanczos algorithm determine eigenvalues iteratively and provide a running time of $O(n)$, with n being the amount of rows or columns in the shot similarity matrix. A detailed overview of this approach can be found in [13]. Current sets of shots C_x will be connected to a temporal graph by connecting them with the following rule: Create an edge between C_n and C_m if there is a shot S_i in set C_n and a shot S_{i+1} in set C_m where i denotes the temporal shot order. Finally we extract the scenes of the temporal graph by computing the shortest path $(S_1, S_2, ..., S_n)$ with the Dijkstra algorithm. The edge from S_i to S_j will be disconnected if $i = j + 1$ pertains. If there is no path from S_i to S_j or vice versa, S_i and S_j do not belong to the same scene.

4   Evaluation

The data set for testing the detection rate of the shot similarity algorithms is a mixture of different genres and varying color environments. We chose videos from a wide range of genres to test our algorithms, because our intended use cases (non-linear e-learning videos, navigation in existing videos) can not be limited to a single type of video or genre. Furthermore we selected the videos according to their potential difficulties for the algorithms. As described in the introduction, we created a scene groundtruth by choosing those parts of the videoclips which present continuous action in one place. Because of the definition of 'a place' being fuzzy, we created three groundtruth versions of every video: coarse, medium and fine. All three properties specify the radius where the action takes place, from small radius (fine) to high radius (coarse). An expert group is used to determine the scenes of the different levels in each videoclip. The best result of an algorithm on a granularity level is compared with the result of another algorithm which might attain its best values on another granularity level.

---

[1] http://code.google.com/p/sofia-ml/

Table 1. Features of the test videos for evaluation

Video	Number of scenes			Number of shots	Length (min)
	coarse	medium	fine		
Advertising	31	68	87	213	06:43
Documentary	21	22	23	69	06:57
News	13	21	23	45	09:31
Sitcom	15	15	15	195	11:25
Film1	24	28	36	124	09:19
Film2	15	18	21	310	07:13
Series	15	17	21	195	08:43

The advertisement film includes 31 commercials with a hardly noticeable story-line. The news clip is a compilation of the attack of 11 September in New York which could be seen as one scene. The advertisement, documentation and news clip have no explicit story line which makes it difficult for scene detection algorithms to decide whether visually different shots belong together. A compilation of 15 scenes with one person is tested with the sitcom Big-Bang-Theory. This clip shows the same background setting throughout the whole clip. Only slight differences of the positioning of the main characters characterize new scenes. Film1 is a compilation of scenes from the action film Heat playing in different locations. Film2 is a compilation of the best fighting scenes from the Matrix trilogy with scenes playing in dark ambiance and fast camera recordings. These action movies contain a lot of fast dolly shots where explosions and other effects hamper the scene detection process. The series test contains the first ten minutes of an episode of Criminal Minds containing the intro and other scenes. Several zooms occur during the intro and the following shots which results in different coloration. Those effects are crucial to shot detection and scene detection algorithms. A precise description of the features of the test data set can be found in Table 1. The scene detection is performed on the results of a shot detection algorithm with high recall values and values for the F1 measure between 78% and 100% (average 91,21%). Our shot detection algorithm can be seen as a strongly modified version of the fuzzy-logic algorithm proposed in [3]. The used keyframe extraction algorithm is a 2-frames-algorithm, which uses one of the first and last keyframes (depending on whether the transition is gradual or not) of the shots. Additionally a third keyframe is extracted in the middle of the respective shot. The corresponding Fig. 1 shows the results of the shot similarity functions χ^2-distance (WCD), χ^2-distance in combination with our new complexity feature (WCD+C), color histogram intersection (CHI) and quadratic form distance (QFD). Nearly all scene detection algorithms use CHI as main similarity function due its simplicity and computability. QFD and WCD have not been used in scene detection so far. For that reason we show the different overall results using these measurements and feature a better similarity metric than CHI. Our printed overall performance index (OPI) [5] is the highest value, which is achieved by one of the granularity stages: coarse, medium or fine. The OPI value may be compared to a modified F1 value, which was adapted to a scene detection

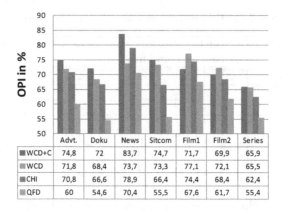

Fig. 1. OPI of the tested shot similarity functions

	Advt.	Doku	News	Sitcom	Film1	Film2	Series
■ WCD+C	74,8	72	83,7	74,7	71,7	69,9	65,9
■ WCD	71,8	68,4	73,7	73,3	77,1	72,1	65,5
■ CHI	70,8	66,6	78,9	66,4	74,4	68,4	62,4
■ QFD	60	54,6	70,4	55,5	67,6	61,7	55,4

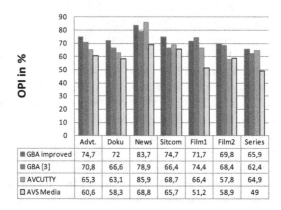

Fig. 2. OPI of the tested shot similarity functions

	Advt.	Doku	News	Sitcom	Film1	Film2	Series
■ GBA improved	74,7	72	83,7	74,7	71,7	69,8	65,9
■ GBA [3]	70,8	66,6	78,9	66,4	74,4	68,4	62,4
■ AVCUTTY	65,3	63,1	85,9	68,7	66,4	57,8	64,9
□ AVS Media	60,6	58,3	68,8	65,7	51,2	58,9	49

evaluation. Anyway a detailed comparison of the WCD and CHI function shows differences of 5 percent in nearly all tested videos. Only the news clip provides better results with the CHI function. After the addition of our complexity comparison, WCD gets excellent results in all videos. Surprisingly the QFD suffers from a poor OPI, although its computational complexity is the highest of all tested functions and its bin comparison process regards not only corresponding bins.

In a second step, we compared our new algorithm (GBA improved) with two scene detection tools, AVCutty [2] (AVC) and AVS Video Editor [3] (AVE), and of course with the approach from [9]. AVC and AVE uses color comparison methods and a 2-frames keyframe algorithm to detect scene boundaries. AVE additionally uses the temporal distance as an indicator for scenes. AVC and AVE both provide

---

[2] http://www.avcutty.de (accessed March 12, 2013)
[3] http://www.avsmedia.com (accessed March 12, 2013)

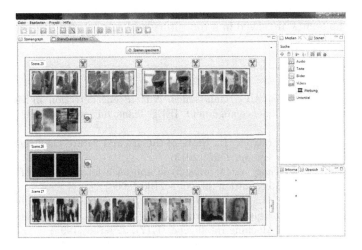

Fig. 3. Screenshot of the user correction GUI

an average OPI value of 67,44 and 58,92 percent contrary to our algorithm with 73,21. The results from GBA [9] with 69,7 percent are second overall. Precise results are shown in Fig. 2. We could optimize our algorithm further with features like the influence of gradual transitions in the scene detection process [11] or further shot similarity features to distinguish shots even better. To sum up, we can say that a combination of χ^2-distance and complexity comparison improves every scene detection which uses CHI as a similarity measure between shots. Additionally our graph-based algorithm attains good results in every tested video clip and outperforms the other algorithms.

5 Conclusion and Future Work

We have presented an improved approach for video scene detection. Our main algorithm shows that detecting scenes mainly depends on the capability to distinguish arbitrary shots. In this context the χ^2-distance combined with our complexity feature improves all scene detection algorithms which use CHI as shot comparison mode. In the future, we will focus our research on combining audio features with visual features to significantly improve the video analysis in general. The shot detection is integrated into a software for interactive video authoring [8]. A small settings window allows the configuration of the parameters of the scene detection. The user is able to activate parallel processing, the detection of fades and dissolves (or just hart cuts), or the storage of the results as an MPEG-7 file as well as the keyframe extraction algorithm and other variable settings for the algorithms. After shot and scene detection, an overview of the results determined by the algorithms is displayed (see Fig. 3), where the user is able to add and delete shot or scene boundaries. The adding of a shot boundary is accomplished by the scene editor described in [8]. The splittings and mergings of scenes is allowed at shot boundaries and can be done directly in the according overview. The scenes are saved in a repository and can be used in a scene graph.

References

1. Chasanis, V.T., Likas, A.C., Galatsanos, N.P.: Scene detection in videos using shot clustering and sequence alignment. IEEE Trans. on Multimedia 11(1), 89–100 (2009)
2. Del Fabro, M., Böszörmenyi, L.: State-of-the-art and future challenges in video scene detection: a survey, pp. 1–28 (2013)
3. Fang, H., Jiang, J., Feng, Y.: A fuzzy logic approach for detection of video shot boundaries. Pattern Recognition 39, 2092–2100 (2006)
4. Hafner, J., Sawhney, H., Equitz, W., Flickner, M., Niblack, W.: Efficient color histogram indexing for quadratic form distance functions. IEEE Transactions on Pattern Analysis and Machine Intelligence 17(7), 729–736 (1995)
5. Hua, X.-S., Zhang, D., Li, M., Zhang, H.-J.: Performance evaluation protocol for video scene detection algorithms. In: Workshop on Multimedia Information Retrieval, in Conjunction with 10th ACM Multimedia (2002)
6. Kyperountas, M., Kotropoulos, C., Pitas, I.: Enhanced eigen-audioframes for audiovisual scene change detection. IEEE Transactions on Multimedia 9(4), 785–797 (2007)
7. Lu, L., Cai, R., Hanjalic, A.: Audio elements based auditory scene segmentation. In: Proceedings of the 2006 IEEE International Conference on Acoustics, Speech and Signal Processing, ICASSP 2008, vol. 5, p. V (2006)
8. Meixner, B., Matusik, K., Grill, C., Kosch, H.: Towards an easy to use authoring tool for interactive non-linear video. Multimedia Tools and Applications, 1–26 (2012) ISSN 1380-7501
9. Ngo, C.-W., Ma, Y.-F., Zhang, H.-J.: Video summarization and scene detection by graph modeling. IEEE Trans. on Circuits and Syst. for Video Technology 15(2), 296–305 (2005)
10. Niu, F., Goela, N., Divakaran, A., Abdel-Mottaleb, M.: Audio scene segmentation for video with generic content. In: Presented at the Society of Photo-Optical Instrumentation Engineers (SPIE) Conference. Society of Photo-Optical Instrumentation Engineers (SPIE) Conference Series, vol. 6820 (2008)
11. Petersohn, C.: Improving scene detection by using gradual shot transitions as cues from film grammar. In: Proc. of Multimedia Content Access: Algorithms and Systems II, vol. 6820, p. 68200D (2008)
12. Rubner, Y., Tomasi, C., Guibas, L.J.: The earth mover's distance as a metric for image retrieval. International Journal of Computer Vision 40(2), 99–121 (2000)
13. Shi, J., Malik, J.: Normalized cuts and image segmentation. IEEE Transactions on Pattern Analysis and Machine Intelligence 22(8), 888–905 (2000)
14. Sidiropoulos, P., Mezaris, V., Kompatsiaris, I., Meinedo, H., Bugalho, M., Trancoso, I.: Temporal video segmentation to scenes using high-level audiovisual features. IEEE Trans. Cir. and Sys. for Video Technol. 21(8), 1163–1177 (2011)
15. Swain, M.J., Ballard, D.H.: Color indexing. International Journal of Computer Vision 7(1), 11–32 (1991)
16. Vendrig, J., Worring, M.: Systematic evaluation of logical story unit segmentation. IEEE Transactions on Multimedia 4(4), 492–499 (2002)

EvoTunes: Crowdsourcing-Based Music Recommendation

Jun-Ho Choi and Jong-Seok Lee

School of Integrated Technology
Yonsei University, Korea
{idearibosome,jong-seok.lee}@yonsei.ac.kr

Abstract. In recent days, there have been many attempts to automatically recommend music clips that are expected to be liked by a listener. In this paper, we present a novel music recommendation system that automatically gathers listeners' direct responses about the satisfaction of playing specific two songs one after the other and evolves accordingly for enhanced music recommendation. Our music streaming web service, called "EvoTunes," is described in detail. Experimental results using the service demonstrate that the success rate of recommendation increases over time through the proposed evolution process.

Keywords: music streaming, recommendation, crowdsourcing.

1   Introduction

Recently, there has been a significant change in the way of listening music due to development of the technology for contents delivery over Internet and increasing availability of digital music sources. Especially, people can obtain a variety of music easily by buying or downloading its digital copy from Internet, while it had to be bought directly from offline music stores in the past. Consequently, gathering data related to patterns of buying and listening, recommending other music that is not played often, and inducing purchase are in spotlight of recent studies.

There are two major approaches of music recommendation that are studied and commercially used. First, some music recommendation services rely on similarity between songs in terms of particular audio/musical properties analyzed by using signal processing techniques [1,2] or by hand [3], including low-level representations like spectral features and higher level properties such as tempo, timbre, genre, and even emotional elements. By these properties, they recommend the song that has the most similar properties to the current one as the next song to be played. This type of recommendation has an assumption that listeners prefer songs that have cognate properties. Second, listeners can input some metadata of specific music such as rating, and the system recommends a song that is liked by other users having similar musical tastes based on collaborative filtering [4,5]. However, most users tend to add metadata only for some popular songs, so less popular music may not be recommended due to the lack of data.

C. Gurrin et al. (Eds.): MMM 2014, Part II, LNCS 8326, pp. 331–338, 2014.

In this study, we take a different, novel approach for music recommendation. We use a crowdsourcing method by gathering listeners' direct feedback about the satisfaction of playing specific two songs in a row. To get experimental data, we created a system named "EvoTunes," which plays songs chosen by our music recommendation algorithm, and at the same time, gathers participants' responses for the recommended music to enhance the recommendation algorithm. As the name implies, the recommendation algorithm *evolves* based on the collected user responses. Distinctly from the previous work, our method can generate playlists where orders of songs are important. Moreover, our system can spontaneously adapt to changes in the trend of music and incorporate new songs in the recommendation system.

2   Proposed System

2.1   Service Platform

"EvoTunes" is a web music player platform to conduct our crowdsourcing-based experiment (Fig. 1). Anyone who has a Facebook account can sign up the website. It plays songs one by one and participants can use two playback control buttons: play/pause and skip (Fig. 2). The appearance of the player is very similar to existing music players, so that listeners would experience our experiment almost in the same way as their usual listening situation. Also, because EvoTunes is distributed as a public website, people can use it anytime.

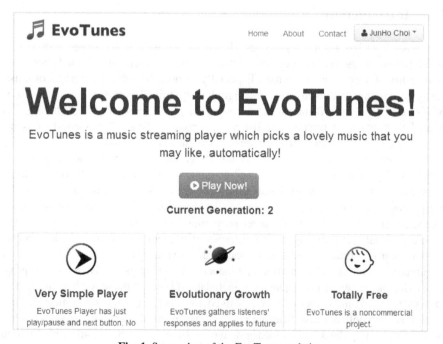

Fig. 1. Screenshot of the EvoTunes website

Fig. 2. Interface of the EvoTunes player

When a participant clicks the skip button or a song finishes, the system server determines the next song based on the currently played song. Information regarding the participant's actions and the played song list are gathered and accumulated in the server, which is exploited to enhance the music recommendation algorithm.

2.2   Selecting Music for Experiment

Although our system can be applied commercially, we needed to avoid using commercial music resources in our experiment due to the copyright issue. Therefore, we collected music clips having Creative Commons License from ccMixter.org, ccMixter Korea, and SoundCloud. Only songs with vocals were kept in our music dataset under the assumption that they are more popular in music streaming than instrumental music. The genre of music was not fixed because one of our goals is to find the relationship between specific two songs regardless of their genres. At the end, the dataset contains 37 songs. The lengths of the songs were between about 2 and 7 minutes.

2.3   Gathering Experimental Data

The following procedure was used to gather and classify experimental data (Fig. 3):

- On start, the server returns an arbitrary song for play because there is no previous music listening history to determine the next song.
- If the subject listened to the song till the end (or sufficiently long), the server determines the next song.
- If the subject also listened to the next song till the end (or sufficiently long), the server not only determines the next song, but also regards the relationship between these two songs as a positive result.

- If the subject skipped the song, the server chooses another song and regards the relationship between the previous two songs as a negative result.
- Repeat the above process for a number of subjects to accumulate preferential relationship between specific two songs.

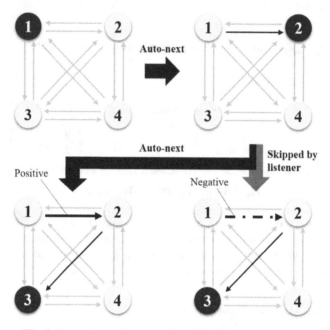

Fig. 3. Illustration of the way of handling listeners' behavior

2.4   Updating Probabilities of Transition

Our experiment is designed not to apply each user's response to update the recommendation algorithm in real time. As shown in Fig. 4, we accumulate experimental data and when a sufficient amount of results are gathered, they are applied to update the probabilities of transition from one song to another, which is referred to as a *generation*.

When updating the transition probabilities by applying accumulated data, each response is classified into two cases. If a user listened a song fully or more than one minute, it is regarded as a positive response and the probability of the transition from the previous song to the current song is increased. Otherwise, it is regarded as a negative response and the probability is decreased. At the beginning of generation 1, all the transition probabilities are set to 0.5. In any case, the minimum transition probability is set to a non-zero small value instead of zero in order to incorporate so-called *long-tail* behavior.

Participants' Responses

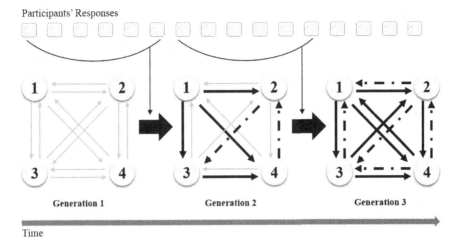

Generation 1   Generation 2   Generation 3

Time

Fig. 4. Updating transition probabilities between songs over generations

2.5   Data Validation

There is a possibility that the subject is not listening to music but the player keeps playing songs, for example, the participant is doing other activities, the volume of the speaker is mute, or the network connection is unstable. Then, data gathered in this situation should not be applied to algorithm enhancement. For this, EvoTunes has the following two data validation mechanisms.

First, the service occasionally chooses a song whose transition probability from the previous song is much lower than other songs. If the participant keeps listening and do not skip despite this "wrong" recommendation, the data gathered from this participant is less reflected to algorithm enhancement.

Second, the service occasionally pauses the current playback and shows a dialog box having a "continue" button, as shown in Fig. 5. The participant has to click the button to continue listening. If the button is not clicked for a long time, it can be assumed that the participant is not properly listening, so the recent data gathered from this participant is excluded in further processing.

Fig. 5. Dialog box to continue listening for data validation

3   Results

486 participants were signed up in the experiment platform – 317 males, 168 females, and 1 unknown. These users logged in the system 585 times, which means that each user logged in about 1.2 times on average. We run five generations, from which the total number of gathered song-to-song transitions (by clicking the skip button or after complete listening) is 3,905.

3.1   Transition Probabilities

Fig. 6 shows probabilities of music recommendation generated by participants' responses gathered from generation 1 to 5. The y-axis is the currently played song and the x-axis is the song that will be played next. A brighter box means a higher transition probability between the two songs.

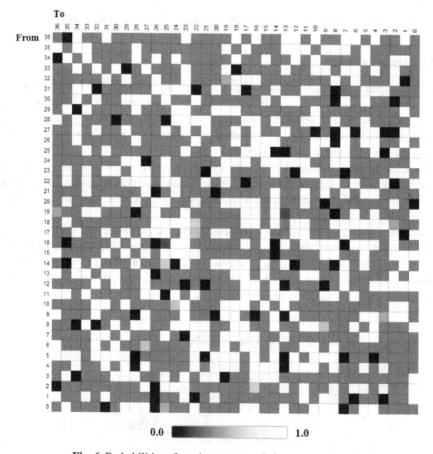

Fig. 6. Probabilities of music recommendation at generation 5

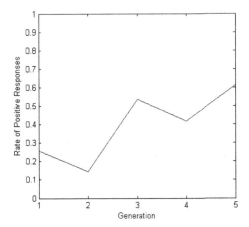

Fig. 7. Rate of positive responses at each generation

As shown in the figure, each transition has a distinct probability as a result of evolution over generations. In addition, it is observed that the probability matrix is not symmetric, which means that there are some difference in preference about the order of playing songs. For example, a transition from music #34 to #36 has a low possibility value, while the reversed transition (from music #36 to #34) has a high probability value.

3.2   User Satisfaction

Fig. 7 shows the rate of positive responses with respect to the generation. The figure confirms that, as time goes, more response data are applied to update the transition probabilities in our algorithm, and thus more satisfactory recommendation can be performed in the next generation.

4   Conclusion

Music recommendation is one of the interesting research topics these days, but requires more studies to achieve users' high satisfaction. Most recent studies related to music recommendation have used similarities about characteristics of songs. However, we focused on the idea that listeners' positive responses for listening to two songs can be also used as useful data for music recommendation. To implement this idea, we created a web service to collect crowdsourced data about preferred playlists, and it was successfully verified from our experiment.

Our system will be still open to collect more data, which will be used for evaluation of longer term effectiveness. Furthermore, we will conduct a larger scale experiment with a larger music database changing dynamically.

Acknowledgment. This work was supported by the Ministry of Science, ICT and Future Planning (MSIP), Korea, under the "IT Consilience Creative Program" (NIPA-2013-H0203-13-1002) supervised by the National IT Industry Promotion Agency (NIPA).

References

1. Cano, P., Koppenberger, M., Wack, N.: An industrial-strength content-based music recommendation system. In: Proceedings of the 28th ACM SIGIR International Conference on Research and Development in Information Retrieval, p. 673. ACM (2005)
2. Eck, D., Lamere, P., Bertin-Mahieux, T., Green, S.: Automatic generation of social tags for music recommendation. In: Advances in Neural Information Processing Systems, pp. 385–392 (2007)
3. Lee, J.H.: Crowdsourcing music similarity judgments using Mechanical Turk. In: Proceedings of the 11th International Society for Music Information Retrieval Conference, pp. 183–188 (2010)
4. Sarwar, B., Karypis, G., Konstan, J., Riedl, J.: Item-based collaborative filtering recommendation algorithms. In: Proceedings of the 10th International Conference on World Wide Web, pp. 285–295 (2001)
5. Barrington, L., Oda, R., Lanckriet, G.R.: Smarter than genius? Human evaluation of music recommender systems. In: Proceedings of the International Society for Music Information Retrieval, vol. 9, pp. 357–362 (2009)

Affect Recognition
Using Magnitude Models of Motion

Oussama Hadjerci, Adel Lablack, Ioan Marius Bilasco, and Chaabane Djeraba

Laboratoire d'Informatique Fondamentale de Lille, Université de Lille 1, France
{oussama.hadjerci,adel.lablack,marius.bilasco,chabane.djeraba}@lifl.fr

Abstract. The analysis of human affective behavior has attracted increasing attention from researchers in psychology, computer science, neuroscience, and related disciplines. We focus on the recognition of the affect state of a single person from video streams. We create a model that allows to estimate the state of four affective dimensions of a person which are arousal, anticipation, power and valence. This sequence model is composed of a magnitude model of motion constructed from a set of point of interest tracked using optical flow. The state of the affective dimension is then predicted using SVM. The experimentation has been performed on a standard dataset and has showed promising results.

Keywords: Dimensional affect recognition, affective state, facial expressions, feature extraction, magnitude model of motion.

1   Introduction

A number of researchers advocate the use of dimensional description of human affect, where an affective state is characterized in terms of a number of latent dimensions. The purpose of the recognition of a dimensional affect is to represent it by continuous values. Unlike categorical description of the emotion (e.g. happy, surprised, angry, etc.) that is more restrictive, the dimensional modeling has the advantage to handle small differences in an affect through the time. It also allows to distinguish the different states in an efficient manner. Because of the importance of face in emotion expression and perception, most of the vision-based affect recognition studies focus on facial expression analysis.

The four dimensions: Arousal, Anticipation, Power, and Valence are all well established in the psychological literature. A recent study [1] argues that these four dimensions account for most of the distinctions between everyday emotion categories. The Arousal (Activity) is the individual's global feeling of dynamism. It subsumes mental activity as well as physical preparedness to act. Anticipation (Expectation) subsumes various concepts that can be separated as expecting, anticipating, being taken unaware. Power (Dominance) subsumes two related concepts, power and control. However, people's sense of their own power is the central issue that emotion is about, and that is relative to what they are facing. Valence is an individual's overall sense of "weal or woe". The Figure 1 shows the values of each dimension for a video sequence.

C. Gurrin et al. (Eds.): MMM 2014, Part II, LNCS 8326, pp. 339–344, 2014.
© Springer International Publishing Switzerland 2014

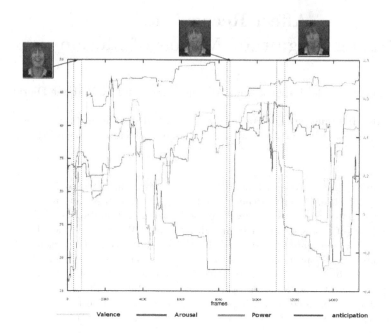

Fig. 1. The four affect dimensions across a video sequence

The goal of this paper is to propose a sequence based approach to recognize the continuous change of the values of each dimensions in terms of positive, negative and neutral state. The approach uses a set of points of interest that are tracked on consecutive frames using optical flow. A magnitude model of motion is then constructed and its state predicted using SVM.

The remainder of this paper is organized as follows. Section 2 briefly reviews some related work in the field of automatic dimensional affect analysis. In Section 3, we describe our mid-level descriptor and the affect recognition procedure. Section 4 presents the experimentation and reports the obtained results. Finally, we give concluding remarks and potential future work in Section 5.

2   Related Work

There is an abundance of literature concerning this topic. Two surveys on affect recognition [2,3] describe recent progress in the field of affective computing by reviewing the models, methods and their applications. Several systems have been proposed and can be categorized into three categories: 1) systems that detect the emotions of the user, 2) systems that express what a human would perceive as an emotion (e.g. an avatar) and 3) systems that actually "feel" an emotion. Some systems use other modalities adding to the facial expression that contribute to the affective state such as speech or body gestures and depend on the environment where the affective behavior occurs. Nicolaou and al. [4] fuses facial expression, shoulder gesture, and audio cues for dimensional and continuous prediction of emotions in valence and arousal dimension. Gunes and Piccardi [5] propose a method to automatically detects the temporal segments or phases of face and body displays, explores whether the detection of the temporal phases can effectively support recognition of affective states, and recognizes affective states based on phase synchronization/alignment. Chen and al. [6] combines both MHI-HOG and Image-HOG through temporal normalization method, to describe the dynamics of face and body gestures for affect recognition.

3   Approach Description

We propose in the following an approach that recognizes the affect state of a single person. Figure 2 shows the main steps divided into two experiments. The first one includes the face only and the second one the body gesture and the face. This approach uses the energy of the scene as a descriptor through the magnitude that encodes the motion features. It has been successfully used by Benabbas et al. [7] to recognize actions performed by a single person. We extract a feature vector from a Region of interest of the image. It's divided in a set of blocks that corresponds the face region or the body region. The first one based only on facial feature allows to estimate the emotional state where the second one uses two modalities to estimate it.

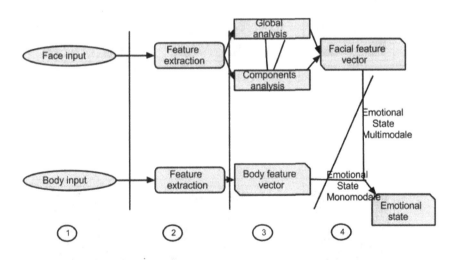

Fig. 2. The proposed approach

In order to create the model of an interval of a video sequence, we start by extracting a set of point of interest from each input frame of a set of consecutive frames. We consider Shi and Tomasi feature detector [8] which finds corners with high eigenvalues in the frame. We also consider that, in our targeted video scenes, camera positions and lighting conditions allow a large number of interest points to be captured and tracked easily. Once we define the set of points of interest, we track them over the next frames using optical flow vectors. The result of the operation of matching features between frames is a set of three-dimensional vectors $V(X_i, Y_i, M_i)$ where X_i and Y_i are the image location coordinates of the feature i, and M_i is the motion magnitude of the feature i and it corresponds to the distance between the position of feature i in the frame f and its corresponding position in the frame f +1. We removed the noise features which correspond to features that have magnitudes that exceed the threshold. In our experiments, we set the minimum motion magnitude to 1 pixel per frame and the maximum to 20 pixels per frame. Then, we divide the scene into a grid of WxH blocks. The grid is applied to the face detected using Viola and Jones [9] algorithm. Then, each vector is allocated to its corresponding block depending on its origin. We cluster the magnitudes of the optical flow vectors for each block using Gaussian mixtures. Each model is represented by an accumulation of magnitude during a predefined interval of time. It is composed of consecutive patterns. The set of estimated Gaussian mixtures constitutes the magnitude model as illustrated in Figure 3. This model was constructed on the arousal dimension of a training video over 30 frames.

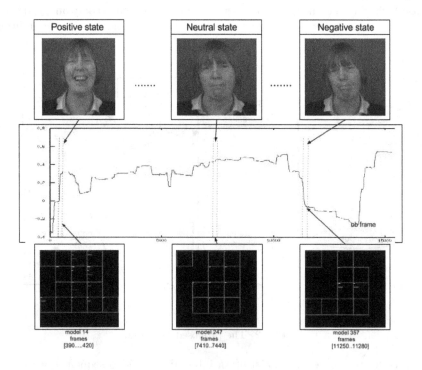

Fig. 3. The construction of the model over 30 frames

4 Experiments and Results

In order to evaluate the described method, we tested it on a publicly available dataset called SEMAINE [10] that has been used in the AVEC challenge [11]. To have better insight of the accuracy of the system, we tested it on a subset of the corpus. The dataset consists of 30 videos. Each video was recorded at 50 frames per second at a spatial resolution of 780x580 pixels. Trace style continuous ratings were made on five core dimensions for all recordings where annotators were instructed to provide ratings for their overall sense of where an individual should be placed along a given dimension. The Figure 4 shows some screenshots extracted from the corpus.

Fig. 4. The SEMAINE Corpus

For the experiment, we find empirically that the best results are obtained by constructing the model over 30 consecutive frames. The prediction has been done using SVM on a model constructed on feature vectors extracted from the face and from the body. The Table 1 presents the obtained results for each dimension using the model constructed from the face only or the body. It is presented in terms of positive, neutral and negative state.

Table 1. Performance results on SEMAINE corpus

Dimension	Face	Body
Arousal	94% (96%, 98%, 88%)	92% (95%, 97%, 84%)
Valence	92% (94%, 88%, 83%)	89% (92%, 96%, 79%)
Power	65% (76%, 51%, 68%)	76% (73%, 85%, 70%)
Anticipation	30% (32%, 31%, 27%)	66% (81%, 64%, 53%)
Total average	70%	80%

The results show that our method has a good performance for all dimensions except for Anticipation. Fusing the face and the body results shows mixed results: for Power and Anticipation we observe a marked improvement. However, for Arousal and Valence the results are quite the same than using the face only.

5   Conclusion

In this paper, we have presented an unconstrained system that estimates the state of the four dimensions of an affect. Evaluation using the SEMAINE corpus proves that joint face and body information results in a better affect estimation on power and anticipation. However, the percentage of the error has been reduced for the two other dimensions. In our future work, we will focus on reducing the information contained in the descriptor and fusing the information that comes from each dimension.

Acknowledgements. This work was conducted in the context of the ITEA2 "Empathic Products" project, ITEA2 1105, and is supported by funding from the French Services, Industry and Competitivity General Direction.

References

1. Fontaine, J.R., Scherer, K.R., Roesch, E.B., Ellsworth, P.C.: The world of emotions is not two-dimensional. Psychological Science 18(2), 1050–1057 (2007)
2. Zeng, Z., Pantic, M., Roisman, G.I., Huang, T.S.: A Survey of Affect Recognition Methods: Audio, Visual, and Spontaneous Expressions. IEEE Transactions on Pattern Analysis and Machine Intelligence 31(1), 39–58 (2009)
3. Calvo, R.A., D'Mello, S.: Affect Detection: An Interdisciplinary Review of Models, Methods, and Their Applications. IEEE Transactions on Affective Computing 1(1), 18–37 (2010)
4. Nicolaou, M.A., Gunes, H., Pantic, M.: Continuous Prediction of Spontaneous Affect from Multiple Cues and Modalities in Valence-Arousal Space. IEEE Transactions on Affective Computing 2(2), 92–105 (2011)
5. Gunes, H., Piccardi, M.: Automatic temporal segment detection and affect recognition from face and body display. IEEE Transactions on Systems, Man, and Cybernetics Part B 39(1), 64–84 (2009)
6. Chen, S., Tian, Y., Liu, Q., Metaxas, D.N.: Recognizing expressions from face and body gesture by temporal normalized motion and appearance features. Image and Vision Computing 31(2), 175–185 (2013)
7. Benabbas, Y., Amir, S., Lablack, A., Djeraba, C.: Human Action Recognition using Direction and Magnitude Models of Motion. In: International Conference on Computer Vision Theory and Applications, pp. 277–285 (2011)
8. Shi, J., Tomasi, C.: Good features to track. In: Internatioal Conference on Computer Vision and Pattern Recognition, pp. 593–600 (1994)
9. Viola, P., Jones, M.J.: Rapid object detection using a boosted cascade of simple features. In: Internatioal Conference on Computer Vision and Pattern Recognition, pp. 511–518 (2001)
10. McKeown, G., Valstar, M.F., Cowie, R., Pantic, M.: The SEMAINE Corpus of emotionally coloured character interactions. In: IEEE International Conference on Multimedia and Expo, pp. 1079–1084 (2010)
11. Schuller, B., Valstar, M., Eyben, F., McKeown, G., Cowie, R., Pantic, M.: AVEC 2011–the first international audio/Visual emotion challenge. In: D'Mello, S., Graesser, A., Schuller, B., Martin, J.-C. (eds.) ACII 2011, Part II. LNCS, vol. 6975, pp. 415–424. Springer, Heidelberg (2011)

Effects of Audio Compression on Chord Recognition

Aiko Uemura, Kazumasa Ishikura, and Jiro Katto

Waseda University, 3-4-1 Okubo, Shinjuku-ku, Tokyo, 169-8555, Japan
{uemura,ishikura,katto}@katto.comm.waseda.ac.jp

Abstract. Feature analysis of audio compression is necessary to achieve high accuracy in musical content recognition and content-based music information retrieval (MIR). Bit rate differences are expected to adversely affect musical content analysis and content-based MIR results because the frequency response might be changed by the encoding. In this paper, we specifically examine its effect on the chroma vector, which is a commonly used feature vector for music signal processing. We analyze sound qualities extracted from encoded music files with different bit rates and compare them with the chroma features of original songs obtained using datasets for chord recognition.

Keywords: chroma vector, audio compression, chord recognition.

1 Introduction

Audio compression technologies such as MPEG-1 Audio Layer-3 (MP3), Advanced Audio Coding (AAC), and Windows Media Audio (WMA) have made important contributions to music distribution services via the Internet. These technologies have enabled efficient compression of music files with high quality. Feature analysis of audio compression with different bit rates is necessary to achieve high accuracy for musical content analysis and content-based music information retrieval (MIR). The bit rate difference has an unexpected adverse effect on musical contents recognition and MIR results because of the frequency response change induced by encoding.

However, examinations of bit rate differences for music content analyses are rarely reported in the literature. We analyzed Mel-Frequency Cepstrum Coefficient (MFCC) differences of compressed audio with various bit rates and their effects on content-based MIR results [1]. Our results show that the MIR results of music data consisting of various bit rates differ significantly from the results of music data, which consist of raw files. A chroma vector might also be influenced by audio compression, as in the case of MFCC. Figure 1 presents an example of the chroma vectors in time-frequency with different bit rates. It is apparent that some chroma power bins differ even for the same music.

A chroma vector, a 12-dimensional vector that represents the intensity of each of the 12 semitones pitch classes of the chromatic scale irrespective of octaves, is typically used for chord recognition. This chroma vector was originally proposed by Fujishima, who presented a real-time chord recognition system using the chroma vector from Discrete Fourier Transform (DFT) [2]. Since then, many approaches have been developed to improve the chroma vector [3–6]. Harte et al. proposed a means to tune the chroma vector and thereby avoid distributing its spectral power to other bins [3].

C. Gurrin et al. (Eds.): MMM 2014, Part II, LNCS 8326, pp. 345–352, 2014.

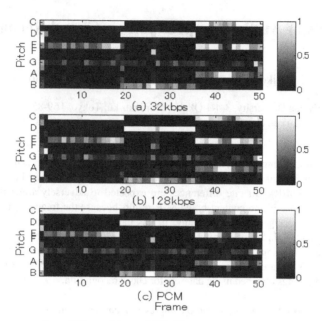

Fig. 1. Chroma vectors from different bit rate MP3 files: extracted from first 50 frames of *"Let It Be"* by The Beatles

Other approaches have been undertaken to combine various musical features to calculate better chroma vectors. Ellis showed a chroma vector that is synchronized with the beat [4]. These features have been used for chord recognition as well as cover song identification. Some cases show means to transform pitch representation. Mauch and Dixon applied Non-Negative Least Squares problems to the spectrogram and obtained clearer chroma vectors [5]. Müller proposed Chroma DCT-Reduced log Pitch (CRP), which is calculated using logarithmic compression and Discrete Cosine Transform (DCT) [6].

The present study analyzes sound qualities of encoded music files with different bit rates and compares them with chroma vectors obtained from the original PCM files. We evaluate audio qualities between encoded files and original ones using the Perceptual Evaluation of Audio Quality (PEAQ) [7] and analyze their effects on chord recognition results. Our results are expected to contribute to assistance of the data format for chord recognition by clarifying the impact of frequency response change induced by audio compression.

This paper is organized as follows. Section 2 presents analysis and evaluation of audio compression effects. Sections 3 and 4 explain effects analysis and discussion. Then section 5 shows our implementation example. Section 6 concludes this report and describes future work.

2   Evaluation of Effects of Audio Compression

2.1   Chroma Vector Calculation

• Audio Compression

We calculate the various chroma vectors extracted from MP3, Low Complexity AAC (AAC LC) and Ogg Vorbis files at different bit rates. The encoding settings are shown in Table 1. All files are based on the Constant Bit Rate (CBR) mode, which uses the same bit rate in every frame throughout the entire file.

LAME (ver. 3.99.5) [8] is used for encoding to obtain MP3 files. Each music file has bit rates of 14 kinds from 32 kbps to 320 kbps. NeroAACEnc (ver. 1.5.1.0) [9] is used to obtain AAC LC files. Each music file has bit rates of 16 kinds from 12 kbps to 320 kbps. Oggenc2.87 [10] is used to obtain Ogg Vorbis files. Ogg Vorbis is quality based encoding and recommended the use of Variable Bit Rate (VBR) mode which the amount of output file varies per time segment. For this study, we set the same parameter value of the maximum and minimum bitrate using the management mode to obtain encoded files similar to CBR mode. Each music file has bit rates of 14 kinds from 32 kbps to 320 kbps.

Table 1. Encoding settings

Codec	MP3	AAC LC	Ogg Vorbis
Extension	mp3	m4a	ogg
Encoder	LAME [8]	NeroAACEnc [9]	oggenc2 [10]
Bit Rate (kbps)	32–320	12–320	32–320

• Chroma Vector

Chroma features introduced by Ellis are based on spectrogram. The instantaneous frequency is used to enhance spectral resolution [4]. Herein, we obtain chroma vectors, which are calculated using the Intelligent Sound Processing (ISP) tool box [11]. The DFT calculation is executed using window frames of length 93 ms and 75% overlap. We align chroma vectors using a spectrum centered on 400 Hz. To obtain the final features, we adjust frames averaged in each beat by reference to beat synchronization [4].

2.2   Perceptual Evaluation of Audio Quality (PEAQ)

PEAQ is a standardized algorithm for objectively measuring perceived audio quality as ITU-R BS.1387-1. ODG (Objective Difference Grade) value represents the measured audio quality of the signal on a continuous scale from -4 (very annoying impairment) to 0 (imperceptible impairment). OGD score is calculated from some variables based on human sound perception. These objective measures are called MOV (Model Output Variable).

Table 2. Model output variables (MOV) of PEAQ basic version

Model Output Variables (MOV)	Interpretation
$WinModDiff_B$	Windowed modulation (envelopes) difference
$AvgModDiff1_B$	Averaged modulation difference
$AvgModDiff2_B$	Averaged modulation difference with emphasis on introduced modulations
$RmsNoiseLoud_B$	Root Means Square (RMS) value of the noise loudness
$BandwidthRef_B$	Bandwidth of the Reference Signal
$BandwidthTest_B$	Bandwidth of the Test Signal
$RelDistFrames_B$	Frequency of audible distortions
Total NMR_B	Averaged Total Noise to Mask Ratio
$MFDP_B$	Detection probability after low pass filtering
ADB_B	Average Distorted Block
EHS_B	Harmonic structure of the error

There are two algorithms of PEAQ, a basic and an advanced version. A basic version is featuring a low complexity approach and an advanced version for higher accuracy at the trade-off of higher complexity. Table 2 shows a list of those MOVs and their interpretation. For audio quality evaluation, we used basic version implemented by Kabal [12].

2.3   Application to Chord Recognition

To evaluate the effect of audio compression for chord recognition, we computed the accuracy using chroma vectors by reference to [13]. One method is a single Gaussian model of Hidden Markov Model (HMM) for its output probability. Subsequently, to decide the chord class, we use maximal gamma values (which are chord likelihood when chroma vectors are observed) from the forward–backward algorithm instead of the Viterbi algorithm. Another is a support vector machine (SVM) model trained by the ground truth data. The structured SVM training package is provided by [14]. The chord sequences are estimated with the chroma vectors and the trained SVM. Both models use two chord types: major and minor chords. We respectively use the 12 cyclic shifts of a 12-dimensional chroma vector to obtain several training data for major and minor chords. We transpose all chroma features to C major or C minor first. Then, the models for two chords are defined. Finally, we obtain the models of 24 chords by transposing the C major and C minor models.

3   Experimental Results

We implement our proposed method using MATLAB and examine experiments using a personal computer. The original music sources are collected from CDs. All audio signals are sampled to 44.1 kHz. We use ground truth chord annotations as the dataset of isophonics [15] and AIST annotations [16] consisting of 307 songs. The dataset provides labels other than major and minor chords. We group triads and other chords with the same root into the same category. For instance, we treat the C minor triad and C minor augment chord as the C minor chord.

3.1   PEAQ Evaluation

First, we compute ODG of 307 songs. Figure 3 shows average ODG results for different bit rates of 32–320 kbps. It is apparent that ODG increases as the bit rate increases, as expected, which implies that the frequencies of signal are easily influenced by compression with lower bit rates but are unsusceptible to higher bit rates.

It is also readily apparent that ODG of encoded files are low under 100 kbps. The sound quality of the encoded file at a lower bit rate differs markedly from that of PCM files. It can be said that ODG has a relation with the recommended bit rate of each codec.

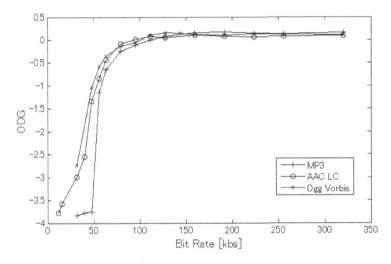

Fig. 2. ODG results

3.2   Chord Recognition Performance

Next, we compute chord recognition accuracy to analyze the effects of audio compression on chord recognition results. For each song, accuracy is calculated as the ratio between the lengths of the correctly analyzed chords and the total length of the song. For our evaluation, we use four-fold cross validation for each bit rate. The final average score is obtained by averaging the scores of all 307 songs for each bit rate.

The accuracies of the HMM method and the SVM one for PCM files were, respectively, 63.5% and 65.7%. The results for each bit rate are expected to be changed to differ significantly for chroma vectors according to Section 3.1 results. However, Figure 3 shows the recognition-rate accuracy in percentage terms. The chroma vectors are not so different among various bit rates in each codec, which shows that the lower bit rate has little or nothing to do with chord recognition using machine learning.

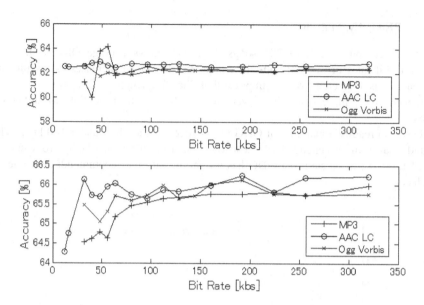

Fig. 3. Chord recognition results with different bit rates (top: HMM, bottom: SVM)

4   Discussions

The audio compression for chord recognition does not strongly affect the power distribution of chroma vectors, which indicates that the frequencies left in compression such as the fundamental frequency is important for chroma vector calculation because a chroma vector is defined as a 12-dimensional vector irrespective of the octave. The frequency response might also be changed by encode algorithm and downsampling. For example, HE-AAC (High-Efficiency Advanced Audio Coding) is of current interest, which is an extension of AAC LC for low bit rate applications. The influence can be analyzed by downsampling on chroma vectors, but it needs further study.

Simple template-based methods [2] can be used instead of HMM-based method for chord recognition. Chord templates are regarded as ideal chroma. Therefore, we should analyze chord recognition accuracy using template method. In addition, the calculation step of chroma vectors remains suspect as a cause because powers of chroma vectors are shifted in the frequency direction.

The power of chroma vectors extracted from original songs is scattered relative to other chroma vectors from MP3, as shown in Figure 1, because original files include higher frequencies than those of encoded files. Signals that consist of noise-like drum and attack sounds have some effect on chroma vectors.

5   Implementation

We implement MATLAB application using GUIDE tool. Figure 4 shows our application for chord recognition. First, chroma vectors are calculated from input source. Then chord labels are estimated with trained models of HMM or SVM. The system includes pre-trained model to reduce training time in advance. This model is trained by all 307 songs. Finally we choose the score type from guitar (tablature) and piano (keyboard), and then the recognition results are displayed at each beat. The results can be also played by MIDI notes. In addition, we can see the difference of chroma vectors between PCM files and input files.

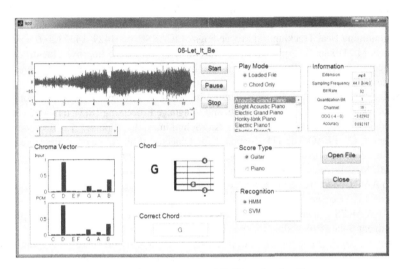

Fig. 4. Chord recognition and PEAQ evaluation System

6   Conclusions

As described in this paper, we analyzed the differences of sound qualities encoded by various bitrates and their effect on chord recognition results. Results indicate that the compression ratio influences the sound quality between compressed and original songs, but low bit rate compression does not strongly affect the chord recognition accuracy.

As the next step, further evaluation is necessary to elucidate the influence on frequency response by encoding algorithm and downsampling. The influence can be analyzed by HE-AAC and downsampling on chroma vectors, but it needs further study.

Acknowledgment. The authors thank Mr. Toshiyuki Nomura of NEC Corporation for his valuable comments on audio compression.

References

1. Hamawaki, S., Funasawa, S., Katto, J., Ishizaki, H., Hoashi, K., Takishima, Y.: Feature Analysis and Normalization Approach for Robust Content-Based Music Retrieval to Encoded Audio with Different Bit Rates. In: Huet, B., Smeaton, A.F., Mayer-Patel, K., Avrithis, Y. (eds.) MMM 2009. LNCS, vol. 5371, pp. 298–309. Springer, Heidelberg (2009)
2. Fujishima, T.: Realtime Chord Recognition of Musical Sound: a System using Common Lisp Music. In: Proceedings of the International Computer Music Association, pp. 464–467 (1999)
3. Harte, C., Sandler, M.: Automatic Chord Identification using a Quantised Chromagram. In: Proceedings of the Audio Engineering Society (2005)
4. Ellis, D., Poliner, G.: Identifying Cover Songs with Chroma Features and Dynamic Programming Beat Tracking. In: Proceedings of ICASSP, pp. 1429–1432 (2007)
5. Mauch, M., Dixon, S.: Approximate Note Transcription for the Improved Identification of Difficult Chords. In: Proceedings of the International Society for Music Information Retrieval Conference (2010)
6. Müller, M., Ewert, S.: Towards Timbre-invariant Audio Features for Harmony-based Music. IEEE Trans. on Audio, Speech, and Language Processing 18(3), 649–662 (2010)
7. Thiede, T., Treurniet, W.C., Bitto, R., Schmidmer, C., Sporer, T., Beerends, J.G., Colomes, C.: PEAQ-The ITU Standard for Objective Measurement of Perceived Audio Quality. Journal of Audio Engineering Society 48(1/2), 3–29 (2000)
8. LAME MP3 Encoder, http://lame.sourceforge.net
9. Nero AAC Codec, http://www.nero.com/enu/company/about-nero/nero-aac-codec.php
10. RAREWARES – oggenc2, http://www.rarewares.org/ogg-oggenc.php
11. Intelligent Sound Processing, http://kom.aau.dk/project/isound/
12. Kabal, P.: An Examination and Interpretation of ITU-R BS.1387: Perceptual Evaluation of Audio Quality. TSP Lab Technical Report, Dept. ECE, McGill University (2002)
13. Supervised Chord Recognition for Music Audio in Matlab, http://labrosa.ee.columbia.edu/projects/chords/
14. Joachims, T.: Sequence Tagging with Structural Support Vector Machines (2008), http://www.cs.cornell.edu/people/tj/svm_light/svm_hmm.html
15. isophonics, http://isophonics.net/
16. Goto, M.: AIST Annotation for the RWC Music Database. In: Proceedings of the International Conference on Music Information Retrieval, pp. 359–360 (2006)

The Perceptual Characteristics of 3D Orientation[*]

Wang Heng[1], Zhang Cong[1], Hu Ruimin[2], Tu Weiping[2], and Wang Xiaochen[2]

[1] School of Mathematic & Computer Science, Wuhan Polytechnic University,Wuhan, China
[2] National Engineering Research Center for Multimedia Software, Wuhan University
{wh825554,hb_wh_zc,hrm1964,echo_tuwp,clowang}@163.com.cn

Abstract. The rapid development of 3D video stimulates demand for 3D audio and promotes the research on perceptual characteristics in 3D sound field. The traditional researches of 3D orientation characteristics were measured sensitivity for some specific locations on site. The test efficiency and results are not good to practical application. This paper first designed a new device to collect optional position in order to establish a 3D sound database and proposed a test method which can be large-scale test. It can ensure the consistency of the experimental data and can't interfere by the experiment personnel. It can improve the efficiency and accuracy of the experiment by using an adaptive method on the computer. It is a significant way to explore the perceptual mechanism of 3D orientation by discovering the relative contribution of frequency and position to selectivity for azimuth and distance. It will provide a theoretical support for the 3D audio acquisition, coding, reconstruction and playback by carrying out the research of 3D orientation characteristics.

Index Terms: 3D orientation, just notice difference, artificial head, perceptual characteristic.

1   Introduction

By the end of 2009, the 3D movie "Avatar" obtained more than $2.7 billion at the box office, it adopted 3d effects production technology bringing people sensory shock effect, so that the industry has assertion of " the film into the 3D era ". 3D movie and TV need to have synchronous 3D sound effects in order to truly achieve the audio-visual feeling in order to be personally on the scene. Early 3D audio system (such as Ambisonics system [1]) because of its structure is complex, the demand of collection and playback equipment is higher, it is difficult to promote practical application. In recent years Dolby and Audyssey launched multichannel playback system with overhead track on the basis of the traditional 5.1 channel. People can experience the sound effects outside level; In 2005, Japan Broadcasting Corporation

---

[*] This work was supported by National Natural Science Foundation of China (Grant No. 61231015, 61272278, 61201340, 61201169), and   Nature Science Foundation of Hubei Province (No. 2012FFB04205).

C. Gurrin et al. (Eds.): MMM 2014, Part II, LNCS 8326, pp. 353–360, 2014.

(Nippon Hōsō Kyō kai, NHK) laboratory introduced 22.2 channel system [2], can reproduce the original 3D scene through the 24 loudspeakers; In 2012, Moving Pictures Experts Group (MPEG) released the latest 3D audio demand and officially launched the collection work of 3D audio technology proposal [3]. It needs to reproduce 3D sound field through the less speakers or headphones at a certain coding efficiency in order to promote the 3D technology to the ordinary home users. Thus 3D audio technology has become the research focus in the multimedia field and the important direction of further development.

Traditional audio technology uses perceptual characteristics to improve efficiency or simplify system in the signal acquisition, coding and playback. The perceptual characteristic of 3D orientation will also play a very important role for the development of 3D audio technology. In 3D audio acquisition and playback, it can guide the spatial arrangement microphones and speakers to acquire the best reconstruction effect of 3D sound field in a certain restriction conditions according to the spatial distribution model of perceptual characteristics for source spatial orientation. In 3D audio content compression, each channel separately coding make the code rate grow linearly. It can increase the compression efficiency by downmixing as a channel and extracting the spatial information parameters, but coding rate is still very high. We can improve parameter coding efficiency through the analysis of statistical distribution of 3D spatial perceptual characteristics and its influence to compression efficiency, removing spatial perceptual redundancy.

So in order to apply and promote the 3D audio technology better, it is necessary to research 3D spatial perceptual mechanism thoroughly, especially to test and analysis 3D spatial azimuth angle and distance perceptual characteristics. The perceptual theory model of 3D orientation will provide theory support for 3D audio acquisition, coding, reconstruction and playback.

The research of 3D spatial azimuth perceptual characteristics has nearly 100 years history, as early as 1936, Stevens studied the horizontal source location. He found that the sound source localization error is 4.6 degrees at 0 degrees in the horizontal plane and the error increases to 16 degrees in 90 degrees horizontal angle. The experimental results showed that the change threshold of source azimuth is different with source position [4].

In 1958, Mills put forward a classical method to measure the human auditory localization sensitivity in Journal of the Acoustical Society of America and defined the human auditory sensitivity as Minimum Audible Angle (MAA). MAA is minimum spatial interval angle that one can distinguish two different locations of the sound sources. The results showed that the minimum is about 1 degree in the horizontal plane, the corresponding frequency is 730 Hz; Maximum Audible Angle is about 3.1 degrees in the horizontal plane, the corresponding frequency is 1800 Hz. MAA can increase to 40 degrees at 90 degrees in the horizontal plane [5].

A large number of relevant researches for auditory azimuth resolution by using Mills' test methods [6, 7, 8], their results showed that: when the source is right ahead horizontal position (horizontal angle is 0 degrees), MAA is minimum, MAA will increase gradually along with the increase of the azimuth; MAA is maximum at ± 90 degrees, the spatial auditory resolution is the worst. But these studies only tested the perceptual resolution of horizontal angle, didn't research perceptual characteristics of vertical angle.

In 2010, Ando of Japan's NHK laboratory tested vertical resolution of sound image reconstructed by speakers through the subjective experimental. The experimental results showed that: when the speaker is in the side, the height perceptual resolution of the sound image gradually increases along with the sound source height; the speaker is in front or behind, the height perceptual resolution of the sound image gradually decrease along with the sound source height as source height increases [9].

At present, the research of perceptual characteristics for spatial position is mainly aimed at certain position for sensitivity measurement, the experiment methods are mostly let people listen live, through adjusting different speakers spacing repeatedly to get different the threshold of spatial positions. Experiment condition of this method is higher, the surrounding environment and some other human factors change will cause large variation of the data, and each test data need huge workload to obtain. Getting the whole 3D spatial perceived sensitivity threshold will be a very difficult thing, so we urgently need a kind of simplified test system to complete this important work. In this urgent demand, this paper designed an experiment device which can simulate any point in spatial source and collect any point source in spatial, also designed a set of adaptive test system, listeners can independently test according to prompt, and then researchers can develop large-scale tests to improve test efficiency and accuracy.

2   Experimental Device Design

This experiment device is designed to collect the source of any position in 3D space by the artificial head. The database of spatial source is established according to a certain rules in order to be called in the test as required. The acquisition by artificial head can guarantee the consistency of the experimental data without interference of surrounding environment and acquisition staff. An adaptive test method can be batch completed on the computer, which greatly shorten the test cycle and workload.

The test device includes: a chassis, half circular arc, an artificial head, a long pole and a wireless speaker. Half circular arc is support above chassis by the two legs, legs and half arc are connected by bearing to rotate the half arc. The artificial head is set above chassis center. The midpoint of artificial head ears is at the centre of half arc bracket. Long pole is set up at the centre of half circular arc and can move along half arc radially. The wireless speaker is set in the long pole close to center of the circle. Figure 1 is schematic diagram of the test device.

Dial is placed in rotation axis of half circular arc and chassis respectively in order to record the rotation angle of half circular arc and artificial head. The scale value is set on the long pole in order to the distance between wireless speaker and the center of circle can be read.

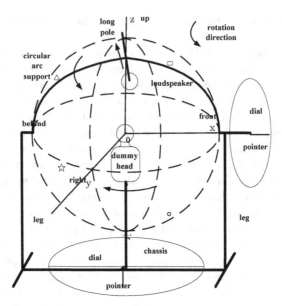

Fig. 1. The device to research perceptual characteristic of 3D orientation

3   Experiment Method

In order to simplify test, this paper also designs a set of 3D spatial location perceptual characteristic adaptive experiment methods, which includes the following steps:

First, collect spatial source signal through change wireless speaker position. The position is rotate chassis, half circular arc and move long pole to proper position. These source signals and corresponding location parameters were saved to database.

The center point of 3D coordinate system XYZ is the center of semi-circular vertices. The distance between the wireless loudspeaker and the center is denoted by the angle referred to as ρ . The angle between the wireless loudspeaker and the plane XOY referred to as α . The angle between the wireless loudspeaker and the plane XOZ referred to as β .The location parameters of the audio signal marked as (ρ, α, β)

Extract spatial source signal as reference sound and test stimuli from the database, the method to measure perceived sensitivity of 3D orientation include following sub steps:

(1) The reference and test stimuli are abstracted from database according to the position parameters of preset sound SP_{ref} and change parameter SP_d . The generation method of sequence is that the test and reference stimuli are synthetized in preset time interval. The sequence is combination of random arrangement, the position parameters of test stimuli is got by the position parameters of reference stimuli SP_{ref} and change parameters SP_d . The change parameters SP_d which is one of

horizontal angle α, height angle β and distance ρ is set to preset initial value at beginning.

(2) The parameter SP_d is changed real-time by a variety of means. When the initial value is large, it can fast approach target value through exponential increase or decrease, in the later, it reverse near the target repeatedly through the linear increase or decrease, whenever a certain number of inversion is reached, linear change step length is reduced, the time test will end until reversal numbers reaches a certain number. The reversal is refers to the parameter changes from increase to reduce or from decrease to increase.

(3) According to the preset reversal numbers threshold L, if the current reversal is L, the average of the reverse values in the last 4-6 times is the goal (Just Noticeable Difference, JND) value of the test, if the current reversal does not meet L, return to (1), abstract corresponding test stimuli according to the current value of the parameter SP_d from a database and produce a new sequence to test.

4   Experiment Results

In order to test the validity of this test method, the perceptual threshold of horizontal angle in horizontal plane will be tested. The source in horizontal plane is already collected and the minimum sampling interval is 0.3 degrees. A simple experimental apparatus used to collect test sources at different horizontal angles in the horizontal plane.

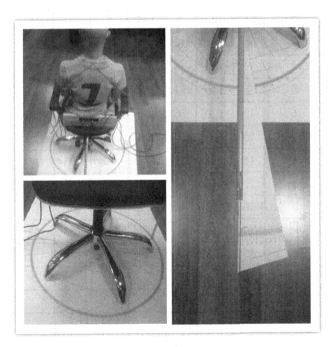

Fig. 2. The device to research perceptual characteristic of 3D orientation

There is eleven listeners in this experiment, eight men and three women, at the age of 19-27 years old, who are undergraduates and graduate students and have professional training and many times subjective listening test before. In this experiment, every person is required to do 12 times audiometry, eight point JND every time, each point costs about 3 minutes to complete, after several test point, listeners need to have a rest for a period of time, each test needs 1 hour or so.

Test source using Gaussian white noise, 16000 Hz low-pass filtering, 48 KHZ sampling, 16 bit quantification, and duration for 300 ms, the sound pressure keep in 70 db or so.

According to the existing literature qualitative or quantitative description on horizontal angle, it is known that the horizontal angle in the front of people is the most sensitive, both sides is the most insensitive, because the human ear perceptual characteristics of sagging face on both sides is roughly symmetrical, only one side data need to be tested, so according to these characteristics, this experiment chooses test point heterogeneously, in order to process data better later. The test angle as follow:

Table 1. Horizontal test point selection table

vertical angle number	0	0	10	10	30	50	75
1	0	70	0	110	0	0	0
2	2	80	3	125	5	7	15
3	4	90	6	140	12	15	35
4	6	100	9	150	20	25	55
5	9	110	12	158	30	40	80
6	12	120	16	166	42	60	110
7	15	130	20	174	55	80	145
8	18	138	25	180	70	105	180
9	22	145	30		90	130	
10	26	152	36		110	150	
11	30	158	42		130	168	
12	35	163	48		145	180	
13	40	168	55		155		
14	45	172	65		165		
15	52	176	75		173		
16	60	180	90		180		

After more than one month test, horizontal angle perceptual threshold value curve in the horizontal plane is obtained as below:

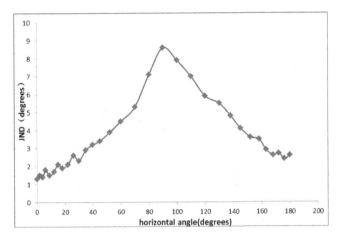

Fig. 3. Horizontal angle perceptual threshold curve

We can see from Figure 3, the final results is consistent with previous qualitative analysis result, horizontal angle in the front of people is the most sensitive, both sides is the most insensitive, at the same time, it also proves the validity of this test method.

To analyze the different heights of the horizontal angle sensing threshold value, the use of cubic spline interpolation fitting surface as shown below:

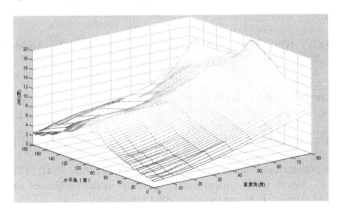

Fig. 4. Horizontal angle perceptual threshold curve surface

As can be seen from the figure, as the elevation angle increases, the horizontal angle sensing threshold value will be a corresponding increase, but at 90 degrees on different height level horizontal orientation sensing threshold value is not very different.

5   Conclusion

Spatial cues involved in 3D audio are mainly horizontal angle, vertical angle and distance. The research of perceptual characteristic for these spatial cues is the important content to explore 3D spatial perceptual mechanism and an effective method to express 3D audio spatial parameters efficiently, so it becomes an important research direction in 3D audio technology. The perceptual characteristics of human ear to the azimuth of sound source charged with the position of the sound source in the 3D spatial. At present, the research of perceptual sensitive characteristic in 3D sound field is just for some specific position such as plane and vertical, it is lack of the perceptual characteristics for all 3D spatial cues; the experiment methods are mostly adjusting the speaker's relative position repeatedly, listener judges whether two sound from different speaker sends out is at the same position in perceptual. Experiment demand of this method is higher, the surrounding environment and some other human factors change will cause large variation of the data, and each test data to obtain all need huge workload. This article designs specific experimental device to acquire source data in 3D spatial, establishes test source database under different environment and different locations; designs a set of 3D spatial position cues subjective perceptual listening test method to study perceptual sensitive characteristic of human ear to spatial orientation and distance. Further, it gets the spatial cues perceptual threshold value of 3D spatial different locations through the large-scale experiment, 3D spatial orientation and distance perceptual model is established from the mathematical analysis of test data, thus it will provide a theoretical support for the 3D audio acquisition, coding and playback.

References

[1] Gerzon, M.A.: With-height sound reproduction. Journal of the Audio Engineering Society 21, 2–10 (1973)
[2] Hamasaki, K., et al.: A 22.2 multichannel sound system for ultrahigh-definition TV (UHDTV). SMPTE Motion Imaging Journal 117, 40–49 (2008)
[3] ISO/IEC JTC1/SC29/WG11 (MPEG), Document N12610, Draft Use Cases, Requirements and Evaluation Procedures for 3D Audio, 99th MPEG Meeting, San Jose, USA (February 2012)
[4] Stevens, S.S., Newman, E.B.: The localization of actual sources of sound. The American Journal of Psychology 48, 297–306 (1936)
[5] Mills, A.W.: On the minimum audible angle. J. Acoust. Soc. Am. 30(4), 237 (1958)
[6] Perrott, D.R., Saberi, K.: Minimum audible angle thresholds for sources varying in both elevation and azimuth. J. Acoust. Soc. Am. 87, 1728 (1990)
[7] Grantham, D.W., Hornsby, B.W.Y., Erpenbeck, E.A.: Erpenbeck Eric A. Auditory spatial resolution in horizontal, vertical, and diagonal planes. The Journal of the Acoustical Society of America 114, 1009–1022 (2003)
[8] Barreto, A., Faller, K.J., Adjouadi M.: 3D Sound for Human-computer interaction: regions with different limitations in elevation localization. In: Proceedings of the 11th International Acm Sigaccess Conference on Computers and Accessibility, pp. 211–212 (2009)
[9] Matsui, K., Ando, A.: Perceptual of sound image elevation in various acoustic environments. In: Audio Engineering Society Conference: 40th International Conference: Spatial Audio: Sense the Sound of Spatial, pp. 8–2 (2010)

Folkioneer: Efficient Browsing of Community Geotagged Images on a Worldwide Scale

Hatem Mousselly-Sergieh[1,2], Daniel Watzinger[1], Bastian Huber[1],
Mario Döller[3], Elöd Egyed-Zsigmond[2], and Harald Kosch[1]

[1] Universität Passau, Innstr. 43, 94032 Passau, Germany
[2] Université Lyon, 20 Av. Albert Einstein, 69621 Villeurbanne, France
[3] FH Kufstein, Andreas Hofer-str. 7, 6330 Kufstein, Austria
firstname.lastname@uni-passau.de, firstname.lastname@insa-lyon.fr,
mario.doeller@fh-kufstein.ac.at, {daniel.watzinger,huber.baste}@gmail.com

Abstract. In this paper, we introduce Folkioneer, a novel approach for browsing and exploring community-contributed geotagged images. Initially, images are clustered based on the embedded geographical information by applying an enhanced version of the CURE algorithm, and characteristic geodesic shapes are derived using Delaunay triangulation. Next, images of each geographical cluster are analyzed and grouped according to visual similarity using SURF and restricted homography estimation. At the same time, LDA is used to extract representative topics from the provided tags. Finally, the extracted information is visualized in an intuitive and user-friendly manner with the help of an interactive map.

1 Introduction

In the era of Web 2 users become able to contribute content to the web by themselves. Online image portals like Flickr[1] is witnessing an explosion in the number of images uploaded everyday. Browsing such huge amount of data is challenging. Therefore, efficient solutions are needed. Currently, an increasing number of the user uploaded images are geotagged beside being annotated by the users. In this paper, we present Folkioneer, a novel system for efficient browsing of community generated images in an intuitive manner on a worldwide scale. Folkioneer exploits geodesic information (longitude and latitude), visual similarity, and user tags to structure visual folksonomies.

The remainder of the paper is organized as follows. In the next section, individual steps of the processing pipeline are discussed briefly. Section 3 provides a description of the system through screenshots. We conclude and discuss future work in Section 4.

2 Folkioneer Architecture

Figure 1 depicts an abstract overview of the entire processing pipeline of Folkioneer. First, a world-scale sample is acquired from a visual folksonomy based on the

---

[1] http://www.flickr.com

C. Gurrin et al. (Eds.): MMM 2014, Part II, LNCS 8326, pp. 361–364, 2014.

small-world phenomenon. After that, the geodesic information of the crawled images is used to extract geographical clusters by applying Clustering Using Representatives (CURE) [1]. In a next step, characteristic polygons describing the shape and the physical extent of each geographical cluster are built by iteratively eroding Delaunay triangulation of the projected image locations [2]. The eroded triangulation is subsequently reused to approximate the area of geodesic polygons in order to prune the cluster hierarchy. Next, the geographical clusters are refined by analyzing and clustering respective images based on visual similarity. Furthermore, Folkioneer also exploits the provided user tags and mines representative tags for each geographical cluster and at different geographical granularity levels. Finally, the acquired information is presented by means of an interactive map.

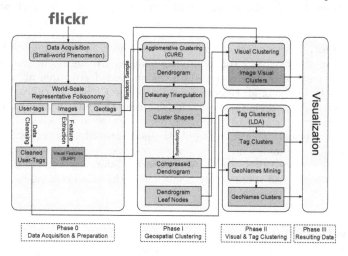

Fig. 1. The processing pipeline followed in Folkioneer

Data Acquisition. In a first step image metadata (identifiers, tags, titles and geodesic information) were obtained from Flickr. To ensure the representativeness of the collected data we applied a crawling strategy that exploits the small-world phenomenon. This was achieved by a breadth-first search on Flickr's friendship graph. The final dataset includes data of 14.1 million images which were uploaded by 211 thousands users in the period from 14/5/2000 to 01/04/2012. The images are annotated with 154 million tags with 6.3 million unique tags.

Geospatial Clustering. In this phase, the crawled images are clustered according to their geographical coordinates. To efficiently handle large amounts of geographical coordinates, we applied a clustering solution based on CURE algorithm for agglomerative clustering [1]. Since spherical coordinates are not suitable for spatial clustering, the geodesic coordinates are projected into the 3-dimensional Cartesian reference system ECEF (Earth-Centered Earth-Fixed). To facilitate nearest-neighbor search CURE is implemented by indexing the 3-dimensional data points using K-D Tree data structure. Furthermore, we used

Fibonacci min-heap to preserve and effectively reuse dissimilarity information across iterations of the clustering algorithm. The output of this a phase is a dendrogram of geographical coordinates. The geospacial clusters are determined by pruning the dendrogram according to a minimum area threshold. Approximation of the geodesic area of geographical clusters is achieved by using Delaunay triangulation [2].

Visual Image Clustering. The geographical clusters are refined by grouping the images of each cluster according to the visual similarity. The main objective is to cluster images representing popular landmarks and events and to determine representative images for each cluster. Visually similar images are identified using the Speeded Up Robust Features (SURF) [3]. SURF is an efficient algorithm for the extraction and matching of local image features. We improved SURF by extending the algorithm with a homography estimation step using LO-RANSAC [4]. Since images depicting the same scene are usually taken from a restricted set of canonical views, the instantiation of new hypotheses for RANSAC is accelerated by restricting the homographies to affine transformations. Homography estimation results in a final number of inliers that obey the generated model. Accordingly, any monotonically decreasing function provides a possible dissimilarity metric. The final visual clusters are then determined by building a complete graph from the pairwise image dissimilarities. To identify the final visual clusters, the graph is fed into a clustering algorithm based on Maximum Standard Deviation Reduction approach (MSDR) [5].

Topic Extraction. After the cleansing step, respective tags of each geographical cluster are analyzed in order to extract representative terms. For this purpose, we used the method of LDA (Latent Dirchlet Allocation) [6] to extract representative topics from larger clusters. For this purpose, tags corresponding to a particular image are regarded as a single document. Subsequently, the document corpus consisting of the respective tags of a geographical cluster is analyzed using LDA to generate possible topics.

3   Demonstration Details

Folkioneer's visualization is based on an interactive map with well-known features like changing the zoom level and panning. In order to visualize contents of the map, the Mercator projection is used to provide a pleasant and familiar viewing experience to the users.

The system is able to manually adjust the granularity of displayed geodesic clusters to meet user preferences by using a simple slider control. Retrieving additional information of particular clusters is initiated by clicking on the respective outlines. Visual clusters are displayed and ordered according to their importance. Additionally, a collection of representative images are displayed with suitable captions for each cluster (Figure 2.a). Furthermore, it is possible to browse through the complete set of images of a given visual clusters as well as

(a)   (b)

Fig. 2. Visual clusters and tag clouds

to view the images in full-size. Finally, Folkioneer supports the visualization of representative tag sets in form of a cloud of tag clouds (Figure 2.b).

4   Conclusion

In this paper, we presented Folkioneer, a system that originally combines different kinds of state-of-the-art algorithms to provide a novel and intuitive browsing experience of community images. In future work, we will consider using the newly gathered information to assist applications that exploit folksonomies such as automatic image annotation.

References

1. Guha, S., Rastogi, R., Shim, K.: Cure: an efficient clustering algorithm for large databases. In: Proceedings of the 1998 ACM SIGMOD International Conference on Management of Data, SIGMOD 1998, pp. 73–84. ACM, New York (1998)
2. Duckham, M., Kulik, L., Worboys, M., Galton, A.: Efficient generation of simple polygons for characterizing the shape of a set of points in the plane. Pattern Recogn. 41(10), 3224–3236 (2008)
3. Bay, H., Ess, A., Tuytelaars, T., Van Gool, L.: Speeded-up robust features (surf). Comput. Vis. Image Underst. 110(3), 346–359 (2008)
4. Chum, O., Matas, J., Kittler, J.: Locally optimized RANSAC. In: Michaelis, B., Krell, G. (eds.) DAGM 2003. LNCS, vol. 2781, pp. 236–243. Springer, Heidelberg (2003)
5. Grygorash, O., Zhou, Y., Jorgensen, Z.: Minimum spanning tree based clustering algorithms. In: Proceedings of the 18th IEEE International Conference on Tools with Artificial Intelligence, ICTAI 2006, pp. 73–81. IEEE Computer Society, Washington, DC (2006)
6. Blei, D.M., Ng, A.Y., Jordan, M.I.: Latent dirichlet allocation. The Journal of Machine Learning Research 3, 993–1022 (2003)

Muithu: A Touch-Based Annotation Interface for Activity Logging in the Norwegian Premier League

Magnus Stenhaug[1], Yang Yang[2], Cathal Gurrin[2], and Dag Johansen[1]

[1] University of Tromsø, Norway
[2] INSIGHT Centre for Data Analytics, Dublin City University, Ireland

Abstract. Annotation of content is a key enabling technology for multimedia systems development. In this demonstration, we present a real-time activity annotation interface designed to be intuitive while imposing minimum effort on the user. Our solution is to use a smartphone and implement a tile-based touch interface. The interface was developed as a part of a larger project in collaboration with Tromsø IL, one of the top soccer teams in Norway. In this demonstration submission we present and evaluate the annotation interface of Muithu.

1 Introduction and Description

Manual and automatic annotation of content are both key enabling technologies for multimedia processing and multimedia information retrieval. In multimedia processing, annotations have been used extensively for both training and evaluation. Meanwhile, the field of information retrieval has a long history of well designed comparative evaluations using document datasets and annotated relevance judgements. The method of gathering annotations has mainly been an ad-hoc activity carried out after the data has been gathered. Outside of the field of Quantified Self analysis using technologies such as Daytum [1], the concept of real-time annotation of activities have received little consideration, yet they are likely to be more accurate than post-hoc annotaitons. The main contribution of this work is to demonstrate a low-cost approach for annotating real-world content in real-time on smartphones, or any handheld device.

Called Muithu [4] and part of a larger project(Bagadus [3]), the annotation interface we present in this work was designed to support easy, real-time annotations of player activities on the field. Muithu was evaluated with football coaches, but could be applied to any real-world annotation task. For this implementation, we found that a typical annotation session yielded 16 events of importance on average. Video sequences corresponding to each annotation were captured from different angles, and presented on a website which enables coaches and players to engage in a constructive dialogue concerning each annotated event.

Our approach to designing the annotation interface was to use the (recently popular) tile interface similar to that of Windows Phone. By using our interface, a user can make high-level semantic annotations with any handheld device.

C. Gurrin et al. (Eds.): MMM 2014, Part II, LNCS 8326, pp. 365–368, 2014.

Although we present a demonstration in the area of sports, this annotation interface is suitable for any real-world annotation, such as those needed for lifelogging [2] or video processing [5].

The possible annotation options are structured as a directed acyclic graph. Each of the nodes in the graph contains a short descriptive text, with edges describing the relation between the nodes. Each of the root nodes are represented with their own tile arranged in a grid as shown in Figure 1 - left. The tiles are coloured squares or images, with the nodes description shown as text labels. Simple annotations can be done by clicking the respective tiles. By pressing and holding a tile, more detailed annotations can be made by dragging a tile into a boundary of a secondary tile (Figure 1 - right). These secondary tiles represent the children of a node in the annotations graph. The use of hierarchical annotations allows the user to enter both coarse and fine grained annotations using the same interface.

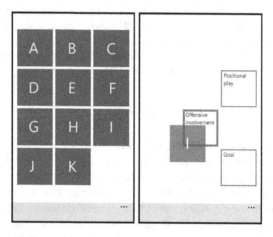

Fig. 1. Two screens of the Muithu interface. Left screen shows the player selection screen and the right shows the options screen for the selected player. The names have been replaced with letters for privacy reasons.

The Muithu annotation interface was designed with the following user experience properties in mind:

Mobile Friendly Interaction. One of our design principles is to ensure that users can accurately make annotations. On mobile devices, button size is generally limited by the size of fingers. Small touch targets will potentially lead to big problems. Thus, we adopt big tile square buttons to increase touch targets, better utilize available screen real estate, and prevent errors. We hide the fine grained annotation screen and only display it when the user employs certain gestures.

Minimise Memory Load. In order to minimise the user's memory load, the user should not have to remember information from one part of the application to another. For instance, when making new annotations, the user's attention may be distracted a few times. Hence, we minimise the user's memory load by making options visible, i.e., display activated grid's information with transparent effect, overlaying it on top of screen (Figure 1 - right).

Learnability. The users interactions should not interfere with the other normal activities, and requires an interface which is easy to learn. We minimise the number of steps it takes for a user to complete an annotation task. Much of the annotation interaction is via touch-and-slide manipulation. Familiarity may come from the fact that it follows standards or that the design follows a metaphor from real world experience.

Adaptive Interface. We designed the Muithu annotation interface to be extensible, and hence useful across a variety of real-world deployments. Annotation can be added or removed on the fly, making this interface easily configurable and essentially use-case agnostic.

2   Evaluation

We present the results of a real-world experiment into the effectiveness of this annotation interface. The user testing the system included a Norwegian Premier League soccer coach (expert) and group of novice users. In [4], we put a soccer coach in charge of operating the system during their national level matches and training sessions. The coach would record high-level important events during the 2012 and 2013 Norwegian Premier League. By facing a camera towards the coach, we measured the time the coach had his attention away from the match in order to annotate. Our experiment showed that the coach would spend around 3.1 seconds on annotating a single event on average.

It is our conjecture that as a result of being able to make an annotation by a single swipe with the thumb, the attention of the user can be maintained on real-world activities. Therefore, we will show that Muithu is 15% faster at annotating than a conventional list-based annotation interface. To evaluate the interface, we measured the time a group of people spent to complete a set of prescribed annotations. A multiple choice drop-down list (two layer hierarchy) provides the baseline for our evaluation. Using a list interface allows us to compareMuithu with a more familiar baseline. Each of the subjects completed a series annotations in succession, altering between the two interfaces, in a real-life environment. Eight users without any prior experience with using the interface completed a series of 20 annotations each, totalling 80 annotations for each interface.

Figure 2 shows the result of our evaluation. The graph shows the average time for annotations one through five and six through ten. The average time used to make an annotation using the list based interface is 4.4 seconds and 3.8 seconds using Muithu's interface. This is a 15% improvement over the list-based interface. This evaluation does not measure the impact of the learnability and attention-maintaining aspects of Muithu, though we are confident that these are improved also.

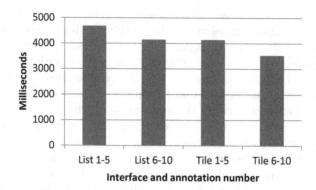

Fig. 2. The average measured time for the tile and list interfaces. The measurments are grouped into the first five and the last five annotations for each interface.

3   Discussions and Conclusions

We have presented our interface for real time annotations. We also compared Muithu with a conventional list-based interface, and showed that our interface improved on the time spent by both experts and normal users to annotate an event. Our experiments shows that the time spent to annotate is around 3.4 seconds from start to finish and that the system was easy to learn. We suggest that this style of user annotation is applicable to many real-world use cases. Aside from use within the sports domain, the use of this type of interface might prove sucessful in other teaching environments such as driver's education.

References

1. Daytum: Collect, categorise and communicate everyday data, http://daytum.com
2. Aizawa, K.: Digitizing personal experiences: Capture and retrieval of life log. In: Proceedings of the 11th International Multimedia Modelling Conference, MMM 2005, pp. 10–15 (2005)
3. Halvorsen, P., Sægrov, S., Mortensen, A., Kristensen, D.K.C., Eichhorn, A., Stenhaug, M., Dahl, S., Stensland, H.K., Gaddam, V.R., Griwodz, C., Johansen, D.: Bagadus: an integrated system for arena sports analytics: a soccer case study. In: Proceedings of the 4th ACM Multimedia Systems Conference, MMSys 2013, pp. 48–59. ACM, New York (2013)
4. Johansen, D., Stenhaug, M., Hansen, R.B.A., Christensen, A., Høgmo, P.-M.: Muithu: Smaller footprint, potentially larger imprint. In: ICDIM, pp. 205–214. IEEE (2012)
5. Silva, J.A., Cabral, D., Fernandes, C., Correia, N.: Real-time annotation of video objects on tablet computers. In: Proceedings of the 11th International Conference on Mobile and Ubiquitous Multimedia, MUM 2012, pp. 19:1–19:9. ACM, New York (2012)

FoodCam: A Real-Time Mobile Food Recognition System Employing Fisher Vector

Yoshiyuki Kawano and Keiji Yanai

Department of Informatics, The University of Electro-Communications
Chofu, Tokyo 182-8585, Japan
{kawano-y,yanai}@mm.cs.uec.ac.jp

Abstract. In the demo, we demonstrate a mobile food recognition system with Fisher Vector and liner one-vs-rest SVMs which enables us to record our food habits easily. In the experiments with 100 kinds of food categories, we have achieved the 79.2% classification rate for the top 5 category candidates when the ground-truth bounding boxes are given. The prototype system is open to the public as an Android-based smartphone application.

1 Introduction

In recent years, food habit recording services for smartphones have become popular. They can awake users' food habit problems such as bad food balance and unhealthy food trend, which is useful for disease prevention and diet. However, most of such services require selecting eaten food items from hierarchical menus by hand, which is too time-consuming and troublesome for most of the people to continue using such services for a long period.

Most of the existing mobile image recognition systems such as Google Goggles need to send images to high-performance servers, which must makes communication delay, requires communication costs, and the availability of which depends on network conditions. On the other hand, image recognition on the client side, that is, on a smartphone is much more promising in terms of availability, communication cost, delay, and server costs. Due to recent rapid progress of smartphones such as iPhone and Android phones, they have obtained enough computational power for real-time image recognition. Then, by taking advantage of rich computational power of recent smartphones as well as recent advanced object recognition techniques, in this demo, we propose a real-time food recognition system which runs on a common smartphone. To boost accuracy and speed of food image recognition, we adopt Fisher Vector and a linear SVM, and implement a system as a multi-threaded system for using quad CPU cores effectively.

Since the recognition process on the proposed system is performed repeatedly about once a second, a user can search for good position of a smartphone camera to recognize foods accurately by moving it continuously without pushing a camera shutter button. This is a big advantage of a real-time image recognition system on a mobile device.

C. Gurrin et al. (Eds.): MMM 2014, Part II, LNCS 8326, pp. 369–373, 2014.
© Springer International Publishing Switzerland 2014

In the experiments, we have achieved the 79.2% classification rate for the top 5 category candidates which outperformed the classification rate of our previous server-side food image recognition system [1].

To summarize novelties of the proposed system, it consists of three folds: (1) An interactive and real-time food recognition and recording system running on a consumer smartphone, (2) using Fisher Vector on a mobile device (3) automatic adjustment of the given bounding box.

2   System Overview

The final objective of the proposed system is to support users to record daily foods and check their food eating habits. To do that easily, we embedded food image recognition engine on the proposed system.

Processing flow of typical usage of the proposed system is as follows (See Figure 1 as well):

1. A user points a smartphone camera toward food items before eating them. The system is continuously acquiring frame images from the camera device in the background.
2. A user draws bounding boxes over food items on the screen. The bounding boxes will be automatically adjusted to the food regions by a GrabCut-based method [2]. More than two bounding boxes can be drawn at the same time.
3. Food recognition is carried out for each of the regions within the bounding boxes using HoG patches and color patches with a linear SVM and Fisher Vector.
4. As results of food recognition and direction estimation, the top five food item candidates and the direction arrows are shown on the screen.
5. A user selects food items from the food candidate list by touching on the screen, if found. Before selecting food items, a user can indicate relative rough volume of selected food item by the slider on the right bottom on the screen for calorie estimation. If not, user moves a smartphone slightly and go back to 3.
6. The name, calorie and nutrition of the recognized food items are shown on the screen and are recorded in the system.

In addition, a user can see his/her own meal record and its detail including calories and nutrition of each food items not only on the screen but also on the Web by sending the records to the server regularly.

3   Implementation and Evaluation

In this section, we describe the implementation of the prototype system we will demonstrate at the conference.

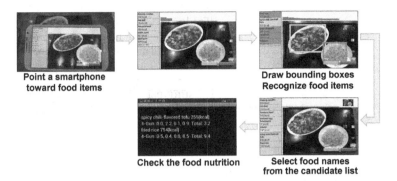

Fig. 1. System process flow

Image Feature. In this paper, we use the following local descriptors for Fisher Vector encoding: HOG Patch and Color Patch.

Histogram of Oriented Gradients(HOG) was proposed by N. Dalal *et al*[3]. Since HOG description is very simple, it is able to describe much faster than popular local descriptor such as SIFT and SURF. This is important characteristic to carry out real-time recognition on a smartphone. In addition, it is able to extract local feature more densely. As a result, it improves recognition accuracy.

We extract HOG features as local features. We divide a local patch into 2×2 blocks, and extract gradient histogram regarding eight orientations from each block. Totally, we extract 32-dim HOG Patch features. Then the 32-dim HOG Patch is L2 normalized to an L2 unit length, not adopt HOG-specific normalization by sliding fusion. PCA is applied to reduce dimensions from 32 to 24.

We use mean and variance of RGB value of pixels as Color Patch feature. We divide a local patch into 2×2 blocks, and extract mean and variance of RGB value of each pixel within each block. Totally, we extract 24-dim Color Patch features. PCA is applied without dimension reduction. The dimension of a Color Patch feature are kept to 24-dim.

Fisher Vector. Fisher Vector [4] can decrease quantization error than BoF[5] by using of a high order statistic and GMM-based soft assignments. Moreover, while BoF needs larger dictionary to improve recognition accuracy, larger dictionary brings increase of computational cost for searching nearest visual words. On the other hand, Fisher Vector is able to achieve high recognition accuracy with even small dictionary, and low computational complexity. This is an advantage for mobile devices. Thus adopting Fisher Vector as encoding method is better in terms of recognition accuracy and processing time for a mobile object recognition. But a system on a smartphone does not exist so far that carries out rapid and high precision image recognition with Fisher Vector. Then we propose a recognition method for a smartphone which is rapid and accurate by making good use of computational resource of the smartphone with Fisher Vector.

In this paper, the number of component of Gaussian is 32 and local descriptors reduced to 24 dimensions by PCA. Thus each feature vector is 1536-dimensional.

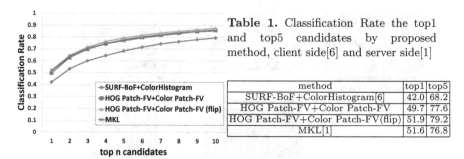

Table 1. Classification Rate the top1 and top5 candidates by proposed method, client side[6] and server side[1]

method	top1	top5
SURF-BoF+ColorHistogram[6]	42.0	68.2
HOG Patch-FV+Color Patch-FV	49.7	77.6
HOG Patch-FV+Color Patch-FV(flip)	51.9	79.2
MKL[1]	51.6	76.8

Fig. 2. Classification rate by proposed method, client side[6] and server side[1]

To improve recognition accuracy, we apply power normalization ($\alpha = 0.5$) and L2 normalization[4].

Classification. As a classifier, we use a linear kernel SVM the computational cost of which is $\mathcal{O}(N)$, and we adopt the one-vs-rest strategy for multi-class classification.

We trained SVMs in off-line. And all the parameter values using recognition steps are loaded on main memory (eigenvalue and eigenvector for PCA, created lookup table, mean of GMM, weight vectors of SVMs). Although all the values can be stored on main memory in advance, Fisher Vector is able to bring better recognition result with even smaller dictionary than that of BoF. We also set the dimension of feature vectors smaller reduced by PCA. As a result, memory space required for Fisher Vector is smaller than the space for codebook for conventional BoF, and in respect of memory Fisher Vector is also superior to BoF.

Experiments. To implement the system, we prepared one-hundred category food dataset which has more than 100 training images per category. All the food items in the dataset photos are marked with bounding boxes. The total number of food images in the dataset is 12,905

In this experiments, we compare with 2 types of the existing food recognition system. One is a client-side recognition system by Kawano et al[6], and the other is a server-side recognition system by Matsuda et al[1].

We compared recognition accuracy with the existing food recognition systems. Our proposed method (HOG Patch-FV+Color Patch-FV) and (HOG Patch-FV+Color Patch-FV(flip)), client side recognition method[6](SURF-BoF+color histogram) and server side recognition method[1]. Note that "flip" means adding horizontally flipped images of training images to training data. [1] adopt total 5 features including hard assignment BoF and global feature based feature and classify with nonlinear $\chi^2 RBF$ kernel MKL-SVM, which is very high computational cost. Figure 2 shows classification rate of each recognition method and Table 1 shows classification rate with top1 and top5. Our HOG Patch-FV+Color Patch-FV achieved 49.7% and 77.6% classification rate within top1 and top5,

Table 2. Average Process Time

	time[sec]	number of food categories
SURF-BoF+ColorHistogram[6]	0.26	50
HOG Patch-FV+Color Patch-FV	0.065	100

respectively. In case of adding flipped images, it achieved 51.9% and 79.2% classification rate within top1 and top5. Our approach is better than Matsuda *et al.*'s food recognition system [1] which was very high cost recognition system running on the server side. system. Therefore we show efficacy our recognition method and it suggests that it is able to carry out high precision image recognition on smartphone.

Next, we measured recognition time by repeating 20 times and averaging them. Table 2 shows the average recognition time. We used Samsung Galaxy NoteⅡ (1.6GHz 4 cores, 4 threads, Android 4.1). We also show the recognition time of [6] for 50 kinds of image recognition. Our approach takes 0.065 seconds for 100 kinds. On the other hand, [6] takes 0.26 seconds for 50 kinds. Thus proposed method is much faster than [6], and this result indicates that the proposed method is more suitable for real-time object recognition.

According to the experiments of recognition accuracy and processing time, we showed efficacy our proposed method. And we achieved better result than server side very high cost recognition method[1]. Therefore we showed rapid image recognition and high precision on smartphone is possible.

Note that Android application of the proposed mobile food recognition system can be downloaded from `http://foodcam.mobi/` .

References

1. Matsuda, Y., Hoashi, H., Yanai, K.: Recognition of multiple-food images by detecting candidate regions. In: Proc. of IEEE International Conference on Multimedia and Expo (2012)
2. Rother, C., Kolmogorov, V., Blake, A.: Grabcut: Interactive foreground extraction using iterated graph cuts. In: SIGGRAPH, pp. 309–314 (2004)
3. Dalal, N., Triggs, B.: Histograms of oriented gradients for human detection. In: CVPR, vol. 1, pp. 886–893. IEEE (2005)
4. Perronnin, F., Sánchez, J., Liu, Y.: Large-scale image categorization with explicit data embedding. In: CVPR, pp. 2297–2304 (2010)
5. Csurka, G., Bray, C., Dance, C., Fan, L.: Visual categorization with bags of keypoints. In: Proc.of ECCV Workshop on Statistical Learning in Computer Vision, pp. 59–74 (2004)
6. Kawano, Y., Yanai, K.: Real-time mobile food recognition system. In: Proc. of IEEE CVPR International Workshop on Mobile Vision, IWMV (2013)

The LIRE Request Handler: A Solr Plug-In for Large Scale Content Based Image Retrieval

Mathias Lux[1] and Glenn Macstravic[2]

[1] Klagenfurt University, Universitätsstrasse 65-67,
Klagenfurt, Austria
mlux@itec.uni-klu.ac.at
[2] World Intellectual Property Organization,
34, chemin des Colombettes,
CH-1211 Geneva 20, Switzerland

Abstract. Big data in the visual domain and large scale image retrieval are current and pressing topics. In this demo paper we describe a specific implementation, that allows for large scale content based image retrieval using the Apache Solr search server. The combination of the robust and well accepted Apache Solr search server with the LIRE content based search framework provides an easy to use and well performing combination of two popular open source tools. We demonstrate the usefulness of our plug-in based the a real life scenario of visual trademark search with more than 1,800,000 images.

Keywords: Content Based Image Retrieval, Retrieval Server.

1 Introduction

Content based image retrieval [1] has been around for years now. Many different local and global features have been proposed, libraries have been published and source code is available for many methods reported in literature. However, actual libraries for use of content based image retrieval in real use cases are rare, some examples are OpenIMAJ [2], the Gnu Image Finding Tool GIFT [3], and LIRE [4]. Services are even more sparse. Google and other big players run their own services, i.e. Google Image Search or Bing Image Search[1]. TinEye[2] on the other hand provides paid content based image retrieval services for customers on their servers.

In this paper we demonstrate a novel open source application that allows for combination of a Apache Solr[3] search server with the LIRE content based image retrieval library and thus provides a content based image retrieval server.

---

[1] http://images.google.com/ and http://www.bing.com/images/
[2] http://www.tineye.com/
[3] http://lucene.apache.org/solr/

C. Gurrin et al. (Eds.): MMM 2014, Part II, LNCS 8326, pp. 374–377, 2014.
© Springer International Publishing Switzerland 2014

2   Implementation

Apache Solr is a text information retrieval server software, which provides a wide range of text retrieval and analysis functions and is well received in real world scenarios. The Apache Solr server can be extended by writing plug-ins. Those plug-ins receive input from requests sent to the server and have access to the underlying Lucene index. LIRE on the other hand is a Java library for content based image retrieval using Lucene as a backend, so combining these two open source projects to build a CBIR server is a natural choice.

For content based image retrieval within Solr we have implemented a bit sampling based hashing function (see e.g. [5]) within LIRE. This hash function $h(v) \in \{0, 1\}$ for a histogram v is defined as $h(v) = sgn(v \cdot r)$, whereas r is a random vector with evenly distributed elements r_i with $-w \leq r_i \leq w$. n hash functions then are – combined as a bit string – one single hash value $H(v) < 2^n$. For indexing m hash values $H_j(v)$, $0 \leq j < m$ were generated. Search then is based on those hashes. Hash values are treated as search terms and Solr retrieves candidate results based on the hashes.

Listing 1.1. XML based document created by LIRE to be added to Solr using the APIs provided by Solr.

```
<doc>
  <field name="id">images01/1/im1005.jpg</field>
  <field name="title">im1005.jpg</field>
  <field name="jc_hi">AQgE6wIRCeoKAuoOA+oKBOoOAuoMBOs=</field>
  <field name="jc_ha">1281 2161 2197 ... </field>
  <field name="eh_hi">
      hqXi1qamxdWl5YfkofTCtcXjteam1cHj1rPV5ZbGlefA18aj5PKUxw==
  </field>
  <field name="eh_ha"> 3271 3649 2274 ... </field>
</doc>
```

Listing 1.1 shows a sample document generated by LIRE to be added to the Solr index with the Solr REST API for image indexing. For each image such a document is created. Each document contains an image ID, a title for displaying purposes and for each feature the feature vector as Base64 encoded data strings as well as the hashes generated from the feature vector. LIRE features more than 20 global descriptors, which can be configured for use with Solr. For search a plug-in has been implemented, called *LireRequestHandler*. This plug-in provides an additional REST API for searching by

- **URL.** In this case the image at the given URL is retrieved and the respective global features are extracted.
- **hashes & histogram** values. If the image is analyzed at the client side the hashes and the histogram can be used to trigger a query.
- **ID.** If a query image is in the index, the image ID can be used for querying.

Search is implemented in a two stage process. After parsing the request, Lucene *IndexSearcher* implementation provided by Solr is used to retrieve m

candidate images based on the hashes with a Boolean term query. This means that the hash values are treated as search terms that should occur in a document, just like a search query string like *"1281 OR 2161 OR 2197 OR ... "*. Based on the Solr similarity computation a document matches the query with a higher score if it contains more of the provided search terms. These m candidates are then re-ranked based on the actual feature vectors and the distance function associated with the feature. The first $n < m$ results are then returned.

3   Demo Application

The Solr server has been used to index 1,873,198 images sampled from registered trademarks in Canada and the USA. A Javascript client allows for search within the index based on already indexed images or query images specified by URL. Fig. 1 shows four different search results, whereas the top left image of each result grid represents the query. Examples (A), (B) and (C) are based on the PHOG descriptor, which captures the gradient orientation of edges within an image. Sample (D) search results for the JCD descriptor, a joint histogram of fuzzy color and edge orientation.

Heuristic evaluation has shown that our application is able to find near duplicates, i.e. logos that have been registered at multiple countries, in different

Fig. 1. Four sample search results from the prototype. The image top left of the individual result grids represents the query.

colors, with different bylines, etc. Fig. 1 gives anecdotal evidence for this use cases. Also very similar logos are returned, whereas similar reflects the notion of similarity of the selected global features. For this use case we selected JCD, PHOG, Edge Histogram and Color Layout (see [6]), which cover a large range of characteristics in both color domain and texture domain. The actual index for this amount of images and the selected features is 1.91 GB and the average response time is 499.67 ms while the median response time is 271.24 (as calculated by the Solr server for 50 sample requests).

4 Conclusion

In this demo paper we have presented a novel combination of LIRE, a content based image retrieval library, with Apache Solr, a popular open source text information analysis and retrieval server. LIRE and Solr can be downloaded and used without restrictions. Note that the plug-in is currently under development and will be merged with the LIRE SVN trunk at the end of the year. A current clone of the git repository is available at `https://bitbucket.org/dermotte/liresolr`.

References

1. Datta, R., Joshi, D., Li, J., Wang, J.Z.: Image retrieval: Ideas, influences, and trends of the new age. ACM Comput. Surv. 40(2), 1–60 (2008)
2. Hare, J.S., Samangooei, S., Dupplaw, D.P.: OpenIMAJ and ImageTerrier: Java libraries and tools for scalable multimedia analysis and indexing of images. In: Proceedings of the 19th ACM MM, MM 2011, pp. 691–694. ACM, New York (2011)
3. Müller, H., Michoux, N., Bandon, D., Geissbuhler, A.: A review of content-based image retrieval systems in medical applications—clinical benefits and future directions. International Journal of Medical Informatics 73(1), 1–23 (2004)
4. Lux, M., Marques, O.: Visual information retrieval using java and lire. Synthesis Lectures on Information Concepts, Retrieval, and Services 5(1), 1–112 (2013)
5. Rüger, S.: Multimedia information retrieval. Synthesis Lectures on Information Concepts, Retrieval, and Services 1(1), 1–171 (2009)
6. Lux, M.: LIRE: Open source image retrieval in java. To appear in Proceedings of the 21st ACM International Conference on Multimedia, MM 2013. ACM, New York (2013)

M³ + P³ + O³ = Multi-D Photo Browsing

Björn Thór Jónsson[1] Áslaug Eiríksdóttir[1], Ólafur Waage[1],
Grímur Tómasson[1], Hlynur Sigurthórsson[1], and Laurent Amsaleg[2]

[1] CRESS and School of Computer Science,
Reykjavík University, Iceland
bjorn@ru.is
[2] IRISA-CNRS,
Rennes, France
laurent.amsaleg@irisa.fr

Abstract. Collections of digital media, personal, professional or social, have been growing ever larger, leaving users overwhelmed with data. Learning from the success of multi-dimensional analysis (MDA) for On-Line Analytical Processing (OLAP) applications, we demonstrate the M³ multi-dimensional model for media browsing and the associated P³ photo browser and O³ media server prototypes.

1 Introduction

With the recent technological changes, multimedia collections have been growing very rapidly and there appears to be no end to this growth. This calls for very effective tools for not only finding media in those collections, but also understanding the collections and analyzing them. Such analysis tools have many applications in such diverse domains as professional media management, online media stores, digital heritage and personal media browsing.

Media search and browsing tools are ubiquitous and to some extent they do help with finding media in collections, although it can be argued that with this growth in media collections the effectiveness of the current search and browsing tools is likely to diminish. They offer no support for understanding and analyzing the collections, however, so what is clearly needed is a new approach to browsing collections that allows for analysis and understanding of the collections.

We have proposed a new model for media browsing, which builds on and combines concepts from a) the multi-dimensional analysis (MDA) model used in on-line analytical processing (OLAP) and b) the faceted search model. We call this model the Multi-dimensional Media Model, or M³; we choose to pronounce this as *emm-cube*, which also refers to the fact that the data model essentially constructs hyper-cubes of media items.

The focus of this demonstration is on presenting the M³ data model as we would like to discuss the data model with the MMM audience. We will do this via prototypes for both a media server (called O³ since it serves hyper-cubes of media objects) and a photo browser (called P³ since it displays hyper-cubes of

C. Gurrin et al. (Eds.): MMM 2014, Part II, LNCS 8326, pp. 378–381, 2014.

photos). In the remainder of this proposal we briefly describe the data model and the software prototypes, before describing the demonstration scenarios.

We have demonstrated previous versions of the prototypes at ICMR 2011 [1] and CBMI 2012 [2]. Those demonstrations focused on the browsing and tagging experiences, respectively, and were not interactive in the sense that conference participants could not play with the system. This demonstration, in contrast, will cater to the MMM audience by focusing on the underlying data model and on allowing conference participants to experiment with the data model and prototypes. Furthermore, we have implemented many new features into our prototypes which significantly enhance the demonstration experience.

2   The M^3 Data Model

For traditional databases, the multi-dimensional analysis (MDA) model used in on-line analytical processing (OLAP) was the key to allowing analysis and understanding of large data collections. The MDA model introduced two key concepts that revolutionized users perception of data, namely *dimensions*, including hierarchies, used for specifying interesting sets of data and *facts*, or numerical attributes, which are aggregated for an easy-to-understand view of the data of interest. These simple concepts put the focus on the value of data items as well as on the relationships that exist between data.

Based on the resounding success of OLAP applications, it is not surprising that multimedia researchers have studied the application of OLAP to multimedia retrieval for some time (e.g., see [3–5]). The fact that the MDA model is geared towards simple numerical attributes, however, is a serious limitation when it comes to multimedia collections where tags and annotations are a very important part of the meta-data.

The use of tags has been studied in faceted search, however. Faceted search uses a single tag-set, but proposes to build multiple hierarchies (or even DAGs) over that tag-set, one for each aspect that could be browsed. These hierarchies are then traversed to interactively narrow the result set, until the user is happy. Item counts or sample queries are typically used to present the result while it is very large; when it is sufficiently small it is presented in a linear fashion. A major drawback of the faceted approach is the use of a single tag-set; although the hierarchies do help users somewhat to disambiguate the different uses of an ambiguous tag, it is more logical to categorize the tags into different tag-sets.

In the M^3 model, media files correspond roughly to OLAP facts, but meta-data items (we call these *tags*, but they can refer to any meta-data, including numerical data, dates and time-stamps, annotations and textual tags) can be associated with the media; one file may be associated with many tags while a particular tag may be associated with many files. Tags are grouped into multiple *concepts*; each concept encapsulates a particular conceptual group of tags, such as people, objects or animals. Borrowing from faceted search, however, *concepts* are then further organized by building (multiple) browsing *dimensions*, typically structured as hierarchies or DAGs, to facilitate browsing.

Fig. 1. Screenshot of P^3 in Cube Mode (left) and Card Mode (right)

The M^3 model thus allows us to define the concepts according to which media items can be grouped and the way tags attached to media files are organized. OLAP operations are transformed into adding or removing selection predicates applied to the dimensions of a media collection. Predicates act as filters on tags, concepts or dimensions, and restrict the set of media to display. A media browsing session thus consists of applying filters and retrieving the media items that pass through all the applied filters.

3   The P^3 Prototype Photo Browser

The P^3 prototype converts the M^3 model to a 3D GUI presentation, which grants users the access to the powerful and flexible browsing operations of the data model, such as the drill-down and roll-up operations, pivoting, and general filtering, via hierarchies and tag-sets. The demonstration version of the P^3 browser is built using Qt and openGL, allowing for three-dimensional browsing of the set of photos of interest [6].

Currently, there are two browsing modes available for P^3. *Cube Mode* is a direct representation of the M^3 data model (see Fig. 1, left) and is the main mode of the browser. In this mode, users can build an image cube of one, two or three dimensions by applying the features supported by M^3, such as filtering, pivoting, drill-down and roll-up. *Card Mode* is used to view images from selected cells in more detail (see Fig. 1, right). Further modes could include slide-shows, web publishing and other similar features.

4   The O^3 Prototype Media Server

The O^3 prototype implements the M^3 model. It is written in C++ and makes extensive use of its standard library, as well as the TR1 C++ library extensions. The current implementation, without plugins, consists of over fifty thousand lines, and runs on Linux. The underlying data-store is MonetDB, an open-source column-store from CWI, which was found to have the best performance of three evaluated data-stores [7].

The O^3 server includes a plug-in subsystem to support automated tagging of images. Plug-ins are called upon to analyze objects during insertion, and may generate new tags or attach objects to existing tags, thus greatly improving tagging effectiveness. The plug-in subsystem loads, manages and isolates plug-ins from the rest of the layers. We already have available plug-ins for EXIF meta-data extraction, color analysis and face recognition [8].

5 Demonstration Scenarios

As mentioned in the introduction, the focus of the demonstration is on allowing conference attendees to experiment with the underlying M^3 data model.

The main demonstration scenario is a photo collection which will be created dynamically during the MMM conference. We will be taking many pictures during the conference and inserting these into the collection, as well as allowing conference attendees to insert photos. Tagging will be performed dynamically by both demonstrators and conference attendees. Tagging may be done according to either predefined concepts and dimensions, e.g., relating to sessions and papers, or by adding tags and augmenting dimensions. This will demonstrate the efficient tagging support of the model and prototypes and the associated positive impact on the browsing experience.

Note that we volunteer to keep an open booth during the entire conference, where conference participants can contribute to the tagging of the conference photos. This would significantly enrich the multimedia experience of the conference. The photos and tags can subsequently be exported to the MMM web-site, or alternatively to a photo sharing site such as Flickr.

For interested parties, we will have on hand other photo collections, including one containing a set of photos from a well-known nature trail in Iceland, tagged with information about participants, events, geography, and other concepts.

References

1. Tómasson, G., Sigurthórsson, H., Jónsson, B.T., Amsaleg, L.: Photocube: Effective and efficient multi-dimensional browsing of personal photo collections (demonstration). In: Proc. ICMR (2011)
2. Tómasson, G., Sigurthórsson, H., Rúnarsson, K., Ólafsson, G.K., Jónsson, B.T., Amsaleg, L.: Using photocube as an extensible demonstration platform for advanced image analysis techniques. In: Proc. CBMI (2012)
3. Zaïane, O.R., Han, J., Li, Z.N., Hou, J.: Mining multimedia data. In: Proc. CASCON (1998)
4. Arigon, A.M., Miquel, M., Tchounikine, A.: Multimedia data warehouses: A multi-version model and a medical application. MTAP 35 (2007)
5. Jin, X., Han, J., Cao, L., Luo, J., Ding, B., Lin, C.X.: Visual cube and on-line analytical processing of images. In: Proc. CIKM (2010)
6. Sigurthórsson, H.: PhotoCube: Multi-dimensional image browsing. Master's thesis, Reykjavik University (2011)
7. Tómasson, G.: ObjectCube – a generic multi-dimensional model for media browsing. Master's thesis, Reykjavik University (2011)
8. Rúnarsson, K.: A face recognition plug-in for the PhotoCube browser. Master's thesis, Reykjavik University (2011)

Tools for User Interaction
in Immersive Environments

Noel E. O'Connor[1], D. Alexiadis[2], K. Apostolakis[2], Petros Daras[2],
E. Izquierdo[3], Y. Li[1], D.S. Monaghan[1], F. Rivera[3],
C. Stevens[4], S. Van Broeck[4], J. Wall[3], and H. Wei[1]

[1] Insight Centre for Data Analytics, Dublin City University, Ireland
[2] Centre for Research and Technology Hellas, Information Technologies Institute,
Greece
[3] Queen Mary University of London, United Kingdom
[4] Alcatel-Lucent Bell Labs, Antwerpen, Belgium

Abstract. REVERIE – REal and Virtual Engagement in Realistic Immersive Environments – is a large scale collaborative project co-funded by the European Commission targeting novel research in the general domain of Networked Media and Search Systems. The project aims to bring about a revolution in 3D media and virtual reality by developing technologies for safe, collaborative, online environments that can enable realistic interpersonal communication and interaction in immersive environments. To date, project partners have been developing component technologies for a variety of functionalities related to the aims of REVERIE prior to integration into an end-to-end system. In this demo submission, we first introduce the project in general terms, outlining the high-level concept and vision before briefly describing the suite of demonstrations that we intend to present at MMM 2014.

1 The REVERIE Project

Within the REVERIE (REal and Virtual Engagement in Realistic Immersive Environments) project we believe that it is time to move online interaction towards the next logical step in its evolution: to immersive collaborative environments that support realistic inter-personal communication. Recent scientific advances in a variety of different research fields mean that it is now possible to integrate research outputs towards technologies that support real-time realistic interaction between humans in online virtual and immersive environments. REVERIE envisages an ambient, content-centric Internet-based environment, highly flexible and secure, where people can work, meet, participate in live events, socialise and share experiences, as they do in real life, but without time, space and affordability limitations. To achieve this we focus on the integration of cutting-edge technologies related to 3D data acquisition and processing, sound processing, autonomous avatars, networking, real-time rendering, and physical interaction and emotional engagement in virtual worlds. The project features two compelling use case scenarios that will be used as the basis for technical integration and

C. Gurrin et al. (Eds.): MMM 2014, Part II, LNCS 8326, pp. 382–385, 2014.

that collectively will allow the project to demonstrate the validity and potential socio-economic benefits of REVERIE's vision of the future of online interaction in immersive environments.

2 Use Cases

The first REVERIE use case is an educational scenario involving teachers and students that experience a virtual school trip. The participants of this use case take part in an immersive 3D educational interactive experience while they have the opportunity to construct their own virtual scenes. The second use case aims at highly realistic visualizations, targeting the look and feel of real physical presence and interaction. It refers to an innovative and highly interactive social experience in a virtual scene based on motion capture techniques and video and audio reconstruction where participants can experience a virtual dance class or take part in a movie inviting their social friends. This is a scenario that involves many participants and therefore a challenge for efficient networking and rendering techniques.

3 Proposed Demonstrations

Currently the project is in a technology development phase whereby individual components are being developed prior to integration into a complete system. In this section, we briefly describe these component technologies. All demos will be live interactive software demonstrations, except for "Navigable 3D video chat" & "Real-time full body 3D reconstruction", due to the impracticality of transporting complex capture set-ups. Illustrations of some of the demos are shown in Figure 1.

3.1 Navigable 3D Videochat

This demonstrates how two users can sit in two different remote locations, each equipped with the REVERIE video chat system, comprising one Kinect, one ordinary computer display, and internet connection. A video will show the characteristics of the virtual view on the remote user and the interactivity features possible[1].

3.2 Gaze Translucency

We refer to being able to capture the information about what other people are looking at as gaze translucency. Current telecommunication tools like Skype do not provide gaze translucency which makes video sessions with several people very difficult and very different from our face to face conversations. Our solution automatically highlights the people that are looking at us and reveals who you are looking at, providing important contextual information in online interactions that we take for granted in our real-world interactions.

---

[1] http://www.reveriefp7.eu/resources/demos/multi-party-3d-visual-and-audio-communication

3.3 Real-Time Full Body 3D Reconstruction

This will demonstrate real-time 3D human reconstruction from multiple Kinect RGB+depth cameras. It will present various textured 3D reconstruction results from various view points, both for still and moving humans[2].

(a) 3D reconstructions of moving humans in various poses

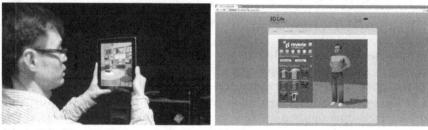

(b) An iPad-based window onto virtual worlds

(c) The REVERIE avatar authoring tool

(d) Puppeting the gaze of an avatar

(e) Real-time head nod and shake detector

Fig. 1. Illustrations of some of the demos to be shown

3.4 RAAT: The REVERIE Avatar Authoring Tool

Here we will demonstrate a powerful online JavaScript library for avatar creation. The demonstration will present how RAAT is employed to power impressive

---

[2] http://www.reveriefp7.eu/resources/demos/real-time-full-body-3d-reconstruction-teleimmesion-applications

online applications that can generate realistic looking 3D characters intended to populate virtual environments[3].

3.5 Puppeting the Gaze of a Virtual Avatar

Increasingly avatars are becoming realistic virtual human characters that exhibit human behavioral traits, body language and eye and head movements. As the interpretation of eye and head movements represents an important part of human communication it is extremely important to accurately reproduce these movements in virtual avatars to avoid falling into the well-known "uncanny valley". In this demo we present a cheap hybrid real-time head and eye tracking system that can allow a human user to robustly puppet a virtual avatar.

3.6 Real-Time Head Nod and Shake Detection

Almost all cultures use subtle head movements to convey meaning. Two of the most common and distinct head gestures are the head nod and head shake. In this demo we present a real-time head nod and shake detection system using Microsoft Kinect.

3.7 Window to Virtual Worlds

Not all REVERIE end users will be active participants, some will wish to attend and observe the events taking place in a virtual environment. This demonstration will show how a tablet can be used as a window on a virtual world. It utilizes the iOS inbuilt sensors to automatically update 3D perspective views for the user in response to their natural motion[4].

4 Conclusion

Although a number of separate demonstrations are proposed, they all address aspects of the REVERIE integrated system which will be ready for demonstration mid-2014. To ease understanding of how the various components fit together, the individual demos will be accompanied by a poster that outlines the architecture of the overall system illustrating how these components fit together.

Acknowledgements. This work was supported by the EU FP7 project REVERIE, ICT-287723.

---

[3] http://www.reveriefp7.eu/resources/demos/reverie-avatar-authoring-tool
[4] http://www.reveriefp7.eu/resources/demos/window-to-virtual-worlds

RESIC: A Tool for Music Stretching Resistance Estimation

Jun Chen and Chaokun Wang

School of Software, Tsinghua University, Beijing 100084, China
Tsinghua National Laboratory for Information Science and Technology
Key Laboratory for Information System Security, Ministry of Education
Key Laboratory of Intelligent Information Processing,
Institute of Computing Technology, Chinese Academy of Sciences
chenjun12@mails.thu.edu.cn, chaokun@tsinghua.edu.cn

Abstract. In this demonstration, we present a useful tool that estimates the stretching **RE**sistance of mu**SIC** (RESIC). The tool takes advantage of both music characteristics and human factors by incorporating audio content features, musical genre and user-tagged music stretching resistance data set to provide reliable estimation. For better understanding of this tool, two front-ends are introduced. Our work fills the gap of music stretching resistance estimation, which aids music resizing techniques in parameter selections, and also expands the user manipulation of music.

Keywords: Music Stretching Resistance, Audio Features, Musical Genre.

1   Introduction

In recent years, music resizing techniques [1–3] have arisen to provide easy manipulation of music tracks on the time length for scenarios like audio-video synchronization, animation products, etc. Among these algorithms, LyDAR [3] deals with track compressions by utilizing the lyrics-density. MUSIZ [2] introduces a music resizing framework consisting of structure analysis and music synthesis by employing segment labeling and audio feature extraction. Besides, the resizing method based on the cropping and insertion is also proposed [1].

However, the evaluation of these resizing algorithms still remains difficult since it contains the subjective feedback of human acceptance on the resized music tracks. Because music resizing can suffer from perceptual artifacts, the acceptance for general audience is not always guaranteed if a music track is overly compressed or overly elongated. To avoid over-stretching, the knowledge of the music stretchability is demanded. Recently, we have focused on the estimation of the stretching ability of music, denoted as music stretching resistance (MSR), also *stretch-resistance* used in [2]. MSR consists of the minimum compressing rate α_{min} and the maximum elongating rate α_{max}, which depicts the human acceptable stretching rate range for a given music piece. As an interplay of music characteristics, human auditory and psychological feedbacks, the estimation of MSR should take both audio and human factors into consideration. Meanwhile,

C. Gurrin et al. (Eds.): MMM 2014, Part II, LNCS 8326, pp. 386–389, 2014.

the estimation of MSR also contributes to more accurate stretching boundaries of music resizing techniques [1–3].

In this demonstration, we showcase the architecture and implementation towards building an applicable MSR estimation tool, **RESIC**, which incorporates audio content features, musical genre and human factors. For better understanding, two front-ends (standalone-client and web-client) have been brought up to show the novelty and utility of this tool in practice.

2 System Overview

RESIC is composed of two parts, the back-end and the front-end. The back-end deals with the audio signal processing and analysis, automatic musical genre classification and MSR estimation, while the front-end interacts with users. Fig. 1 illustrates the back-end architecture and workflow of RESIC. Basically, it contains the following four components.

Audio Feature Extraction. We have extracted overall mean of MFCC (Mel-frequency cepstral coefficients), Chroma, Spectral features (Centroid, Rolloff, Flux), and ZeroCrossing rate with Marsyas[1] for the input music track (as well as all the tracks in our music data set in advance). These features outperformed other features like pitch histogram and tempo in our comparative experiments of feature selection for MSR estimation.

Normalization. Normalization on the features into $[0.0, 1.0]$ range is required because, on the one hand, the value ranges of the extracted features vary a lot, and on the other hand, the weights of different dimensions provided by the metric learner in the following step are calculated on the normalized features.

Musical Genre Setting. As a comprehensive description of music tracks with cultural backgrounds, the genre information is also required to get a more precise MSR estimation. It can either be set manually or set by automatic musical genre classification algorithms. Considering that automatic musical genre classification may induce extra errors, manual setting is more recommended if users have knowledge about musical genres. RESIC supports musical genre classification by simply using KNN classifier with the extracted features.

Estimation. We proposed an MSR estimation method which employs the metric learning technique [4] to adjust the weights of both the normalized audio features and the genre difference in the dissimilarity measurement between music tracks. The metric learner was trained using our MSR-tagged music data set which is built in advance by inviting volunteers to conduct the rewarding listening experiments and judge MSR for music tracks in the data set. The judging criterion of MSR is basically the human acceptance of the stretched tracks, e.g., acceptable singing speed, reasonable lyrics-density. There are in total 894 songs from 11 musical genres in the experiments. The music tracks are stretched using SoundTouch[2]. The volunteers' tagged MSR forms the baseline reference when

---

[1] http://sourceforge.net/projects/marsyas/
[2] http://www.surina.net/soundtouch/

estimating MSR values for a given music track. It also represents the human knowledge in MSR estimation. RESIC provides MSR estimation by performing KNN classification with dissimilarity measurement adjusted by the metric learner.

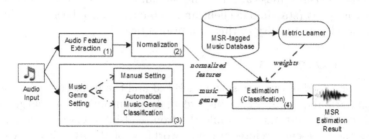

Fig. 1. The back-end architecture and workflow of RESIC

The workflow of the back-end is also numbered in Fig. 1, in which music files or music segments in tracks are the input, while MSR estimation in the form, $[\alpha_{min_{low}}, \alpha_{min_{high}})$ and $[\alpha_{max_{low}}, \alpha_{max_{high}})$ for music pieces or segments, is the output. The MSR values, α_{min} and α_{max}, are presented in ranges considering the estimation errors and subjective difference of human. Besides the back-end, the front-end interacts with users and presents the estimation results in solutions including the web-based and the standalone clients. RESIC focuses on the visualization of MSR and the easy manipulation for general users.

3   Demonstration

Our demonstration tool allows users to have full experience with how much the given music piece can be stretched by visualizing the MSR estimation. Fig. 2(a) illustrates the standalone client which was designed to provide MSR estimation off-line where processing time is guaranteed. The decoding, audio feature extraction and normalization are automatically executed once a music piece is loaded. Manual setting of musical genre is optional and the automatic musical genre classification is a substitute when users are not sure which genre fits best. The estimation results will be presented intuitively through the color change of the slider bar and also the display of MSR values. The red bars represent the "dangerous" and non-suggested ranges while the green bars mean the "safe" ranges. The red-to-green and green-to-red gradual bars are the transition ranges from "dangerous" to "safe" and from "safe" to "dangerous", respectively. Users can experience what over-compressed or over-elongated music tracks sound like when resizing at rates within red bars, and also what acceptable stretched music tracks sound like when resizing within green bars. Thus, once they have a music track to resize, RESIC will alert to them what is the minimum (or maximum) rate it can be compressed (or elongated).

(a) Standalone client   (b) Web client

Fig. 2. Screenshots of two implementations of RESIC front-ends

For easy access to RESIC, the web-client (Fig. 2(b)) takes full advantage of the remote interaction through a web browser with a friendly user interface, related audio, video samples and readings.

The main contribution of this demonstration is to provide a tool which is capable of estimating the MSR values of music pieces. Also, it can be used to estimate MSR values for different segments of the same music track and resize it non-homogeneously. RESIC provides users with general human acceptable music resizing, which on the one hand, aids music resizing algorithms to guarantee the acceptance of results, and on the other hand, expands the user manipulation of music pieces for scenarios like editing or resizing her/his favorite audio clips or even singing/humming tracks recorded on site.

Acknowledgments. The work is supported by the National Natural Science Foundation of China (No. 61373023, No. 61170064, No. 60803016), National High Technology Research and Development Program of China (No. 2013AA013204), and the Opening Foundation of Key Laboratory of Intelligent Information Processing, Institute of Computing Technology, the Chinese Academy of Sciences under Grant (No. IIP 2012-5).

References

1. Liu, Z., Wang, C., Wang, J., Wang, H., Bai, Y.: Adaptive music resizing with stretching, cropping and insertion. Multimedia Systems 19(4), 359–380 (2013)
2. Liu, Z., Wang, C., Bai, Y., Wang, H., Wang, J.: MUSIZ: a generic framework for music resizing with stretching and cropping. ACM Multimedia, 523–532 (2011)
3. Liu, Z., Wang, C., Guo, L., Bai, Y., Wang, J.: Lydar: a lyrics density based approach to non-homogeneous music resizing. In: IEEE ICME, pp. 310–315 (2010)
4. Xing, E., Ng, A., Jordan, M., Russell, S.: Distance metric learning, with application to clustering with side-information. In: NIPS, pp. 505–512 (2002)

A Visual Information Retrieval System for Radiology Reports and the Medical Literature

Dimitrios Markonis[1], René Donner[2], Markus Holzer[2], Thomas Schlegl[2], Sebastian Dungs[3], Sascha Kriewel[3], Georg Langs[2], and Henning Müller[1]

[1] University of Applied Sciences, Western Switzerland, Switzerland
dimitrios.markonis@hevs.ch
[2] Medical University of Vienna, Vienna, Austria
[3] University of Duisburg, Duisburg, Germany

Abstract. The enormous amount of visual data in Picture Archival and Communication Systems (PACS) and in the medical literature is growing exponentially. In the proposed demo, the medical image search of the KHRESMOI project is presented to solve some of the challenges of medical data management and retrieval. The system allows searching for visual information by combining content–based image retrieval (CBIR) and text retrieval in several languages using semantic concepts. 3D visual retrieval in internal hospital sources is supported by marking volumes of interest (VOI) in the data and connection to the medical literature are established to allow further investigating interesting cases. The system is demonstrated on 5TB of radiology reports with associated images and articles of the biomedical literature with over 1.7M images.

1   Introduction

Search for visual information is a common task in the radiology workflow, clinical work, teaching and research activities. Recent studies have shown that radiologists often fail when searching for images using the current information retrieval tools such as Google [1]. Search in hospital records is mostly patient–based, while external sources are often accessed by general search engines that return information of questionable quality. Conventional search by keywords cannot fully cope with all medical tasks, such as for differential diagnosis or in cases where a pattern found is not known. CBIR has been promising in medical applications [2]. However, few applications have reached the clinicians [3] as they were mostly technology driven. User–oriented design [4] has been proposed for creating applications with a real impact in medical information search.

The KHRESMOI project[1] aims at creating a multi–modal search system for biomedical information. One of the target user groups are radiologists and their tasks are strongly related to visual information search. Text–based search and CBIR in hospital databases and in the biomedical literature are supported (see Fig. 1 for an overview of the system). The development of the system followed

---

[1] http://khresmoi.eu/

C. Gurrin et al. (Eds.): MMM 2014, Part II, LNCS 8326, pp. 390–393, 2014.

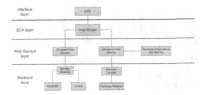

Fig. 1. Overview of the system architecture

a user–centered approach. First, investigation of the image search behavior of radiologists [1] was conducted, followed by user tests on the first version of the system [5]. The resulting prototype is presented in this paper and will be demonstrated at the conference.

2   Clinical Radiology 3D Image Search

The 3D image search is aimed at supporting clinical radiologists during the assessment of radiology data. To query, the radiologist marks a region of interest in the imaging data of a case under assessment. The system then searches for similar image content in a large data base of radiological imaging data. The search result consists of a ranked list of cases, each with markers for regions that have appearance similar to the query region. For each case the radiology report is displayed, and terms relating to anatomical regions, and pathological observations in the image are highlighted.

During indexing, the prototype uses a fragment registration based localization engine to map anatomy labels to each volume [6,7]. Then feature vocabularies are built for each anatomical structure [8]. They form the basis for anatomy specific pathology retrieval. Finally indices are built for individual anatomical structures. During a query, first, the anatomy label of the query region is identified, then closest neighbors in the corresponding index are found, and displayed.

The retrieval process is based on interviews with clinical radioligists, and aimed at fitting well into their workflow. The primary objective is to provide radiologists with efficient access to information in the hospital image storage (PACS) together with corresponding reports even before they make a diagnosis, or specify an observation.

3   2D Medical Document Image Search

In this section, the features, the architecture and the methods used for accessing the images and articles in the medical literature are described. The frontend is based on the ezDL interface [9] but in principle any other type of interface can be used. The system allows querying by keywords and/or image examples and returns a list of images or articles (Fig. 3). Images can be viewed in full size along with information such as the caption and corresponding articles. Filtering

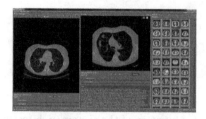

Fig. 2. Screenshot of the 3D search prototype. A query image is displayed on the left. A VOI can be marked and a result list is displayed on the right. Details can be viewed in the middle along with the volumes matching the VOI. Bellow is the radiology report.

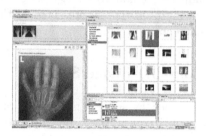

Fig. 3. Screenshot of the 2D search prototype. Main tools are the Query zone (top left) to supply keywords and/or image examples as queries, the result list (top right) inspect results and the details view (bottom left) to view details of selected results.

results by specific imaging modality is supported, as well as using relevance feedback.

The architecture uses RESTful web services for distributing the services and allowing easy connections. The text search is based on Lucene[2]. The visual search uses the Parallel Distributed Image Search Engine (ParaDISE) [10]. ParaDISE uses the Hadoop implementation [11] of the MapReduce framework for parallelizing the indexing of the visual features. The plug–in like design allows easy addition of features and indexing techniques. Like this updates of the system with novel image retrieval methods are easy.

The prototype will be demonstrated on PubmedCentral data containing 1.7M images of 700'000 biomedical articles. For modeling the content of images for CBIR, the features Bag–of–Visual–Words (BoVW) and Bag–of–Colors (BoC), evaluated in [10], were used. Efficient online retrieval was obtained using locality sensitive hashing (E2LSH) [12] for approximate nearest neighbour search.

4   Conclusions

New decision support tools are required to assist radiologists in their daily work and help them cope with the increasing amount of data that they need to analyze daily. Medical CBIR is a promising technique that can be used for a large variety

---

[2] http://lucene.apache.org/

of scenarios from keyword search to visual search of full images and regions or volumes of interest. The demo will show the combination of visual search with text search, the mix connection between the 2D and 3D search functionalities but also the possibilities of an adaptive user interface and other functionalities such as translation of found texts, spelling corrections and links of the keywords to standard terminologies.

References

1. Markonis, D., Holzer, M., Dung, S., Vargas, A., Langs, G., Kriewel, S., Müller, H.: A survey on visual information search behavior and requirements of radiologists. Methods of Information in Medicine 51(6), 539–548 (2012)
2. Müller, H., Michoux, N., Bandon, D., Geissbuhler, A.: A review of content–based image retrieval systems in medicine–clinical benefits and future directions. International Journal of Medical Informatics 73(1), 1–23 (2004)
3. Aisen, A.M., Broderick, L.S., Winer-Muram, H., Brodley, C.E., Kak, A.C., Pavlopoulou, C., Dy, J., Shyu, C.R., Marchiori, A.: Automated storage and retrieval of thin–section CT images to assist diagnosis: System description and preliminary assessment. Radiology 228(1), 265–270 (2003)
4. Vredenburg, K., Mao, J., Smith, P., Carey, T.: A survey of user-centered design practice. In: Proceedings of the SIGCHI Conference on Human Factors in Computing Systems: Changing Our World, Changing Ourselves, pp. 471–478 (2002)
5. Markonis, D., Baroz, F., Ruiz de Castaneda, R.L., Boyer, C., Müller, H.: User tests for assessing a medical image retrieval system: A pilot study. In: MEDINFO 2013 (2013)
6. Donner, R., Menze, B.H., Bischof, H., Langs, G.: Global localization of 3d anatomical structures by pre-filtered hough forests and discrete optimization. Med. Image Anal. (March 2013)
7. Dorfer, M., Donner, R., Langs, G.: Constructing an un-biased whole body atlas from clinical imaging data by fragment bundling. In: Mori, K., Sakuma, I., Sato, Y., Barillot, C., Navab, N. (eds.) MICCAI 2013, Part I. LNCS, vol. 8149, pp. 219–226. Springer, Heidelberg (2013)
8. Burner, A., Donner, R., Mayerhoefer, M., Holzer, M., Kainberger, F., Langs, G.: Texture bags: Anomaly retrieval in medical images based on local 3D-texture similarity. In: Müller, H., Greenspan, H., Syeda-Mahmood, T. (eds.) MCBR-CDS 2011. LNCS, vol. 7075, pp. 116–127. Springer, Heidelberg (2012)
9. Beckers, T., Dungs, S., Fuhr, N., Jordan, M., Kriewel, S., Tran, V.: An interactive search and evaluation system. Open Source Information Retrieval 9 (2012)
10. García Seco de Herrera, A., Markonis, D., Eggel, I., Müller, H.: The medGIFT group in ImageCLEFmed 2012. In: Working Notes of CLEF 2012 (2012)
11. White, T.: Hadoop: The Definitive Guide. O'Reilly Media, Inc. (2010)
12. Andony, A., Indyk, P.: Near-optimal hashing algorithms for approximate nearest neighbor in high dimensions. In: 47th Annual IEEE Symposium on Foundations of Computer Science, FOCS 2006, pp. 459–468 (2006)

Eolas: Video Retrieval Application for Helping Tourists

Zhenxing Zhang, Yang Yang, Ran Cui, and Cathal Gurrin

School of Computing, Dublin City University
Glasnevin, Dublin 9, Dublin, Ireland
{zzhang,cgurrin,yang.yang}@computing.dcu.ie,
cuiran1991@sina.com

Abstract. In this paper, a video retrieval application for the Android mobile platform is described. The application utilises computer vision technologies that, given a photo of a landmark of interest, will automatically locate online videos about that landmark. Content-based video retrieval technologies are adopted to find the most relevant videos based on visual similarity of video content. The system has been evaluated using a custom test collection with human annotated ground truth. We show that our system is effective, both in terms of speed and accuracy. This application is proposed for demonstration at MMM2014 and we are sure that this application would benefit tourists either planning travel or while travelling in real-time.

Keywords: Multimedia Information Retrieval, Video Processing, Exemplar-SVMs, Visual Similarity.

1 Introduction

The motivation of this work is to help tourists automatically finding documentary videos concerning a landmark of interest. The proposed query mechanism is via a photograph, either captured in the moment, or chosen from the photo-album. This application would be extremely helpful especially when they travelled in foreign countries with different languages. Recent advances in content-based video retrieval research suggests that is now possible to apply robust and efficient techniques to solve some real world problems; in this case, the problem of 'finding out about' certain landmarks. There has been some prior work in this area, such as using a photo to identify certain classes of objects [4], location recognition from captured images with a mobile device [5], or automatically identifying a sculpture [1] and labelling it and so on. In this demo paper, our purpose is not only to develop a novel application for tourists, but also to bring state-of-the-art video retrieval technologies beyond desktop environment into a real-time mobile devices usage.

In this demonstration we present video retrieval system, named Eolas (the Irish word for 'eyes'), that we implemented using exemplar-SVMs based object detection technologies [3]. The focus of this implementation is on linking famous landmarks/attractions with documentary videos from online sources such as YouTube.

C. Gurrin et al. (Eds.): MMM 2014, Part II, LNCS 8326, pp. 394–397, 2014.
© Springer International Publishing Switzerland 2014

By compare the visual similarity between a selected query image and keyframes extracted from a video archive, the system is able to present a list of documentary videos which are most closely visually-related to the query image.

In the rest of this demo paper we first provide an overview of retrieval system, then present the implementation details and system performance, and finally we present some concluding remarks and suggestions.

2   System Overview

Eolas is composed of two main components, a smartphone application and a online web service. Even though smartphones have evaluated dramatically with both computational and storage capabilities, they are still not ideal for heavy processing tasks such as video processing and retrieval. Hence, we use the online web service based on a remote server to accomplish efficient content analysis and retrieval, with the smartphone simply being the user interaction tool.

Online Search Architecture. The online query processing engine is triggered when the smartphone transmits a photo (from the Eolas applicaiton via the camera or the photo-album). Prior to transmission, there is phase of initial filtering and encoding which optimises the photo for upload. The video retrieval service that received the query has three main components: *a)* Query Parsing Module which responsible for parsing a query request, extracting feature representation of query image and passing the result to retrieval engine. *b)* Retrieval Module which performs searching operation based on the pre-indexed dataset, and return the ranked results to client side in JSON format. *c)* Server Log Module which can save the service operations related to each query.

Mobile Application GUI. The screenshots of smartphone application are displayed in Figure 1. Users can take a photo to query. A ranked results will be displayed and more details can be presented after click any result.

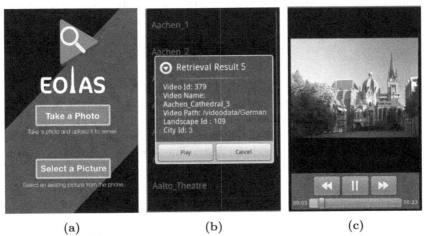

     (a)   (b)   (c)

Fig. 1. Snapshot of smartphone application GUI

3   System Implementation and Performance Evaluation

Based on the previously described architecture, we choose the tourist application as an initial use-case. We constructed a dataset of famous German tourist attractions. 310 documentary videos were downloaded from Flickr website under the Creative Commons license. The videos average at two minutes in duration and each one focuses on one attraction. The server application indexed these videos as follows.

3.1   Offline Video Processing Pipeline

There are three main steps in the offline video processing pipeline. Firstly, in order to reduce the complexity, each video has been segmented into a series of shots using a conventional approach to shot boundary detection (background colour change) and each shot is represented by one (central) keyframe. This gives 2,610 keyframes. Secondly, Histogram of Oriented Gradients (HOG) are picked to describe each keyframe, due to its good performance for object detection, robustness, and its speed. More importantly, it could offer stable performance even when there are many local changes, such as illumination changes, where local descriptors like SIFT would normally fail to match. Finally, a relational database has been used to index the meta data and the feature representation.

3.2   Exemplar-SVM Based Scoring Scheme

Different from approach of [7] using a bag of visual word representation and a text-based indexing algorithm, we implemented a linear discriminative object classifier for each query image, as was done by [3], [6]. There are two major benefits of this approach, firstly unlike [7], there are no quantization error because we are using the Hog descriptors directly, and secondly, a unique weighting score can be learned by using this data-driven learning method. This allows us to determine the most discriminative visual features according to one positive query example and many negative examples. After obtaining a weighting vector $\vec{w}$ for a query image, each video can be sorted using the following technique:

$$S(I_q, I_i) = \vec{w}^T X_i \qquad (1)$$

where I_q is the query image, I_i is the ith video and X_i is the feature vector. The top ranked videos are returned to the user. About 10,000 random images have been downloaded from Flickr website matching common topics like human, tree, parties. These images are checked and then used as negative examples for every training process which can provide for fast online retrieval. *LIBLINEAR* library [2] has been employed to achieve fast online training.

3.3   Performance Evaluation

Ten topics have been manually annotated as ground truth and in a user study, we calculated the mean average precision (mAP) and average query time across all topics (shown in Table 1). Eolas is shown to return 69% true positive results in less then 15 millisecond.

Table 1. System Evaluation Performance

Dataset Size		Performance	
Videos	Keyframes	mAP	Query time
311	2610	0.69	0.014 (s)

4   Conclusion

This demonstration paper presents a real-world application of a content based video retrieval engine using machine learning technologies. We describe the system and provide evaluation results in a small user study. Further work would be to expand the data to include different locations, to include GPS data from the image EXIF header to filter potential videos to a region, and to extend this work to different use-cases.

References

1. Arandjelović, R., Zisserman, A.: Name that sculpture. In: ACM International Conference on Multimedia Retrieval (2012)
2. Fan, R.-E., Chang, K.-W., Hsieh, C.-J., Wang, X.-R., Lin, C.-J.: LIBLINEAR: A library for large linear classification. Journal of Machine Learning Research 9, 1871–1874 (2008)
3. Malisiewicz, T., Gupta, A., Efros, A.A.: Ensemble of exemplar-svms for object detection and beyond. In: ICCV (2011)
4. Google Mobile. Open your eyes: Google goggles now available on iphone in google mobile app. (2010), http://googlemobile.blogspot.ie/2010/10/open-your-eyes-google-goggles-now.html/
5. Schroth, G., Huitl, R., Chen, D., Abu-Alqumsan, M., Al-Nuaimi, A., Steinbach, E.: Mobile visual location recognition. IEEE Signal Processing Magazine, Special Issue on Mobile Media Search 28(4), 77–89 (2011)
6. Shrivastava, A., Malisiewicz, T., Gupta, A., Efros, A.A.: Data-driven visual similarity for cross-domain image matching. ACM Transaction of Graphics (TOG) (Proceedings of ACM SIGGRAPH ASIA) 30(6) (2011)
7. Sivic, J., Zisserman, A.: Video Google: A text retrieval approach to object matching in videos. In: Proceedings of the International Conference on Computer Vision, vol. 2, pp. 1470–1477 (October 2003)

Audio-Visual Classification Video Browser

David Scott[1], Zhenxing Zhang[1], Rami Albatal[1], Kevin McGuinness[1],
Esra Acar[2], Frank Hopfgartner[2], Cathal Gurrin[1], Noel E. O'Connor[1],
and Alan F. Smeaton[1]

[1] Dublin City University, Glasnevin, Dublin 9, Ireland
[2] Technische Universität Berlin, Ernst-Reuter-Platz 7, 10587 Berlin, Germany

Abstract. This paper presents our third participation in the Video Browser Showdown. Building on the experience that we gained while participating in this event, we compete in the 2014 showdown with a more advanced browsing system based on incorporating several audio-visual retrieval techniques. This paper provides a short overview of the features and functionality of our new system.

1 Introduction

In recent years, interactive audio-visual search techniques are playing a paramount role within video retrieval systems. In order to enhance the user search experience and outperform the current state-of-the-art, we embrace five multi-modal retrieval methodologies in an advanced system which does not rely on text retrieval approaches. In this paper we present our VBS Multimedia Retrieval system incorporating HTML5 interface, that communicates with a middle-ware layer utilizing the audio-visual search elements which are described in the following sections.

2 System Features

2.1 Video Segmentation

We segment the video into shots based on techniques outlined by Pickering et al.[5]. Further, we extract multiple key frames for each shot to represent the shot in the graphical user interface. A segmentation into shots allows us to provide a quick overview over different scenes within a video, a strategy which proved successful in the TRECVid known item search task. It is likely that some key frames will be very similar to other key frames of the same shot. To avoid including these similar frames, we remove duplicates by comparing the global color layout of all key frames within each shot.

2.2 Visual Concept Classification

We use trained models to classify the visual content based on concepts such as person, landscape or buildings. We use the judgements from the classification to

C. Gurrin et al. (Eds.): MMM 2014, Part II, LNCS 8326, pp. 398–401, 2014.

act as ranked list when used alone. We also use the concept lists to either filter or boost content when used in tandem with other searches. The initial classifiers were trained on TrecVid 2013 Semantic Indexing task (SIN) training set [7]. In this task 60 concepts were requested to be identified in the shots of the SIN video dataset. Three visual content descriptors were used: two types of Opponent SIFT Bag of Visual words, the first is based on dense sampling, the second on sparse sampling; the third descriptor is a concatenation of a normalized RGB Histogram and a normalized Gabor transform. A Support Vector Machine (RBF-euclidean distance kernel)[1] was trained for each of the three descriptor and each of the 60 concepts. To provide a judgement about the existence of a particular concept in a shot, the correspondent three classifiers of that concept were used to evaluate the visual features extracted from the shot, then a weighted sum of the three judgement scores generated by the classifiers is performed to provide a final score. An initial framework for feature extraction and classification parameter evaluation and optimisation is developed and tested in this task; the framework is designed to be extendible to work on a large scale data, it is installed on the machines of the Irish Centre for High-End Computing (ICHEC)[1].

2.3 Audio Concept Classification

We generate models to detect audio concepts such as explosions, gunshots and screams. In a manner similar to the visual concept classification, we use audio concept lists to either filter or boost content when used in tandem with other searches. The classifiers are trained on the MediaEval 2013 Violent Scenes Detection task (VSD) training set using high-level audio concept annotations provided in the VSD dataset. We employ Mel-Frequency Cepstral Coefficients (MFCC) features as low-level audio features. For the representation of video shots, we use a Bag-of-Audio Words (BoAW) approach based on MFCC with a sparse coding scheme. We adopt the dictionary learning technique presented in [4]. In the coding phase, we construct the sparse representations of audio signals by using the LARS algorithm [2]. In order to generate the final sparse representation of video shots which are a set of MFCC feature vectors, we apply the *max-pooling* technique. A Support Vector Machine (SVM) with an RBF kernel is trained for each audio concept using sparse audio representations. In order to provide a judgment about the existence of a particular concept in a shot, the probability estimates of SVM models are used. Normally, in a basic SVM, only class labels or scores are output. The class label results from thresholding the score, which is not a probability measure. The scores output by the SVM are converted into probability estimates using the method explained in [9].

2.4 Visual Similarity Search

The aim of visual similarity search is to offer our system the ability to find the most visual relevant shots to a given shot query and then provide the most

---

[1] http://www.ichec.ie/

possible shots for users to identify. Supposing the searching topic is happened in front of a special scene, we could easy filter out the shots which did not happened in that scene.

Different from the approach of our previous participation using locally aggregation descriptors (VLAD) and nearest neighbours searching, we followed the route of [6] and built a linear discriminative object classifier for each query image. The main benefit of this approach is that a unique weighting score will be learnt to determine the most discriminative visual features for retrieval from training data which contains one positive data and many negative data.

For each keyframe, local feature descriptors are extracted and an aggregation descriptor is generated to represent it. A large size of negative training set is created and reused for every classifier training. In the online process, the same dimensional descriptor is extracted to the search query image and a linear classifier is trained by using the open source library [3]. Finally, each video shot from dataset can be sorted by calculating the inner product of weighting vector from classifier and their feature vectors.

2.5   Face Browsing and Search

We anticipate that providing functionality to allow users to get a high-level overview of all the human faces appearing in a video will be useful for queries involving people. To this end, we provide a face view that shows all the faces found in the video, and allows users to quickly navigate to the locations in the video in which selected faces appear. We use the Viola-Jones face detector [8] to first locate faces in the videos, and then to cluster these faces by using agglomerative techniques. This clustering also allows face-based search to be easily implemented: when the user chooses to search for similar faces on a given key frame, all images associated with the clusters containing any faces that appear in the key frame can be retrieved and displayed.

3   User Interface

Our user interface features a standard multi-modal platform powered by the python based Django framework. Users will be presented with metrics for accessing each of the system features explained above and will allow for multiple users to issue collaborative search for items of interest in the collection. As an optional module to the system we will use a mobile device to capture live screenshots of the example videos to use as input to a similarity based query. Our interface will attempt to maximise the usable canvas to enable users to find items as quickly as possible.

4   Conclusion

In this paper we present a technical overview of the system which will be presented at MMM Video browser showdown 2014. In this paper we outline the technologies used and the interaction with the system. Having come in second place overall last year, we have made some modifications in order to have a more rounded and better system which will challenge for the top spot this year.

References

1. Cortes, C., Vapnik, V.: Support-vector networks. In: Machine Learning, pp. 273–297 (1995)
2. Efron, B., Hastie, T., Johnstone, I., Tibshirani, R.: Least angle regression. The Annals of Statistics 32(2), 407–499 (2004)
3. Fan, R.-E., Chang, K.-W., Hsieh, C.-J., Wang, X.-R., Lin, C.-J.: LIBLINEAR: A library for large linear classification. Journal of Machine Learning Research 9, 1871–1874 (2008)
4. Mairal, J., Bach, F., Ponce, J., Sapiro, G.: Online learning for matrix factorization and sparse coding. The Journal of Machine Learning Research 11, 19–60 (2010)
5. Pickering, M.J., Rüger, S.M.: Evaluation of key frame-based retrieval techniques for video. Computer Vision and Image Understanding 92(2-3), 217–235 (2003)
6. Shrivastava, A., Malisiewicz, T., Gupta, A., Efros, A.A.: Data-driven visual similarity for cross-domain image matching. ACM Transaction of Graphics (TOG) (Proceedings of ACM SIGGRAPH ASIA) 30(6) (2011)
7. Smeaton, A.F., Over, P., Kraaij, W.: Evaluation campaigns and trecvid. In: MIR 2006: Proceedings of the 8th ACM International Workshop on Multimedia Information Retrieval, pp. 321–330. ACM Press, New York (2006)
8. Viola, P., Jones, M.: Rapid object detection using a boosted cascade of simple features, pp. 511–518 (2001)
9. Wu, T.-F., Lin, C.-J., Weng, R.C.: Probability estimates for multi-class classification by pairwise coupling. The Journal of Machine Learning Research 5, 975–1005 (2004)

Content-Based Video Browsing with Collaborating Mobile Clients

Claudiu Cobârzan, Marco A. Hudelist, and Manfred Del Fabro

Alpen-Adria-Universität Klagenfurt
9020 Klagenfurt, Austria
{claudiu,marco,manfred}@itec.aau.at

Abstract. A system comprised of collaborating mobile clients and a server is introduced in order to solve known item search tasks. The clients query the server for small video sequences according to some search criteria and are kept informed about the actions of collaborating participants (viewed frames, queries submitted, bookmarks set). The results are browsed and refined through a GUI that takes advantage of modern tablets' capabilities.

1 Introduction

Users often face the daunting task of finding a specific video segment they have previously seen in a large video collection that is either personal or public. Those tasks are often performed alone or with little help from others and usually employ a desktop computer and a tool that is either a specialized one or, more often, a simple video player.

The system we propose aims at taking advantage of the current high availability, popularity and ease of use of mobile clients (in our case, tablets) in solving such a task. Our approach is a collaborative one: multiple clients are querying a server for specific segments according to some criteria. Once the results are returned, they are inspected through a combination of viewing and browsing of key frames and smaller video segments. The clients can refine their initial search by issuing subsequent queries and can bookmark certain frames in order to ask the opinion of the other participants. The queries of the participants are all public and can be consulted, modified and resubmitted. The system keeps all clients informed about set or dismissed bookmarks as well as already viewed frames.

2 System Architecture

The system employs a 3-tier architecture: mobile clients, a server and a database. The communication is performed wirelessly and a proprietary TCP based protocol is applied. It combines pull and push techniques on both the clients and the server in order to pass relevant information: queries being made, responses and status information (viewed frames, set/canceled bookmarks, frame to be submitted as result of a search task, etc.). The following subsections provide an overview of the server and client components without details regarding the communication and the underlying data model and database implementation.

C. Gurrin et al. (Eds.): MMM 2014, Part II, LNCS 8326, pp. 402–406, 2014.

2.1   Server

The main tasks of the server are coordination of participants (mobile clients) and handling client queries. The server collects status information (queries, bookmarks, visited frames) from each client and makes it available in real-time to all other clients. New clients entering the system, register at the server and receive an update of the current state. Whenever a client issues a query, the server selects all segments from its internal database that match the filtering options set. A sequence consisting of pairs of start and end frames of the matching segments is returned to the client.

The content analysis is performed offline. Several content-based analysis tools are applied to detect different types of segments in the available videos. We use a color layout tool to identify segments where one color dominates the scene, e.g. the background color of a scene in a TV studio. A face detector is applied in order to index interview or anchorman scenes. Motion estimation is performed to identify foreground and background motion in video scenes. An audio classifier is used to detect speech and music segments in the audio stream of a video.

The color analysis is based on the MPEG-7 color layout descriptor [3] and the motion histogram introduced by Schoeffmann et al. [6] is applied for the motion analysis. For each frame of a video the color layout and the motion histogram are calculated and coherent frames with similar colors or respectively similar motion are detected. A sliding window of length 30 is used to iterate over all frames. If the difference between a frame and the average of the 30 preceding frames exceeds an empirically estimated threshold a segment boundary is detected.

This segment detection can be compared to shot detection, as the resulting segments often correspond to video shots. The segments found are mapped to predefined colors and motion directions that can be selected by the user during the search process. Color-based and motion-based segment detection was already successfully applied to the search tasks of the Video Browser Showdown 2013 [1].

The face detection is based on boosted cascades of Haar-like features [8]. We did not train a classifier on our own, but relied on classifiers shipped with the OpenCV[1] library. In each frame of a video, faces are detected and tracked among consecutive frames using a sliding window of 30 frames again. If a face is present at least in 25 frames within that sliding window, a face segment is detected, otherwise it is discarded. This procedure should help avoiding the detection of false positives.

We implemented the Continuous Frequency Activation (CFA) feature [7] for the speech/music discrimination. It basically works by looking for stationary parts in the audio spectrum. The advantage of CFA is that music can be detected even if it is played in the background at low volume with louder signals being present in the foreground.

---

[1] http://opencv.org/ (last access 2013-10-16).

2.2   Mobile Clients

The clients issue search requests within the available movie collection according to various criteria and have the possibility to refine the process by taking into consideration the segments already visited (by oneself or by others) as well as frames suggested for submission as result of the search process (bookmarked frames). The interface is divided into multiple task specific areas:

- *query target selection* (see Figure 1A): allows the user to choose the video(s) in which the search task will be performed. The presented video list also relays information regarding the sections which were already explored by the user in contrast to other current participants;
- *query construction* (see Figure 1F): enables setting various options regarding the desired scenes (sound, music, faces, dominant color, background and

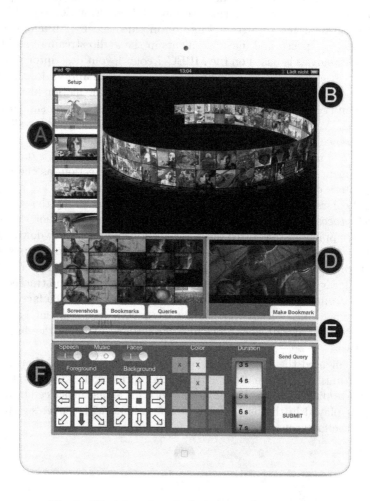

Fig. 1. Client interface during a typical search task

foreground movement, scene duration). The user submits the query to the server by tapping on the 'Send Query'-button;

- *results overview presentation* (see Figure 1B): displays the results of a query in the form of selectable video thumbnails arranged as a rotating 3D ring. Such an arrangement had positive effects on search time for image browsing ([2], [4]). Tapping a video thumbnail on the ring provides further exploration options (like the ones in Figure 1C and Figure 1D);
- *results exploration*: the selected video is presented as a sequence of consecutive frames (Figure 1C) as well as in a preview player (Figure 1D). The granularity of the displayed consecutive frames can be controlled by the user. The users can also navigate inside the video segment by using a large enhanced seeker bar (Figure 1E) that incorporates information on parts already inspected by oneself and others (green respectively blue markings). Segments of interest, on which the user would value a second opinion, can be bookmarked. All the participants are notified whenever a bookmark is set. All the submitted queries can be inspected and reused in full or with modified options. The currently selected frame, respectively, the current playback position of the preview player can be submitted by tapping on the 'Submit"-button in the lower right corner (see Figure 1 downright).

Figure 1 provides a snapshot of a search task in which the user searches for video segments that contain faces and speech, with a downwards moving foreground, a steady background, a combination of brown and orange colors and a minimum duration of five seconds (details specified according to the options set in Figure 1F).

3   Conclusion

We introduced a collaborative system for video browsing within the Video Browser Showdown (http://www.videobrowsershowdown.org) competition. It employs mobile clients and aims at taking advantage at the client side of the interaction capabilities provided by last generation tablets. It focuses on client and server features and organization that support and have the potential to enhance collaborative video browsing.

Acknowledgments. This work was funded by the Federal Ministry for Transport, Innovation and Technology (bmvit) and the Austrian Science Fund (FWF): TRP 273-N15 and by Lakeside Labs GmbH, Klagenfurt, Austria, and funding from the European Regional Development Fund (ERDF) and the Carinthian Economic Promotion Fund (KWF) under grant KWF-20214/22573/33955.

References

1. Del Fabro, M., Münzer, B., Böszörmenyi, L.: Smart Video Browsing with Augmented Navigation Bars. In: Li, S., El Saddik, A., Wang, M., Mei, T., Sebe, N., Yan, S., Hong, R., Gurrin, C. (eds.) MMM 2013, Part II. LNCS, vol. 7733, pp. 88–98. Springer, Heidelberg (2013)
2. Hudelist, M.A., Schoeffmann, K., Ahlström, D.: Evaluation of Image Browsing Interfaces for Smartphones and Tablets. Accepted for publication in: Proceedings of the 9th International Symposium on Multimedia (ISM). IEEE, Anaheim (2013)
3. Kasutani, E., Yamada, A.: The MPEG-7 color layout descriptor: a compact image feature description for high-speed image/video segment retrieval. In: Proceedings of the 2001 International Conference on Image Processing. IEEE, Thessaloniki (2001)
4. Schoeffmann, K., Ahlström, D.: Using a 3D cylindrical interface for image browsing to improve visual search performance. In: 13th International Workshop on Image Analysis for Multimedia Interactive Services (WIAMIS). IEEE, Dublin (2012)
5. Schoeffmann, K., Cobârzan, C.: An Evaluation of Interactive Search with Modern Video Players. In: IEEE International Conference on Multimedia and Expo (ICME). IEEE, San Jose (2013)
6. Schoeffmann, K., Lux, M., Taschwer, M., Böszörmenyi, L.: Visualization of Video Motion in Context of Video Browsing. In: Proceedings of the IEEE International Conference on Multimedia and Expo (ICME). IEEE, New York (2009)
7. Seyerlehner, K., Pohle, T., Schedl, M., Widmer, G.: Automatic music detection in television productions. In: Proceedings of the 10th International Conference on Digital Audio Effects (DAFx), Bordeaux (2007)
8. Viola, P., Jones, M.: Rapid object detection using a boosted cascade of simple features. In: Proceedings of the 2001 IEEE Computer Society Conference on Computer Vision and Pattern Recognition (CVPR). IEEE, Kauai (2001)

Browsing Linked Video Collections for Media Production

Werner Bailer, Wolfgang Weiss, Christian Schober, and Georg Thallinger

JOANNEUM RESEARCH Forschungsgesellschaft mbH
DIGITAL – Institute for Information and Communication Technologies
Steyrergasse 17, 8010 Graz, Austria
firstname.lastname@joanneum.at

Abstract. This paper describes a video browsing tool for media (post-) production, enabling users to efficiently find relevant media items for redundant and sparsely annotated content collections. Users can iteratively cluster the content set by different features, and restrict the content set by selecting a subset of clusters. In addition to clustering by features, similarity search by different features is supported, and a set of linked video segments can be explored for a segment of interest. Desktop and Web-based variants of the user interface, including temporal preview functionality, are available.

1 Introduction

The proposed video browsing tool has originally been designed for applications in the (post-) production phase of movie and broadcast production. In this application scenario, users typically deal with large amounts of audiovisual material with a high degree of redundancy and need to select a small subset for use in a production. Newly shot material is typically sparsely annotated, thus the browsing tool has to rely on automatically extracted features. The content sets in this application are typically much larger than those used in the Video Browser Showdown, reaching about 100 hours.

Automatic content analysis is performed during the ingest of content. Currently, camera motion estimation, visual activity estimation, extraction of global color features and estimation of object trajectories are performed. The extracted features are represented using the MPEG-7 Audiovisual Description Profile [1] and are indexed in an SQLite database.

In order to select content, the user follows an iterative selection process, consisting of alternating steps of clustering and selecting subsets of the current data set. The user clusters content by one of the automatically extracted features and can then select relevant clusters to reduce the content set. Further clustering by the same or other features can then be applied to the reduced set. In addition, items similar to one of the items in the cluster can be retrieved. A more detailed description of the tool and the browsing process can be found in [2]. An earlier version of the tool was used in the first two editions of the Video Browser Showdown (VBS) in 2012 and 2013. In addition to similarity search, the functionality

C. Gurrin et al. (Eds.): MMM 2014, Part II, LNCS 8326, pp. 407–410, 2014.
© Springer International Publishing Switzerland 2014

to explore video segments related to a segment of interest (e.g., a potential result segment) has been added.

2   Browsing User Interface

The central component of the video browsing tool's user interface is a light table (cf. Figure 1). The light table (A) shows the current content set and cluster structure using a number of representative key frames for each of the clusters. The clusters are visualized by colored areas around the images. The size of the images in the light table view can be changed dynamically so that the user can choose between the level of detail and the number of visible images without scrolling. By clicking on a key frame (B) in the light table view, a video player is opened and plays the segment of the video that is represented by that image. The temporal context of a key frame is shown by a time line (C) of temporally adjacent key frames that appears when the user moves the mouse over a frame. This time line shows one line of key frames which is limited by the width of the screen. If the user wants to get a broader range of temporally adjacent key frames, it is possible to zoom in. Therefore, a larger list of key frames is shown in the light table.

The tool provides support for performing similarity search (D) based on pre-calculated features, such as: camera motion, motion activity or color layout. To execute a similarity search, the user drags a key frame into the similarity search area, selects a similarity search option and executes the search (see Figure 2). In addition, segments from the temporal proximity of a result segment can be retrieved. The latter supports the user in cases where a presented item is topically similar to the wanted item and might thus be nearby in the programme, but is not similar in terms of visual features.

On the left side of the application window the history (E) and the result list (F) are displayed. The history window automatically records all clustering and selection actions done by the user. By clicking on one of the entries in the history, the user can go back to a previous point. Then users can choose to discard the subsequent steps and use other cluster/selection operations, or to branch the browsing history and explore the content using alternative cluster features. The result list can be used to memorize video segments and to extract segments of videos for further video editing, e.g. as edit decision list (EDL). Users can drag relevant key frames into the result list at any time, thus adding the corresponding segment of the content to it.

A new feature allows the exploration of relations and links within the video collection. In the ingest step, we extract SIFT descriptors [4] from the videos and match between the pairs of segments in the content set. We also consider partial and local matches, as those might indicate backgrounds or objects shared between the segments. The approach for determining these segments is described in [3], with only the part of the method based on visual features being employed. In order to efficiently handle larger content collections, the similarities are pre-calculated and not determined at browsing time. However, the feature extraction

Fig. 1. Screenshot of the Web-based video browsing tool, using parts of the VBS 2013 data set (videos kindly provided by the Flemish public broadcasting organization VRT)

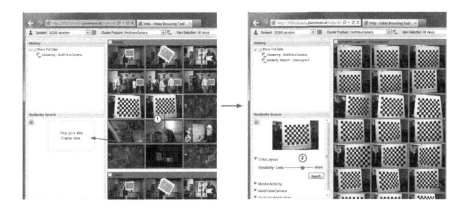

Fig. 2. Executing the similarity search

and matching has been implemented on the GPU using NVIDIA CUDA, so pre-computing is quite efficient. The results of feature matching are then presented as a set of linked segments related to a result segment. Thus the user can use the paths defined by links between segments to explore the content collection. This feature is seamlessly integrated with other browsing capabilities of the tool, so that the user can switch between different ways of exploring the content.

The user interface is available both as desktop and as a Web-based version, offering the same functionality. Both use the same backend implementation, which is accessible as a SOAP Web service for the Web-based client.

Acknowledgements. The research leading to these results has received funding from the European Union's Seventh Framework Programme (FP7/2007-2013) under grant agreements n° 215475, "2020 3D Media – Spatial Sound and Vision" (http://www.20203dmedia.eu/) and n° 287532, "TOSCA-MP - Task-oriented search and content annotation for media production" (http://www.tosca-mp.eu).

References

1. Information technology - multimedia content description interface - part 9: Profiles and levels, amendment 1: Extensions to profiles and levels. ISO/IEC 15938-9:2005/Amd1:2012 (2012)
2. Bailer, W., Weiss, W., Kienast, G., Thallinger, G., Haas, W.: A video browsing tool for content management in post-production. International Journal of Digital Multimedia Broadcasting (March 2010)
3. Lokaj, M., Stiegler, H., Bailer, W.: The Search and Hyperlinking Task at MediaEval 2013. In: MediaEval 2013 Workshop, Barcelona, Spain, October 18-19 (2013)
4. Lowe, D.: Distinctive image features from scale-invariant keypoints. International Journal of Computer Vision 60(2), 91–110 (2004)

VERGE: An Interactive Search Engine for Browsing Video Collections

Anastasia Moumtzidou[1], Konstantinos Avgerinakis[1], Evlampios Apostolidis[1],
Vera Aleksić[2], Fotini Markatopoulou[1], Christina Papagiannopoulou[1],
Stefanos Vrochidis[1], Vasileios Mezaris[1], Reinhard Busch[2], and Ioannis Kompatsiaris[1]

[1]Information Technologies Institute/Centre for Research and Technology Hellas,
6th Km. Xarilaou - Thermi Road, 57001 Thermi-Thessaloniki, Greece
{moumtzid,koafgeri,apostolid,markatopoulou,cppapagi,stefanos,
bmezaris,ikom}@iti.gr
[2]Linguatec Sprachtechnologien GmbH, Gottfried-Keller-Str. 12, 81245 München
{v.aleksic,r.busch}@linguatec.de

Abstract. This paper presents VERGE interactive video retrieval engine, which is capable of searching and browsing video content. The system integrates several content-based analysis and retrieval modules such as video shot segmentation and scene detection, concept detection, clustering and visual similarity search into a user friendly interface that supports the user in browsing through the collection, in order to retrieve the desired clip.

1 Introduction

This paper describes VERGE interactive video search engine[1], which is capable of retrieving and browsing video by integrating different indexing and retrieval modules. VERGE supports the Known Item Search task, which requires the incorporation of techniques for browsing and navigation within a video collection. VERGE was evaluated with participation in workshops and showcases such as TRECVID and VideOlympics, where it was shown to significantly improve user search experience over single or fewer search modalities. Specifically, VERGE demonstrated the best results in the interactive known item search of TRECVID 2011 by achieving a Mean Inverted Rank of 0.56, while the concept detectors of VERGE achieved good balance between detection accuracy (e.g. 15.8% MXinfAP for TRECVID 2013) and low computational complexity.

The proposed version of VERGE aims at participating to the KIS task of the Video Browser Showdown (VBS) Competition [1]. In this context, VERGE supports interactive searching of a known video clip in a large video collection by incorporating content-based analysis and interactive retrieval techniques.

---

[1] More information and demos of VERGE are available at: http://mklab.iti.gr/verge/

C. Gurrin et al. (Eds.): MMM 2014, Part II, LNCS 8326, pp. 411–414, 2014.
© Springer International Publishing Switzerland 2014

In the next sections we present the content-based analysis and retrieval techniques supported by VERGE, as well as the interaction with the user.

2   Video Retrieval System

VERGE is a retrieval system, which combines advanced retrieval functionalities with a user-friendly interface. The following basic modules are integrated: a) Shot and Scene Segmentation; b) Textual Information Processing Module; c) Visual Similarity Search; d) High Level Concept Detection; e) Clustering.

The **shot segmentation module** performs shot segmentation by extracting visual features, namely color coherence, Macbeth color histogram and luminance center of gravity, and forming a corresponding feature vector per frame [2]. Then, given a pair of frames, the distances between their vectors are computed, composing distance vectors that are finally evaluated using one or more SVM classifiers, resulting to the detection of both abrupt and gradual transitions between the shots of the video.

Scene segmentation is based on the previous analysis and groups shots into sets that correspond to individual scenes of the video. The algorithm [3] introduces and combines two extensions of the Scene Transition Graph (STG); the first one aims to reduce the computational cost by considering shot linking transitivity, while the second one constructs a probabilistic framework towards combining multiple STGs.

The **textual information processing module** applies Automatic Speech Recognition (ASR) on videos. We employ the VPE (Voice Pro Enterprise) framework, which is based on RWTH-ASR technology [4]. Finally, each shot is described by a set of words, which are used to create a taxonomy to facilitate browsing of the collection.

The **visual similarity search module** performs content-based retrieval based on global and local information. To deal with global information, MPEG-7 descriptors are extracted from each keyframe and they are concatenated into a single feature vector. Efficient retrieval is achieved by employing the r-tree indexing structure. In the case of local information SURF features are extracted. We apply two Bag of Visual Words techniques for representing and retrieving images efficiently. On the first, we calculate visual vocabularies via hierarchical k-means clustering [5], while on the second, we follow K-Means clustering and VLAD encoding for representing images.

The **high level concept retrieval module** indexes the video shots based on 346 high level concepts (e.g. water, aircraft). For each keyframe we employ up to 25 feature extraction procedures [6]. For learning these concepts, a bag of linear Support Vector Machines (LSVM) is trained for each feature extraction procedure and each concept. A sampling strategy is applied to partition the dataset into 5 subclasses and for each subset a LSVM is trained. During the classification phase, a new unlabeled video shot is given to the trained LSVMs, each of them returns the degree of confidence that the concept is depicted in the image, and late fusion is used for combining these scores.

Finally, the **clustering module** incorporates an agglomerative hierarchical clustering process [7], which provides a hierarchical view of the keyframes. In addition to the feature vectors used as input to the high level concept retrieval module, we extract vectors consisting of the responses of the trained concept detectors for each video shot. The clustering algorithm is then applied to these representations in order to group the keyframes into clusters, each of which consists of keyframes having visually or semantically similar content.

VERGE is built on Apache server, PHP, JavaScript and mySQL database. Besides the aforementioned basic modules, VERGE integrates the following complementary functionalities: a) basic temporal queries, b) shot storage structure and c) history bin.

3   Interaction Modes

The aforementioned modules aid the user to interact with the system through a user-friendly interface (Figure 1), in order to discover the desired video clip during known item search tasks. The interface comprises of three main components: a) the central component, b) the left side and c) the upper panel.

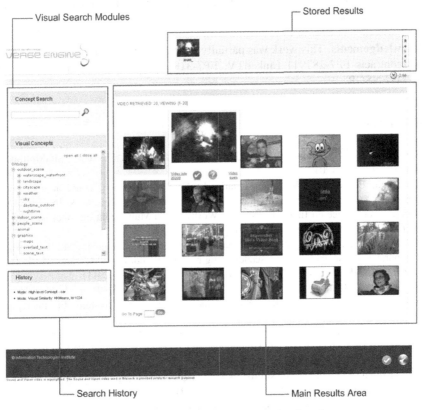

Fig. 1. Screenshot of VERGE video retrieval engine

The central component of the interface includes a shot-based representation of the video in a grid-like interface. In this grid each video shot is visualized by a representative key frame. When the user hovers over an image, a pop-up frame appears that contains a larger preview of the image in order to allow for better inspection of its content, as well as several links that support:

- browsing the temporally adjacent shots and all shots of the specific video
- registering an image as relevant for a specific topic or query
- searching for visually similar images to the given (query) image

On the left side of the interface, the search history, as well as additional search and browsing options are displayed. The history module automatically records all searching actions done by the user, while the search and browsing options include the taxonomy based on the ASR transcriptions, the high level visual concepts and the hierarchical clustering. Using the aforementioned functionalities, the user can browse the dataset at shot and scene level taking also into account ASR and concept taxonomies.

Finally, the upper panel is a storage structure that mimics the functionality of the shopping cart found in electronic commerce sites and holds the shots selected by the user throughout the session.

Acknowledgements. This work was partially supported by the European Commission under contracts FP7-287911 LinkedTV, FP7-318101 MediaMixer and FP7-610411 MULTISENSOR.

References

1. Schoeffmann, K., Bailer, W.: Video Browser Showdown. ACM SIGMultimedia Records 4(2), 1–2 (2012)
2. Tsamoura, E., Mezaris, V., Kompatsiaris, I.: Gradual transition detection using color coherence and other criteria in a video shot meta-segmentation framework. In: IEEE International Conference on Image Processing, Workshop on Multimedia Information Retrieval (ICIP-MIR 2008), San Diego, CA, USA, pp. 45–48 (2008)
3. Sidiropoulos, P., Mezaris, V., Kompatsiaris, I., Meinedo, H., Bugalho, M., Trancoso, I.: Temporal video segmentation to scenes using high-level audiovisual features. IEEE Trans. on Circuits and Systems for Video Technology 21(8), 1163–1177 (2011)
4. Rybach, D., Gollan, C., Heigold, G., Hoffmeister, B., Lööf, J., Schlüter, R., Ney, H.: The RWTH Aachen University Open Source Speech Recognition System. In: Interspeech, Brighton, UK, pp. 2111–2114 (2009)
5. Nister, D., Stewenius, H.: Scalable recognition with a vocabulary tree. In: Proceedings of the 2006 IEEE Computer Society Conference on Computer Vision and Pattern Recognition (CVPR 2006), vol. 2 (2006)
6. Moumtzidou, A., Gkalelis, N., Sidiropoulos, P., Dimopoulos, M., Nikolopoulos, S., Vrochidis, S., Mezaris, V., Kompatsiaris, I.: ITI-CERTH participation to TRECVID 2012. In: Proc. TRECVID 2012 Workshop, Gaithersburg, MD, USA (2012)
7. Johnson, S.C.: Hierarchical Clustering Schemes. Psychometrika 2, 241–254 (1967)

Signature-Based Video Browser

Jakub Lokoč, Adam Blažek, and Tomáš Skopal

SIRET Research Group, Department of Software Engineering,
Faculty of Mathematics and Physics, Charles University in Prague
{lokoc,skopal}@ksi.mff.cuni.cz, blazekada@gmail.com

Abstract. In this paper, we present a new signature-based video browser tool relying on the natural human ability to perceive and memorize visual stimuli of color regions in video frames. The tool utilizes feature signatures based on color and position extracted from the key frames in the preprocessing phase. Such content representation facilitates users in drawing simple query sketches and enables also effective and efficient processing of the query sketches. Besides user drawn simple sketches of desired scenes, the tool supports also several additional automatic content-based analysis techniques enabling restrictions to various concepts like faces or shapes.

Keywords: Content-based Video Retrieval, Feature Signatures, Known Item Search, Sketch Search.

1 Introduction

The results of the Video Browser Showdown 2013 have demonstrated that known item video search tasks can be successfully performed using the combination of automatic content-based analysis and user friendly human interaction video browsers. For example, the winner of Video Browser Showdown 2013 [2] has used a tool supporting face detection techniques, that have significantly improved the efficiency and effectiveness of the last year retrieval tasks. However, the performance of the tools relying on a particular concept depends on whether or not the concept appears in the video clip. Therefore, we have focused on a new signature-based video browser tool employing the query by sketch approach, utilizing more general (but also more frequent) low level concepts like simple color regions and their position. More specifically, we utilize feature signatures based on color and position that can flexibly represent each key frame by the set of simple color regions as depicted in Figure 1. We may observe, that unlike the traditional spatial color sketch techniques employing fixed-grid color maps (for more details see [4]), the feature signatures [3] flexibly adjust to the content of the key frames, thus allowing users to specify and locate less dominant color regions (e.g., the yellow sun in 1b). Furthermore, such simple representation (color circles) can be trivially matched to the user drawn sketches, where the users can also record the observed color stimuli by simple circles.

Since the videos often contain several high-level concepts, our tool supports also some more specific automatic content-based concept detectors (e.g., faces or

C. Gurrin et al. (Eds.): MMM 2014, Part II, LNCS 8326, pp. 415–418, 2014.

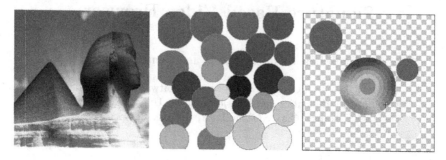

Fig. 1. a) The original image, b) the corresponding feature signature and c) the sketch drawing component with an enhanced color picker

shapes) implemented by the OpenCV library, however, their usage is not the key contribution of this paper and so we mainly focus on our new signature-based approach in the following text.

2   Signature-Based Video Browser

Our signature-based video browser utilizes a preprocessing phase, where we select a set of significant key frames[1]. For each selected key frame we extract one feature signature comprising set of centroids (w, x, y, L, a, b), where w represents weight (depicted as radius in Figure 1b), x, y denotes the position of the centroid and L, a, b corresponds to the color coordinates in the CIE LAB color space. The extraction algorithm uses an adaptive variant of the k-means algorithm resulting in feature signatures of various sizes depending on the complexity of the key frames. For more information about the coordinates and the feature extraction process see [1]. During the preprocessing step, we also detect additional meta information for each key frame, typically a probability that the frame contains a specific object (e.g., a human face).

Whereas the key frames are preprocessed to feature signatures comprising tens of centroids[2], we assume the query sketches will consist only of few memorized color regions (e.g., a brown monument and the blue sky at the top). Such regions can be simply drawn by few color circles (i.e., user defined centroids), which more or less correspond to the centroids of the searched feature signatures (as depicted in Figure 1c). As the searched video clip may contain more visually distinct scenes, we expect the users may want to quickly draw a sequence of consecutive sketches in order to improve the filtering power of the query. For these reasons, we have designed a user friendly sketch drawing component focusing on simple, yet sufficiently precise reproduction of the significant color stimuli to the

---

[1] First we select every k^{th} frame and then we remove frames similar to some extent (considering color distribution) to their antecedent neighbor frames.

[2] The number of the centroids correlates with the complexity of the scene in the key frame.

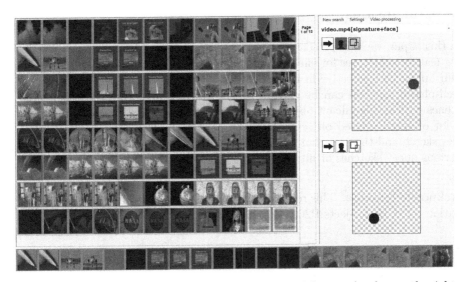

Fig. 2. Result list of the sketch-based query consisting of the two sketches on the right. The bottom line depicts an extended version of a user selected result item.

query sketch. Let us denote, we assume the observed video clip contains scenes with suitable color stimuli, while for more complicated scenes (a lot of widely distributed colors) we plan to implement additional filters.

For our approach, the color selection is the most crucial task, and thus we have implemented a special color picker (depicted in Figure 1c), where the colors are organized in the HSV color space and displayed in the concentric circles around the actually edited centroid. Thus a user can quickly specify a desired hue and saturation, while different value levels can be simply changed by the mouse wheel. For a more comfortable editing, the drawn circles can be moved, deleted, enlarged or their color can be altered at any time with the mouse, the undo and redo operations are also implemented. Beside drawing of the sketches, a user can specify sketches with additional filters which use the preprocessed meta information (depicted over each sketch in Figure 2).

The search algorithm works for a drawn sketch as a top-k operator finding the most corresponding feature signatures. More specifically, for each drawn centroid in the sketch the algorithm finds the most similar centroids from all the feature signatures (using the Euclidean distance) and improves the ranking of the corresponding feature signatures (i.e., key frames). Finally, the frames are sorted according to the ranking and for each frame the line sequence of static frames surrounding the matched frame(s) is displayed (see Figure 2). If a user draws two sketches, then the algorithm selects such pairs of matched frames that are both within a user-defined time interval. By clicking any presented static frame, its signature is loaded as a sketch in order to browse to similar scenes which helps in cases the video consists of similar or repetitive scenes (e.g., news studio).

3 Conclusions

In this paper, we have presented a new signature-based video browser tool utilizing feature signatures for effective and efficient filtering of the video key frames. Our approach is effective in finding scenes with unique color regions where one well-placed centroid can be enough to find desired scene. On the other hand, scenes with no significant object or color (e.g., desaturated desert landscape or a lot of small colored objects) might be hard to find only by simple few circles sketch and thus we plan to extend our tool to support more ways to define various query sketches or filters.

Acknowledgments. This research has been supported by Czech Science Foundation (GAČR) projects P202/11/0968 and P202/12/P297.

References

1. Kruliš, M., Lokoč, J., Skopal, T.: Efficient extraction of feature signatures using multi-GPU architecture. In: Li, S., El Saddik, A., Wang, M., Mei, T., Sebe, N., Yan, S., Hong, R., Gurrin, C. (eds.) MMM 2013, Part II. LNCS, vol. 7733, pp. 446–456. Springer, Heidelberg (2013)
2. Le, D.-D., Lam, V., Ngo, T.D., Tran, V.Q., Nguyen, V.H., Duong, D.A., Satoh, S.: NII-UIT-VBS: A video browsing tool for known item search. In: Li, S., El Saddik, A., Wang, M., Mei, T., Sebe, N., Yan, S., Hong, R., Gurrin, C. (eds.) MMM 2013, Part II. LNCS, vol. 7733, pp. 547–549. Springer, Heidelberg (2013)
3. Rubner, Y., Tomasi, C.: Perceptual Metrics for Image Database Navigation. Kluwer Academic Publishers, Norwell (2001)
4. Wang, J., Hua, X.-S.: Interactive image search by color map. ACM Trans. Intell. Syst. Technol. 3(1), 12:1–12:23 (2011)

NII-UIT: A Tool for Known Item Search by Sequential Pattern Filtering

Thanh Duc Ngo[1,2], Vu Hoang Nguyen[1], Vu Lam[1,2], Sang Phan[4],
Duy-Dinh Le[3], Duc Anh Duong[1], and Shin'ichi Satoh[3]

[1] University of Information Technology - VNU HCMC, Vietnam
{thanhnd,vunh,ducda}@uit.edu.vn
[2] University of Science - VNU HCMC, Vietnam
lqvu@fit.hcmus.edu.vn
[3] National Institute of Informatics, Japan
{ledduy,satoh}@nii.ac.jp
[4] The Graduate University for Advanced Studies, Japan
{plsang}@nii.ac.jp

Abstract. This paper presents an interactive tool for searching a known item in a video or a video archive. To rapidly select the relevant segment, we use query patterns formulated by users for filtering. The patterns can be formulated by drawing color sketches or selecting predefined concepts. Especially, our tool support users to define patterns for sequences of consecutive segments, for instance, sequences of occurrences of concepts. Such patterns are called sequential patterns, which are more powerful to describe users' search intention. Besides that, the user interface is organized following a coarse-to-fine manner, so that users can quickly scan the set of candidate segments. By using color-based and concept-based filters, our tool can deal with both visual and descriptive known item search.

Keywords: Video Browser Showdown, Known Item Search, Concept Filtering, Sequential Pattern.

1 Introduction

The exponential growth of video data creates a growing demand for tools which can help users to rapidly search for a known item or a specific video segment. The challenging issues in building such a tool, especially with an interactive tool, include query formulation, selecting candidate video segments for users' judgment and navigating users through a huge pool of candidate segments.

Regarding existing systems [1,2,3], using filters has been shown as a promising solution for query formulation and candidate segment selection. Users first formulate a query pattern by drawing a color sketch or selecting a predefined concept. The pattern is then used as a filter to quickly remove all irrelevant video segments. However, a limitation of such systems is that their filters are designed and applied to individual video segments. By doing that, video segments

C. Gurrin et al. (Eds.): MMM 2014, Part II, LNCS 8326, pp. 419–422, 2014.

are considered independent each other. Thus, the correlations between consecutive video segments are disregarded. For example, given two consecutive video segments in which the first has the *outdoor* concept detected and the second has the *interview* concept detected, the sequential transition from an *outdoor* scene to an *interview* scene between the segments can not be described and used for filtering while such information is extremely informative. In this work, we propose to take such information into account by extending existing filtering methodology. Particularly, we support users to formulate patterns for sequences of video segments. The formulated patterns are called sequential patterns.

On the other hand, we keep using our coarse-to-fine display strategy as in our former system [1] to help users efficiently scan through the pool of candidate video segments. In general, the key techniques of our tool are as following:

- *Coarse-to-fine display.* At the coarsest level, one key-frame is used to represent content of each 2-minute segment. By looking at these key-frames, some irrelevant segments can be eliminated quickly. At the finest level, 5 key-frames are used to represent content of each 15-second segment. By looking at these key-frames, the segment that is relevant to the target segment can be identified and confirmed.

- *Using filters to reduce the number of video segments to be judged by users.* The basic video segments for filtering are 15 second-segments. Query patterns are of 3 types: i) patterns based on scene concepts: *indoor, outdoor, daytime, nighttime, sport, performance, interview*; ii) patterns based on object concepts: face, pedestrian; and, iii) color-sketch patterns, which are formulated by selecting colors for cells of 3×4 grids. More importantly, for each type, users can use two individual patterns to form a sequential pattern. For example, using *outdoor-interview* as a pattern to find a pair of video segments having a transition from an outdoor scene to an interview scene.

2   System Overview

2.1   Preprocessing

For each video, a set of 15-second segments will be obtained by equally dividing the video. The middle frame of a segment is used for thumbnail display, since frames within a segment are mostly similar. By using segments of such length, we expect to achieve a balance between recall and efficiency. The relevant segment (typically with a duration of 20 seconds) will not be missed and there are not too many segments for a user or the system to process. With an one-hour video, the total number of such segments is 240. Five key-frames of each segment are selected. Features of each frame (i.e. Bag-of-Visual-Words, Histogram of Oriented Gradients, Haar-like) are extracted and then used as inputs for concept classifiers and object detectors.

The concept classifiers are trained using Support Vector Machine (SVM) classifiers. Training data is crawled from the Internet using Google Image Search

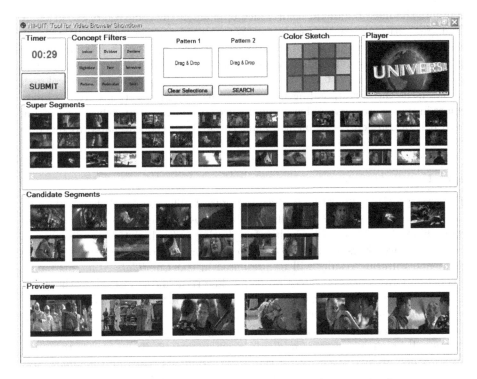

Fig. 1. The user interface of NII-UIT video browsing tool

with queries are names of the concepts. Top images of each concept are annotated as positive samples. Meanwhile, images of the other concepts are used as negative samples. With object detectors, we use a robust approach introduced by Viola et al. [4] for *face* and Deformable Part Models (DPM) [5] for *pedestrian*. Although many object and scene concepts can be employed using the same framework, we only select those which can be detected or recognized with high accuracy (more than 80% accuracy on our development datasets).

2.2 Indexing and Retrieval

Given classification and detection results as binary labels indicating occurrences of concepts in segments, we index the videos by using inverted index technique for fast retrieval and filtering. With each concept, we store a list of video segments in which the concept is detected. This list will be instantly returned when a user select the concept at the retrieval stage. Similarly, for each pair of concepts, we store a list of pairs of consecutive segments in which the concepts appear sequentially. Note that we consider pairs of concepts with different orders as different pairs for indexing.

On the other hand, we also index dominant colors in video segments. All key-frames of a video segment are divided into 3 × 4 grids. For each cell, we compute the most dominant color which appears at the same cell of the grids. A final grid with dominant colors in cells represent the whole segment. With each cell of the grid and each color, we store a list of all video segments having the color appear at the cell. When a user choose a color at a cell, video segments of the corresponding list will be returned. If more than one cells are colored, we intersect the corresponding lists and return the intersection only. To index colors in two consecutive video segments, we extend the approach by processing each pair of consecutive segments. Their two grids are concatenated to generate a unified 3×8 grid.

2.3   User Interface

The user interface of our tool is given in Figure 1. The top panel shows all filters including scene concept-based filters, object-based filters and a color sketch-based filter. Users can use one filter for filtering individual 15-second segments or coupling two filters of the same type for filtering pairs of consecutive 15-second segments. Filtering results will be presented in the *Candidate Segments* panel. Their temporal orders in videos are kept. The right most position of the panel is a video player to play a selected segment in fast forward mode (2x).

The *Super Segments* panel shows key-frames that represent for 2-minute segments. Once users select a 2-minute segment, its corresponding 15-second segments will be shown in the *Candidate Segments* panel. If the task is to search over a video archive, each row of the *Super Segments* panel or *Candidate Segments* panel display segments of one video. The bottom *Preview* panel is to show key-frames of a selected 15-second segment in the *Candidate Segments* panel. It provides the finest overview of a segment for final relevance confirmation. All panels can be expanded so that more representative thumbnails can be displayed at the same time.

References

1. Le, D.-D., Lam, V., Ngo, T.D., Tran, V.Q., Nguyen, V.H., Duong, D.A., Satoh, S.: NII-UIT-VBS: A Video Browsing Tool for Known Item Search. In: Li, S., El Saddik, A., Wang, M., Mei, T., Sebe, N., Yan, S., Hong, R., Gurrin, C. (eds.) MMM 2013, Part II. LNCS, vol. 7733, pp. 547–549. Springer, Heidelberg (2013)
2. Scott, D., et al.: DCU at MMM 2013 Video Browser Showdown. In: Li, S., El Saddik, A., Wang, M., Mei, T., Sebe, N., Yan, S., Hong, R., Gurrin, C. (eds.) MMM 2013, Part II. LNCS, vol. 7733, pp. 541–543. Springer, Heidelberg (2013)
3. Bai, H., Wang, L., Dong, Y., Tao, K.: Interactive Video Retrieval Using Combination of Semantic Index and Instance Search. In: Li, S., El Saddik, A., Wang, M., Mei, T., Sebe, N., Yan, S., Hong, R., Gurrin, C. (eds.) MMM 2013, Part II. LNCS, vol. 7733, pp. 554–556. Springer, Heidelberg (2013)
4. Viola, P., Jones, M.: Robust Real-time Face Detection. International Journal of Computer Vision (IJCV), 137–154 (2004)
5. Felzenszwalb, P., Girshick, R., McAllester, D., Ramanan, D.: Object Detection with Discriminatively Trained Part Based Models. IEEE Transactions on Pattern Analysis and Machine Intelligence (TPAMI), 1627–1645 (2010)

Author Index